Semantic Models in IoT and eHealth Applications

Intelligent Data-Centric Systems

Semantic Models in IoT and eHealth Applications

Edited by

Sanju Tiwari
Autonomous University of Tamaulipas,
Victoria, Tamaulipas, Mexico

Fernando Ortiz Rodriguez
Autonomous University of Tamaulipas,
Victoria, Tamaulipas, Mexico

M.A. Jabbar
Department of Computer Science and Engineering,
Vardhaman College of Engineering,
Hyderabad, Telangana, India

Series Editor
Fatos Xhafa
UPC-BarcelonaTech, Barcelona, Spain

Academic Press is an imprint of Elsevier
125 London Wall, London EC2Y 5AS, United Kingdom
525 B Street, Suite 1650, San Diego, CA 92101, United States
50 Hampshire Street, 5th Floor, Cambridge, MA 02139, United States
The Boulevard, Langford Lane, Kidlington, Oxford OX5 1GB, United Kingdom

Notices

Knowledge and best practice in this field are constantly changing. As new research and experience broaden our understanding, changes in research methods, professional practices, or medical treatment may become necessary.

Practitioners and researchers must always rely on their own experience and knowledge in evaluating and using any information, methods, compounds, or experiments described herein. In using such information or methods they should be mindful of their own safety and the safety of others, including parties for whom they have a professional responsibility.

To the fullest extent of the law, neither the Publisher nor the authors, contributors, or editors, assume any liability for any injury and/or damage to persons or property as a matter of products liability, negligence or otherwise, or from any use or operation of any methods, products, instructions, or ideas contained in the material herein.

ISBN: 978-0-323-91773-5

For information on all Academic Press publications
visit our website at https://www.elsevier.com/books-and-journals

Publisher: Mara Conner
Editorial Project Manager: Emily Thomson
Production Project Manager: Prasanna Kalyanaraman
Cover Designer: Miles Hitchen

Typeset by VTeX

Working together
to grow libraries in
developing countries

www.elsevier.com • www.bookaid.org

Contents

Contributors

Karima Boudaoud
University of Nice Sophia Antipolis, Sophia Antipolis, France

Ozgu Can
Ege University, Department of Computer Engineering, Bornova-Izmir, Turkey

Inaldo Capistrano Costa
Division of Computer Science, Aeronautics Institute of Technology (ITA), São José dos Campos, SP, Brazil

Antonella Carbonaro
Department of Computer Science and Engineering – DISI, University of Bologna, Cesena, Italy

Gustavo de Assis Costa
Department of Computer Science, Federal Institute of Education, Science, and Technology of Goiás, Jataí, GO, Brazil

Gerard Deepak
Manipal Institute of Technology Bengaluru, Manipal Academy of Higher Education, Manipal, India
National Institute of Technology, Tiruchirappalli, Tiruchirappalli, India

Antonio De Nicola
ENEA, Centro Ricerche Casaccia, Rome, Italy

Moses E. Ekpenyong
Department of Computer Science, University of Uyo, Uyo, Nigeria
Centre for Research and Development, University of Uyo, Uyo, Nigeria

R.R. Rubia Gandhi
Department of EEE, Sri Ramakrishna Engineering College, Coimbatore, India

Manas Gaur
Ohio Center of Excellence in Knowledge-enabled Computing (Kno.e.sis), Wright State University, Dayton, OH, United States
University of South Carolina, Columbia, SC, United States

Ayush Goyal
Department of Electrical Engineering and Computer Science, Texas A&M University, Kingsville, TX, United States

Amelie Gyrard

M3 (Machine-to-Machine Measurement), Paris, France

Trialog, Paris, France

Ohio Center of Excellence in Knowledge-enabled Computing (Kno.e.sis), Wright State University, Dayton, OH, United States

Funebi F. Ijebu

Department of Computer Science, University of Uyo, Uyo, Nigeria

Abinaya Inbamani

Department of EEE, Sri Ramakrishna Engineering College, Coimbatore, India

M.A. Jabbar

Department of Computer Science and Engineering, Vardhaman College of Engineering, Hyderabad, Telangana, India

Utkarshani Jaimini

Ohio Center of Excellence in Knowledge-enabled Computing (Kno.e.sis), Wright State University, Dayton, OH, United States

University of South Carolina, Columbia, SC, United States

Normunds Kante

Latvian Biomedical Research and Study Centre, Riga, Latvia

Janis Klovins

Latvian Biomedical Research and Study Centre, Riga, Latvia

Antonio Kung

Trialog, Paris, France

Thirumeni Mariammal

Rajalakshmi Engineering College, Chennai, India

Anastasija Nikiforova

University of Tartu, Tartu, Estonia

Latvian Biomedical Research and Study Centre, Riga, Latvia

Fernando Ortiz-Rodriguez

Autonomous University of Tamaulipas, Victoria, Tamaulipas, Mexico

M. Preethi

Department of IT, Sri Ramakrishna Engineering College, Coimbatore, India

R. Rajalakshmi

Department of IT, Sri Ramakrishna Engineering College, Coimbatore, India

Vita Rovite

Latvian Biomedical Research and Study Centre, Riga, Latvia

A. Siva Sakthi
Department of BME, Sri Ramakrishna Engineering College, Coimbatore, India

Saeedeh Shekharpour
University of Dayton, Dayton, OH, United States

Amit Sheth
Ohio Center of Excellence in Knowledge-enabled Computing (Kno.e.sis), Wright State University, Dayton, OH, United States
University of South Carolina, Columbia, SC, United States

Deepak Surya S.
Manipal Institute of Technology Bengaluru, Manipal Academy of Higher Education, Manipal, India
National Institute of Technology, Tiruchirappalli, Tiruchirappalli, India

Francesco Taglino
IASI-CNR, Rome, Italy

Krishnaprasad Thirunarayan
Ohio Center of Excellence in Knowledge-enabled Computing (Kno.e.sis), Wright State University, Dayton, OH, United States

Sanju Tiwari
Autonomous University of Tamaulipas, Victoria, Tamaulipas, Mexico

Kommomo J. Usang
Department of Computer Science, University of Uyo, Uyo, Nigeria

Patience U. Usip
Department of Computer Science, University of Uyo, Uyo, Nigeria

Veerapandi Veerasamy
Dept. of Electrical Engineering, University of Putra Malaysia, Serdang, Selangor, Malaysia

Rita Zgheib
Canadian University Dubai, Dubai, United Arab Emirates

Acknowledgments

At the outset, it gives me a great pleasure in expressing my sincere regards and thanks to my co-editors, Professor Fernando Ortiz-Rodriguez and Dr. M.A. Jabbar and the potential authors for making this a successful project. I had a great experience during the whole time with the series editor, Professor Fatos Xhafa and the Elsevier team. I gratefully acknowledge the attention, the suggestions, and friendly treatment that they gave me through the communication that it takes between all of us. In fact, thanks are not enough to express my gratitude and appreciation toward all. I am grateful to my family and friends for their endless support, motivation, and encouragement. I shall fail in my duty if I do not mention the name of my lovely son, Saransh, for his patience throughout the entire work. As it is my first edited book, I would like to dedicate this book to my dear mother, Mrs. Prabha Mishra (Sr. Retired Teacher of the Govt School), who always inspired me to do the hard work and taught the lesson to never give up.

Sanju Tiwari
(PhD, Post-Doc), SMIEEE

Semantic modeling for healthcare applications: an introduction

Sanju Tiwari[a], Fernando Ortiz-Rodriguez[a], and M.A. Jabbar[b]

[a]*Autonomous University of Tamaulipas, Victoria, Tamaulipas, Mexico*
[b]*Department of Computer Science and Engineering, Vardhaman College of Engineering, Hyderabad, Telangana, India*

Contents

1.1 Introduction

The Healthcare domain needs many health devices linked to the internet due to the emerging growth of Internet of Things (IoT) technologies. For the initial diagnosis and prevention, the Internet of Things plays an essential role in analyzing health issues and generating alerts with the help of IoT devices. IoT technologies play a significant role in providing efficient and effective solutions in health information system areas. The Healthcare domain is crucial for people and organizations to get the services and the delivery of services simultaneously. In this area, a massive collection of domain knowledge is gathered by the medical devices that efficiently conducts the activities for remote patient monitoring in a dynamic way. Different devices can be used to collect the patient data, such as monitoring wellness, elderly care, chronic diseases, fitness activities, and assuring sustainability. These devices (diagnostic, medical imaging, and sensor devices) are significant and effectively used for diagnosing, medicating, and treatment. Different types of sensing devices such as blood pressure monitors, glucose meters, thermometers, and heart rate sensors can be the best examples of IoT devices in health care. The stored and analyzed health data gathered from devices are later used for querying and generate alerts by defining rules under emergency conditions. These applications provide IoT-based medical services for reducing the cost, monitoring the patients, reducing healthcare providers' time, and increasing patient guidance.

These sensing devices can exchange the collected data (user information, health sensing data, history, device information, or other domain data). During the interchange of data by IoT devices, it is necessary to ensure the confidentiality and privacy of the data and exchange the data without losing its meaning. Semantic web technologies such as ontologies and rules are the best effort to maintain the consistency of data. These technologies have proven to be a challenging task for the exploitation and integration of data from different sources. IoT-based applications are included in semantic web technologies for providing sensor-based services and remote monitoring by linking the end-users. There is a need to present the semantic modeling of data with IoT devices description. IoT devices generate massive data every day by observations and actuations; every device has its own information and it becomes challenging to access the information by different applications. This problem can be resolved by involving semantics to reduce the ambiguity and increase the reusability of data for a significant decision [1].

Semantic modeling generalizes the entities with their description and depicts the relationship among entities to structure the data with its meaning. It is presented as a conceptual model to include semantic data representation with its meaning and the possible relationship between them. In the semantic web approach, data is modeled and organized to advance the knowledge base and maintain data consistency as information is updated remotely. A single entity, word, or data cannot interpret the context or meaning to humans, but linked entities can always provide the meaning with a context. For example, the context of data in a database is expressed mainly by its structure but not focused on its meaning, while in semantic modeling, the structure is described inherently with the meaning and description of data. The hierarchy of classes and subclasses can describe semantic models as data classification, and classes are related to the properties to generalize the structure. This abstraction of classification provides strong readability and interpretation to both humans and machines, that was the semantic web's central aspect. Ontologies are presented as the backbone to achieve the goal of the semantic web to define the concepts explicitly and presented as a semantic model. Abstractions can be presented in a semantic data model as

- Classification – "instance_of" relations
- Aggregation – "has_a" relation
- Generalization – "is_a" relation

Recently, various ontology-driven healthcare systems have been leveraged by the IoT technologies that provide the opportunities to enhance patient monitoring and detection of the abnormal situation with the help of medical wearables and cloud infrastructure. It has been observed that elderly people and some patients require tele-homecare; hence, homecare systems are emerging in daily life. This chapter will discuss the semantic modeling of data with IoT techniques in healthcare with sensing devices such as sensors, meters, etc. to monitor patients. These devices are helpful to collect the data and pre-process the collected data for extracting knowledge; the extracted knowledge later is semantically modeled in a knowledge base. Semantic modeling of the knowledge base is further helpful for semantic representation and semantic reasoning. In healthcare, the physician and patients are connected with heterogeneous IoT devices to provide semantic interoperability. The patient data set is semantically modeled with the metadata information using semantic technologies like RDF, RDFs, OWL, SPARQL, etc., and the data of any patient can be extracted any time from any place by the IoT devices. Semantic modeling with IoT has presented a new approach known as the Semantic Web of Things (SWoT). Ontologies play a significant role in dealing with the massive amount of data by a semantic representation of the domain knowledge. Several ontologies with IoT-based applications in healthcare have been proposed recently.

The rest of the paper has been organized as follows: Section 1.2 discusses the related work on existing semantic models in healthcare, Section 1.3 focuses on the background to highlight the semantic web technologies, web-of-things, semantic web-of-things, etc. Section 1.4 discusses the semantic modeling of data with different phases and examples. Finally, Section 1.5 concludes the chapter.

1.2 Literature review

The Internet of Things became a growing technology in healthcare for monitoring patients remotely and discovering new drugs. The quality and efficiency of healthcare have robust features such as adaptability, flexibility, cost shrinkage, affinity, and high speed. Considerable work has been done with semantic interoperability in healthcare and is discussed in this section.

Titi et al. [2] presents a fuzzy-ontology-based system with the IoT for the remote monitoring of diabetic patients. The primary focus of this work is to propose the ontology-based model and the semantic fuzzy decision-making mechanism. They evaluated the proposed model by query answering and it resulted in the effective monitoring for patients.

Saad et al. [3] presented an ontology-based framework with IoT technologies for the continuous monitoring of the patient's status remotely.

Chiang and Liang [4] proposed a context-aware, smart, homecare system for the tele-homecare of persons under care. Under supervision, the proposed system keeps the essential contexts in knowledge ontology, including the person's environmental and physiological information. The sensing devices enable the person under care to interact with the system through gestures.

Selvan et al. [5] proposed a fuzzy ontology-based recommender system using Type-2 fuzzy logic to advise drugs and diet for chronic disease patients. Sensing devices like wearable sensors and IoT-based medical records are used to extract the patients' risk factors and is connected with Linked Open Data (LOD) to generate a public knowledge base.

Reda et al. [6] proposed a semantic data model to evaluate the information from different data sources with domain-specific or generic data sets and merged them into an interconnected data space with IoT health data sources. This model enables logical reasoning and automatic inferencing and significantly provides the reusability of existing sources.

Subramaniyaswamy et al. [7] proposed a health-centric recommender system, ProTrip, that can recommend food availability by taking climate attributes based on the user's choice and the nutritious value of food. The proposed recommender system has been designed over the ontological knowledge base by the semantic modeling of heterogeneous user profiles and their descriptions. The proposed model has been evaluated for real-time IoT-based healthcare systems.

Moreira et al. [8] proposed SAREF4health, an ETSI Smart Appliances REFerence (SAREF) IoT ontology for real-time electrocardiography (ECG) to describe the need of representing time series. It follows ontology-driven semantic modeling to develop the ECG ontology in the Unified Foundational

Ontology (UFO) as a reference model. SAREF4health is efficient in endowing semantic interoperability of IoT techniques that require handling frequency-based time series.

Jabbar et al. [9] proposed an IoT-based Semantic Interoperability Model (IoT-SIM) to present semantic interoperability between heterogeneous IoT devices in the healthcare domain. A physician can remotely monitor a patient's health status with the help of heterogeneous IoT devices. All entities' descriptions have been semantically annotated and interpreted with its meaning by users and devices. They have used RDF and RDFs to design the semantic model of healthcare entities and relate them as triple (subject, predicate, object) to make a semantic representation. SPARQL queries are used to extract the information.

Kim et al. [10] designed a medical lifelog ontology (MELLO) by extracting the lifelog concepts and their relationships among these concepts. It follows the ontology-development methodologies and is semantically annotated to provide a clear definition of each concept. The main aim of MELLO is to support the categorization and semantic mapping of lifelog data from heterogeneous health self-tracking devices.

Rhayem et al. [11] have designed a HealthIoT ontology to model the information of connected devices and knowledge of medical things on the internet. This semantic model also offers semantic interoperability with healthcare devices and medical equipment. This semantic model has been evaluated with SWRL rules and inbuilt reasoners.

Jin and Kim [12] presented a semantic model to promote semantic interoperability with the server, e-Health objects, and clients. This model also reused Semantic Sensor Network (SSN) to deal with the sensing devices' interoperability issues. This ontology has been created to organize the e-Health information by reusing SSN and assessed only by the reasoner at modeling time.

Linked Health Resource (LHR) [13] ontology is designed using semantic web technologies to describe the healthcare information. This ontology has reused the SSN/SOSA IoT-based model to connect with IoT devices. The reasoning and SPARQL queries have evaluated the proposed ontology to analyze the completeness of modeled knowledge.

1.3 Background

This section discusses semantic web technologies and basic terminologies, Internet-of-Things (IoT), Semantic Web-of-Things (SWoT), and IoT-based general-purpose and domain-specific semantic models.

1.3.1 Semantic web and terminology

The semantic web has been introduced as Web 3.0, an extension of WWW. The primary aim of the semantic web is to present the web of data rather than documents. It enables machines to interpret the meaning and link one resource to other resources with their unique address. The semantic web contains some specific standards as its technology, such as RDF, RDFs, OWL, and SPARQL. The Resource Description Framework (RDF) was presented as a fundamental element of the semantic web to create a statement for an entity or resource. The RDF statement is described with three parts, subject-predicate-object (SPO). A collection of RDF statements is also known as RDF triples [40]. The RDF became a W3C standard in 2004 to present labeled, directed, and typed graphs. RDF provides the facility for linking and publishing the data [41]. RDF is found appropriate with OWL. RDFs are an extension over RDF to offer more power to increase the expressivity of attributes of the classes. RDF and OWL, both standards empower the web for sharing data and documents with machines as well as humans, which provides efficient searching and makes it easy to reuse information all over the web [40]. SPARQL is the way to query, extract, and update the stored data in the RDF format on the semantic web. SPARQL plays a vital role in navigating different databases and explores the relationships among data. A few query languages that include SeRQL [39] and RQL [38] have been presented for RDF throughout the years, but SPARQL became the standard language for querying with RDF data [37]. SPARQL is the most commonly used graph-based standard query language to extract and update data stored in the RDF format.

1.3.2 Web-of-Things

The Internet of Things (IoT) enables connecting the devices to provide autonomous behavior and guarantee privacy, trust, and security. The IoT presents as a robust research area with distributed methods from different aspects for gaining popularity and promoting IoT development methods. The expansion of the IoT is the Web of Things (WoT) that is presented as the open web standards to support machine interoperability and information sharing. The conventional web services are empowered with physical world services by carrying IoT into the present web [36]. This aspect of WoT presents a new practice of bridging the gap between physical and virtual worlds. This is possible only with the help of Hyper Text Transfer Protocol (HTTP) and Constrained Application Protocol (CoAP) and application programming interfaces (API). Hence, applications have the advantage with the HTTP protocol to facilitate the platform to publish data updates into machines for fetching data updates by the machines and exchanging the information.

1.3.3 Semantic Web-of-Things

The Semantic Web of Things (SWoT) is the current research to assimilate the WoT to the semantic web. The main motive of SWoT is to offer the convergence of IoT that is presented as the addition of semantics into Web of Things (WoT) that provide the sharing of ontologies and data with machine interoperability. There are many issues in scalability; heterogeneity and interoperability arise from various interconnected machines [33]. The SWoT assures coherent extension to integrate the IoT with the digital and physical world and concentrates on providing robust level interoperability that permits to reuse and share of things with semantics [34]. Ontologies are a powerful tool to give semantic interoperability among systems. They make able machines to provide full interoperability with the semantics of data to be shared for all devices. Ontology can be managed by machines and express the concept definitions and restrictions with possible understandings to design a rich structure of the specified domain [35].

The semantic analytics is a growing initiative to present the reasoning on Linked Open Data (LOD), it is a way to derive meaningful information from IoT data that helps to process the data in an interoperable manner to derive new knowledge from existing facts [55]. Semantic analytics combines multiple semantics technologies and analytic tools such as machine learning, LOD, and logic-based reasoning. The main aim of semantic analytics is to interpret data into actionable knowledge and provide reasoning along with reasoning approaches and semantic web technologies. SPARQLStream was presented as a novel approach for querying and accessing streaming data sources [56]. SPARQLStream has been presented as an extension to the SPARQL 1.1 query language to manage the real-time sensor data [57].

1.3.4 IoT-based semantic models

IoT is a vast area where devices are deployed to observe the event or action of interest. The existing IoT-based system is generally restricted to the use of sensor data to a single application. The integration of IoT in semantic models provides the reusability of sensor data in different applications and raises productivity with strong heterogeneity and interoperability. Several semantic models are designed to deal with different forms of sensor data and apply the reasoning with gathered sensor data. In this section, general-purpose and domain-specific IoT-based semantic models [32,1] have been discussed to highlight their features. Semantic technologies have been frequently applied in several domains, specifically to deal with the heterogeneity challenges such as association of data, derive new knowledge to design smart applications and interoperability at data processing, storage, and management. The healthcare domain requires serious attention to apply the semantic interoperability in smart healthcare services. Various frameworks have been presented that use fuzzy semantic models and contain different layers to conduct different activities such as heterogeneous health records storage, mapping of local semantic models to global ones by using algorithms, and interface to interact with the domain experts.

1.3.4.1 General-purpose IoT semantic models

Recently, several IoT-based semantic models have been presented for different purposes such as observations and actuations, context modeling, and construction and engineering. Some of the most widely used semantic models are discussed: om-lite [31] is a model-based semantic model on presenting the observation schema. The om-lite allows integrating data explicitly and indicating to observations framed remotely. These observations help to estimate the data quality and are also significant for data discovery. It does not describe any class for the feature of interest or observed qualities.

SSN/SOSA

The SSN/SOSA [30,29] model has been designed by W3C Spatial Data on the Web Working Group and later it was considered as a W3C recommendation. SOSA is an extension to include (sensor, observation, sample, and actuator) concepts. It follows the ODP-based design principle for sensor-based classes to allow multiple ways of creating observable properties.

SAREF ontology

SAREF (Smart Appliances REFerence) [28] is a shared semantic model for the smart appliances domain. This model describes the core concept of the "Device" class, designed for actions, functions, and states. SAREF also describes different categories of devices like actuators, sensors, HVAC systems, etc. SAREF is perfect for refining the standard semantics acquired in the ontology and frame new concepts.

Time

The time semantic model [27] is the most widely used model for temporal entities. It is presented as a W3C recommendation for describing the temporal resources. This model represents the entities about topological relations between time intervals and also date-time information.

QUDT

QUDT semantic model [26] is proposed by NASA to organize the concepts of units of measure, quantities, dimensions, and types. Quantity is the core concept of the QUDT model to measure a specific event, object, or physical device observation.

Table 1.1 Domain-specific semantic models.

Ref.	Domain	Semantic models	Online status	Reused models
[11]	Healthcare	HealthIoT	No	NA
[8]	Healthcare	SAREF4health	Yes	SSN, SAREF
[12]	Healthcare	e-Health	No	SSN
[6]	Healthcare	IFO	No	NA
[13]	Healthcare	Linked Health Resource (LHR)	No	SSN, SOSA
[21]	Healthcare	SHCO	Yes	SSN/SOSA, SAREF
[43]	Building and Construction	BOT	Yes	NA
[44]	Building and Construction	ThinkHome	Yes	NA
[45]	Building and Construction	IoT-O	Yes	SSN
[46]	Building and Construction	saref4bldg	Yes	SSN/SOSA, SAREF
[47]	Water	saref4watr	Yes	SSN/SOSA, SAREF, GeoSparql
[49]	Water	WaterNexus Ontology	Yes	SAREF
[48]	Water	SWIM	No	NA
[50]	Water	xLMINWS.owl	Yes	dbpedia, time, ssn
[51]	Agriculture	saref4agri	Yes	SAREF, SOSA, TIME, GEOSP, SSN
[25]	Agriculture	Agri-IoT	No	SSN, AGROVOC

1.3.4.2 Domain-specific IoT-based semantic models

IoT plays an essential role in providing semantic interoperability with the events and physical devices in a specific domain. Exciting work has already been done to make several domain-specific IoT-based semantic models. Table 1.1 has presented several domain-specific semantic models such as healthcare, waste, water, agriculture, building and construction, etc.

1.4 Semantic modeling of data

Semantic modeling is a process that organizes the knowledge more explicitly by using RDF and other W3C standards; this knowledge can be queried and visualized for multiple decision-making tasks. This section will discover main ideas such as semantic annotation, semantic linking, semantic representation for semantic data modeling.

1.4.1 Semantic annotation

An annotation is a form of metadata that is attached to a part of a document such as a named entity, a paragraph, or the complete document [15]. HTML documents are organized only for the representation of text to form a syntactic web that is document-oriented in nature. A semantic annotation provides

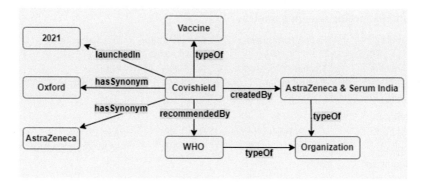

FIGURE 1.1

Semantic annotation of data.

linking some ontology-based metadata to define the semantics in a document and appears as a data-oriented web called the semantic web. Ontologies are considered as formal conceptualizations of a particular domain [53] and referred to as a semantic model to present the meaningful data.

Semantic annotation [54] is a process of assigning relevant information to the concepts and their relationships. Semantic annotation enriched the documents with metadata information of each entity to describe the meaningful data. The main motive to annotate the entities with metadata information is to interpret by both humans and machines, reduce the ambiguity, increase reusability rather than be created from scratch. The integrated metadata information helps improve the accuracy of the classification of the machine learning algorithm during the classification process. An integrated approach of the feature selection and semantic annotation is helpful to deal with the heterogeneous and extensive healthcare data in biomedical AI systems. Semantic annotation provides a unified structure of data extracted from different sources for promoting data integration. According to Pacha et al. [23,22], semantic annotation provides a more detailed description, and a semantic context supports data analytics and intelligent querying.

For example, "Covishield" is a type of Vaccine while "WHO" is an instance of an Organization, and it will be described in the semantic model by the annotation of concepts. Both concepts will be associated with the relationships like "typeOf", "recommendedBy", "createdBy". Fig. 1.1 shows the complete description of semantic annotation and all named entities are annotated with meaningful information such as "Covishield" is a type of Vaccine and recommended by WHO and launched in 2021. The machines will easily interpret this annotation to write the inferences for reasoning purposes and provide richer expressivity to the end-users.

1.4.2 Semantic linking

Semantic linking is the process to hold the relationship between two different resources and find the information of a concept available in various resources [15,24]. Semantic links help identify the description of two objects that refer to the common entity like the same Person or Place. For example, the "Sensor" concept is available in different semantic models; saref1, saref4bldg2, ssn3, which is se-

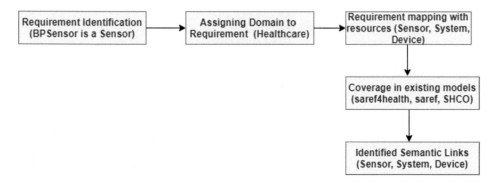

FIGURE 1.2

Flow of semantic linking.

mantic linking and is found among all three models and can be reused for the same purpose. Semantic linking provides strong reusability of existing concepts and reduces the ambiguity of resources as each resource has a unique URI, and it can be easily reused rather than be created from scratch. Semantic linking can be performed by specific steps such as requirement identification, finding domains, mapping requirements with existing resources, coverage of requirements in resources, and identifying the semantic links among different resources in a specific domain. Semantic linking aims to identify the association between existing semantic models and needs to analyze the availability of resources before developing any new semantic model. For example, in the healthcare domain, if you need to have concepts like BloodPressureMeter, GlucoseMeter, BPSensor, etc., "Sensor" or "System" can be reused as a superclass of all types of measurement concepts. A "BPSensor" can be a subClassOf the "Sensor" concept, and Sensor already exists in several models such as, saref,[1] saref4bldg,[2] and ssn.[3] Fig. 1.2 shows the flow of semantic linking.

In Fig. 1.2, the semantic linking process is discussed in five steps. In the first step, a requirement has identified "BPSensors is a Sensor" where "BPSensor" is a device to measure blood pressure, and it is related to healthcare; hence, in the second step it needs to assign the domain "Healthcare" to the identified requirement. In step 3, it needs to map the requirement with existing resources such as "BPSensor" that is a device, and it can be a subclass of Sensor, System, or Device class that may be already created in existing semantic models (saref, sref4health, SHCO, etc.). It is found that there is a semantic linking between three models (saref, saref4health, SHCO) for a particular concept, "BPSensor," as a subclass of Sensor, System, or Device that are available in existing models and can be reused to model any new subclass of it.

---

[1] https://saref.etsi.org/.
[2] https://saref.etsi.org/saref4bldg/Building.
[3] https://www.w3.org/TR/vocab-ssn/.

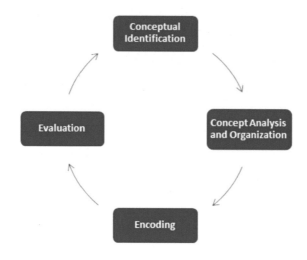

FIGURE 1.3

Construction methodology.

1.4.2.1 Construction of a semantic model

After finding semantic linking and creating a semantic annotation, a semantic model can be easily designed to model the meaningful information of a particular domain. There are several methodologies [17] such as On-To-Knowledge [19], DILIGENT [18], METHONTOLOGY [52], NeOn [16], and tools (Protégé, OntoEdit) to create the semantic model by using the W3C standard. By following these methodologies, Mishra et al. [20] has proposed a construction methodology to create a semantic model for the military domain. This methodology can be applied to construct a new semantic model that generally follows the four major steps: conceptual identification, concept analysis and organization, encoding, and evaluation. In this model, military concepts are identified and organized in a hierarchy to model it semantically using OWL, RDF, RDFs, and later evaluated by querying SPARQL queries with the protégé tool. OOPs! also evaluates this model tool to find the pitfalls [42]. Fig. 1.3 shows the flow of the construction process.

Tiwari et al. [21] extended the work to construct a smart healthcare ontology (SHCO) as a semantic model for organizing the healthcare information and IoT devices. This model has proposed a combined approach to evaluate the designed model. SHCO is modeled by using the Protégé tool and evaluated four different tools (Test-Driven Development (TDD)onto, Themis, Protégé, and OOPs!) has used to validate and verify the designed model. For verification, TDDonto and Themis are used to evaluate the test-cases or requirements while OOPs! and Protégé supported to validate the modeled knowledge in ontology. This model is available for reuse as it has been published online, and anyone can use it. SAREF4health (Moreira et al., 2020) is a healthcare semantic model and leveraged by an iterative and interactive approach in a transparent way. Requirement collection is the first step to implement and validate the ontology as a semantic model to create this model. Several sources are analyzed to collect requirements such as standards, data sets, APIs, specifications, data formats, and expert ideas in the healthcare domain. In the second step, use cases are comprised for finding related concepts and

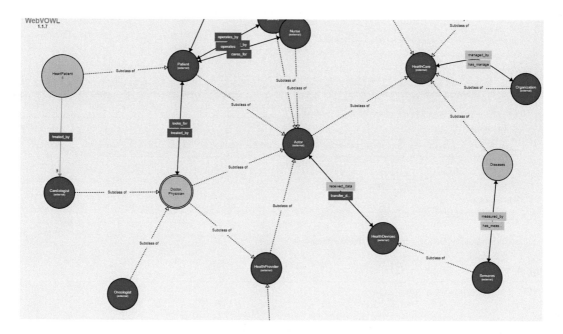

FIGURE 1.4

Semantic representation on WebVOWL.

their relationships. In the third step, the scope and purpose of the ontology have been defined for the specific-use cases by writing competency questions.

1.4.3 Semantic representation

The World Wide Web as a syntactic representation of text is offered by HTML, and later extended by the semantic web as a semantic representation of data with its meaning and a unique address. Semantic representation [14] does not present text as text only, while it is represented as a thing with an individual address and meaningful information. It follows the data-centric approach to represent the real-world entities and conceptualize the entities explicitly. After modeling data in ontology, it needs to represent semantically. Semantic models can be represented online and offline with different tools like OWLGrEd, Gruff, WebVOWL, and online publication using Widoco. Some of the representations are discussed here for the SHCO model; it is uploaded on WebVOWL and represented as a graph in Fig. 1.4.

WebVOWL is presented as a web application for the interactive representation of ontologies in the online mode. Ontologies can be uploaded or a URI link can be provided for the semantic representation of entities. It presents a visual representation for OWL ontologies as a graph representation for ontology entities associated with a force-directed graph structure. The WebVOWL representations are generated from JSON files automatically.

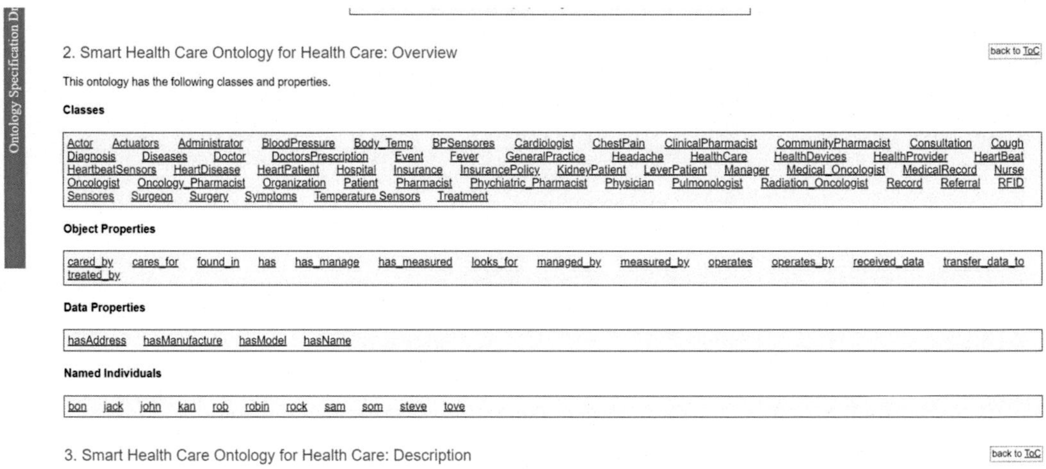

FIGURE 1.5

SHCO representation as a specification.

The SHCO is published online and has an ontology specification draft on the web (http://w3id.org/ def/SmartHealthCare). Every concept and relationship of this model has defined by a unique address such as "Actor," a class defined explicitly with a unique IRI (https://w3id.org/def/SmartHealthCare# Actor) and presented as a resource on the web. Object and data properties are also defined in the same way as classes and have a unique and explicit IRI for object property "cared_by" (https://w3id. org/def/SmartHealthCare#cared_by) and data property "*hasAddress*" (https://w3id.org/def/sanjutiwari# hasAddress).

In this semantic representation, every attribute has its own space and address on the web, so it is easy to reuse any existing attribute rather than create a new one. This representation is called ontology publications and is shown in Fig. 1.5. This representation shows classes, object properties, data properties, and instances as an ontology specification. All entities have a unique address as a link to represent it, and this link can be reused anywhere with its IRI.

The SHCO model has also been presented hierarchically for classes and their subclasses. The protégé tool is a standard tool to use for semantic modeling and semantic representation of data. It also provides inferencing with different reasoners such as Hermit++, Racer, etc. Fig. 1.6 shows a hierarchical representation of the SHCO. This model shows the classes and their subclasses of the healthcare domain that has two major classes "HealthCare" and "HealthDevices." HealthCare further has the subclasses of Actor, Diseases, Event, Organization, Record, and HealthDevices also has subclasses Actuators, Radio Frequency Identification (RFID), and Sensors.

OntoGraph plugin is good to represent in graph structure. Fig. 1.7 shows graph representation using OntoGraph plugin on protégé.

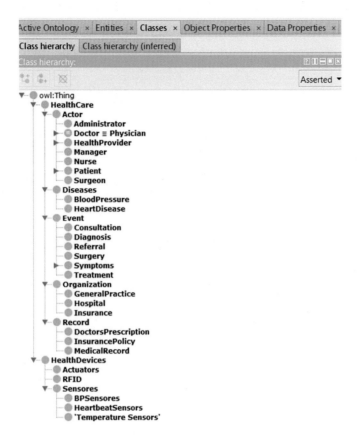

FIGURE 1.6

Hierarchical representation of the SHCO semantic model.

1.5 **Conclusions**

Various semantic-driven healthcare systems have been designed with devices interoperability by leveraging IoT technologies. This chapter has discussed several existing semantic models such as Saref4health, HealthIoT, IoT-SIM, MELLO, the LHR model, SHCO, and many more. We are also concerned about the semantic web technologies (RDF, RDFs, OWL, SPARQL, etc.), web-of-things, semantic web-of-things, and IoT-based semantic models (general-purpose and domain-specific models). Semantic modeling of data is the central aspect of this chapter, and it covers several phases such as semantic annotation, semantic linking, constructing semantic models, and semantic representation. Several case studies and examples have been discussed to present semantic annotation, semantic linking, construction methodology, and semantic representation. Two types of semantic representation (online and offline) have been presented to show different semantic models' hierarchical and graph representation. The SHCO (Semantic Healthcare Model Ontology) model shows the various representation forms such as a graph, hierarchical, and an ontology specification draft.

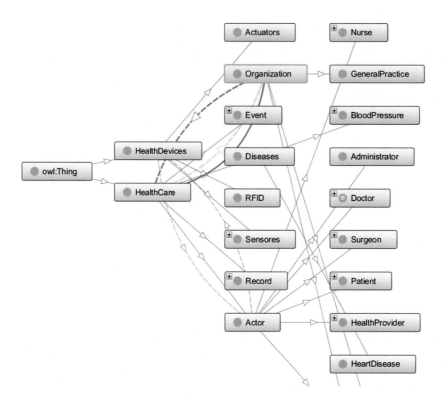

FIGURE 1.7

Graph representation of the SHCO semantic model.

References

[1] Iker Esnaola-Gonzalez, Jesús Bermúdez, Izaskun Fernandez, Aitor Arnaiz, Ontologies for observations and actuations in buildings: a survey, Semantic Web Journal 1 (29) (2020), Preprint.

[2] Sondes Titi, Hadda Ben Elhadj, Lamia Chaari Fourati, A fuzzy-ontology based diabetes monitoring system using Internet of Things, in: International Conference on Smart Homes and Health Telematics, Springer, 2020, pp. 287–295.

[3] Sabah Saad, Bassam A. Zafar, Ahmed Mueen, Developing a framework for e-healthcare applications using the semantic internet of things, International Journal of Computer Applications 182 (34) (2018) 25–33.

[4] Tzu-Chiang Chiang, Wen-Hua Liang, A context-aware interactive health care system based on ontology and fuzzy inference, Journal of Medical Systems 39 (9) (2015) 1–25.

[5] N. Senthil Selvan, Subramaniyaswamy Vairavasundaram, Logesh Ravi, Fuzzy ontology-based personalized recommendation for internet of medical things with linked open data, Journal of Intelligent & Fuzzy Systems 36 (5) (2019) 4065–4075.

[6] Roberto Reda, Filippo Piccinini, Antonella Carbonaro, Semantic modelling of smart healthcare data, in: Proceedings of SAI Intelligent Systems Conference, Springer, Cham, 2018.

[7] V. Subramaniyaswamy, Gunasekaran Manogaran, R. Logesh, V. Vijayakumar, Naveen Chilamkurti, D. Malathi, N. Senthilselvan, An ontology-driven personalized food recommendation in IoT-based healthcare

system, Journal of Supercomputing 75 (6) (2019) 3184–3216.

[8] Moreira João, Luís Ferreira Pires, Marten Van Sinderen, Laura Daniele, Marc Girod-Genet, SAREF4health: towards IoT standard-based ontology-driven cardiac e-health systems, Applied Ontology 15 (3) (2020) 385–410.

[9] Sohail Jabbar, Farhan Ullah, Shehzad Khalid, Murad Khan, Kijun Han, Semantic interoperability in heterogeneous IoT infrastructure for healthcare, Wireless Communications and Mobile Computing (2017), 10 pages.

[10] Hye Hyeon Kim, Soo Youn Lee, Su Youn Baika, Ju Han Kim, MELLO: medical lifelog ontology for data terms from self-tracking and lifelog devices, International Journal of Medical Informatics 84 (12) (2015) 1099–1110.

[11] Ahlem Rhayem, Mohamed Ben, Ahmed Mhiri, Faiez Gargouri, HealthIoT ontology for data semantic representation and interpretation obtained from medical connected objects, in: 2017 IEEE/ACS 14th International Conference on Computer Systems and Applications (AICCSA), IEEE, 2017, pp. 1470–1477.

[12] Wenquan Jin, Hyeun Do Kim, Design and implementation of e-health system based on semantic sensor network using IETF YANG, Sensors 18 (2) (2018) 629.

[13] Cong Peng, Prashant Goswami, Meaningful integration of data from heterogeneous health services and home environment based on ontology, Sensors 19 (8) (2019) 1747.

[14] Kabilan Vandana, Paul Johannesson, Semantic representation of contract knowledge using multi tier ontology, in: SWDB, 2003, pp. 395–414.

[15] Nathalie Pernelle, Fatiha Saïs, Daniel Mercier, Sujeeban Thuraisamy, RDF data evolution: efficient detection and semantic representation of changes, in: Semantic Systems – SEMANTiCS2016, 4, 2016.

[16] Asunción Gómez-Pérez, Mari Carmen Suárez-Figueroa, NeOn Methodology for Building Ontology Networks: A Scenario-Based Methodology, Demetra EOOD, 2009.

[17] Mariano Fernández López, Asunción Gómez-Pérez, Juan Pazos Sierra, Alejandro Pazos Sierra, Building a chemical ontology using methontology and the ontology design environment, IEEE Intelligent Systems & Their Applications 14 (1) (1999) 37–46.

[18] Helena Sofia Pinto, Steffen Staab, Christoph Tempich, DILIGENT: towards a fine-grained methodology for DIstributed, loosely-controlled and evolvInG Engineering of oNTologies, in: ECAI, 16, 2004, p. 393.

[19] Steffen Staab, Rudi Studer, H-P. Schnurr, York Sure, Knowledge processes and ontologies, IEEE Intelligent Systems 16 (1) (2001) 26–34.

[20] Sanju Mishra, Sarika Jain, An intelligent knowledge treasure for military decision support, in: Research Anthology on Military and Defense Applications, Utilization, Education, and Ethics, IGI Global, 2021, pp. 498–521.

[21] Sanju Tiwari, Ajith Abraham, Semantic assessment of smart healthcare ontology, International Journal of Web Information Systems (2020).

[22] M. Manonmani, Sarojini Balakrishnan, A review of semantic annotation models for analysis of healthcare data based on data mining techniques, in: Emerging Research in Data Engineering Systems and Computer Communications, Springer, 2020, pp. 231–238.

[23] Shobharani Pacha, Suresh Ramalingam Murugan, R. Sethukarasi, Semantic annotation of summarized sensor data stream for effective query processing, Journal of Supercomputing 76 (6) (2020) 4017–4039.

[24] Nathalie Pernelle, Semantic enrichment of data: annotation and data linking, PhD diss., Université Paris Sud, 2016.

[25] Andreas Kamilaris, Feng Gao, Francesc X. Prenafeta-Boldu, Muhammad Intizar Ali, Agri-IoT: a semantic framework for Internet of Things-enabled smart farming applications, in: 2016 IEEE 3rd World Forum on Internet of Things (WF-IoT), 2016, pp. 442–447.

[26] Hajo Rijgersberg, Mark Van Assem, Jan Top, Ontology of units of measure and related concepts, Semantic Web 4 (1) (2013) 3–13.

[27] Jerry R. Hobbs, Feng Pan, Time ontology in OWL, W3C working draft, 27, 133, 2006.

[28] Laura Daniele, Frank den Hartog, Jasper Roes, Created in close interaction with the industry: the smart appliances reference (SAREF) ontology, in: International Workshop Formal Ontologies Meet Industries, Springer, Cham, 2015, pp. 100–112.

[29] Krzysztof Janowicz, Armin Haller, Simon J.D. Cox, Danh Le Phuoc, Maxime Lefrançois, SOSA: a lightweight ontology for sensors, observations, samples, and actuators, Journal of Web Semantics 56 (2019) 1–10.

[30] Armin Haller, Krzysztof Janowicz, Simon J.D. Cox, Maxime Lefrançois, Kerry Taylor, Danh Le Phuoc, Joshua Lieberman, Raúl García-Castro, Rob Atkinson, Claus Stadler, The modular SSN ontology: a joint W3C and OGC standard specifying the semantics of sensors, observations, sampling, and actuation, Semantic Web 10 (1) (2019) 9–32.

[31] Simon J.D. Cox, Ontology for observations and sampling features, with alignments to existing models, Semantic Web 8 (3) (2017) 453–470.

[32] Garvita Bajaj, Rachit Agarwal, Pushpendra Singh, Nikolaos Georgantas, Valérie Issarny, 4W1H in IoT semantics, IEEE Access 6 (2018) 65488–65506.

[33] Sanju Mishra, Sarika Jain, Chhiteesh Rai, Niketa Gandhi, Security challenges in semantic web of things, in: International Conference on Innovations in Bio-Inspired Computing and Applications, Springer, Cham, 2018, pp. 162–169.

[34] J. Antonio Jara, Alex C. Olivieri, Yann Bocchi, Markus Jung, Wolfgang Kastner, Antonio F. Skarmeta, Semantic web of things: an analysis of the application semantics for the iot moving towards the iot convergence, International Journal of Web and Grid Services 10 (2–3) (2014) 244–272.

[35] E. Sezer, Okan Bursa, O. Can, M.O. Unalir, Semantic Web Technologies for IoT-Based Health Care Information Systems, SEMAPRO, 2016.

[36] Deze Zeng, Song Guo, Zixue Cheng, The web of things: a survey, Journal of Communications 6 (6) (2011) 424–438, Citeseer.

[37] Waqas Ali, Muhammad Saleem, Bin Yao, Aidan Hogan, Axel-Cyrille Ngonga Ngomo, A survey of RDF stores & SPARQL engines for querying knowledge graphs, arXiv preprint, arXiv:2102.13027, 2021.

[38] Gregory Karvounarakis, Aimilia Magganaraki, Sofia Alexaki, Vassilis Christophides, Dimitris Plexousakis, Michel Scholl, Karsten Tolle, Querying the semantic web with RQL, Computer Networks 42 (5) (2003) 617–640.

[39] Heiner Stuckenschmidt, Richard Vdovjak, Jeen Broekstra, Geert-Jan Houben, Towards distributed processing of RDF path queries, International Journal of Web Engineering and Technology 2 (2–3) (2005) 207–230.

[40] Ora Lassila, Ralph R. Swick, Resource description framework (RDF) model and syntax specification, Citeseer, 1998.

[41] Jorge Cardoso, Amit Sheth, The Semantic Web and its applications, in: Semantic Web Services, Processes and Applications, Springer, Boston, MA, 2006, pp. 3–33.

[42] Sanju Mishra, Sarika Jain, Ontologies as a semantic model in IoT, International Journal of Computers and Applications 42 (3) (2020) 233–243.

[43] Mads Holten Rasmussen, Maxime Lefrançois, Georg Ferdinand Schneider, Pieter Pauwels, BOT: the building topology ontology of the W3C linked building data group, Semantic Web (2021) 1–19, Preprint.

[44] Christian Reinisch, Mario J. Kofler, Félix Iglesias, Wolfgang Kastner, Thinkhome energy efficiency in future smart homes, EURASIP Journal on Embedded Systems (2011) 1–18.

[45] Nicolas Seydoux, Khalil Drira, Nathalie Hernandez, Thierry Monteil, IoT-O, a core-domain IoT ontology to represent connected devices networks, in: European Knowledge Acquisition Workshop, Springer, Cham, 2016, pp. 561–576.

[46] María Poveda-Villalón, Raúl García-Castro, Extending the SAREF ontology for building devices and topology, in: Proceedings of the 6th Linked Data in Architecture and Construction Workshop (LDAC 2018), vol. CEUR-WS, 2159, 2018, pp. 16–23.

[47] Raúl García-Castro, SAREF extension for water, https://saref.etsi.org/saref4watr/v1.1.1/, 2020. (Accessed August 2021).

[48] Laurie Reynolds, Semantic water interoperability model community group, https://www.w3.org/community/swim/, 2016. (Accessed August 2021).

[49] Aitor Corchero, Eugene Westerhof, Lluis Echeverria, Water Nexus Ontology to support generation of policies, https://rioter-project.github.io/rioter-nexus-variables-ontology/, 2019. (Accessed August 2021).

[50] Ahmedi Lule, The InWaterSense Ontologies, https://raw.githubusercontent.com/lule-ahmedi/InWaterSense/master/Ontologies/lminws.owl/, 2015. (Accessed August 2021).

[51] Maria Poveda-Villalon, Quang-Duy Nguyen, Catherine Roussey, Christophe de Vaulx, Jean-Pierre Chanet, Ontological requirement specification for smart irrigation systems: a SOSA/SSN and SAREF comparison, in: 9th International Semantic Sensor Networks Workshop (SSN 2018), CEUR Workshop Proceedings, vol. 2213, 2018, p. 16.

[52] Mariano Fernández-López, Asunción Gómez-Pérez, Natalia Juristo, Methontology: From Ontological Art Towards Ontological Engineering, American Association for Artificial Intelligence, 1997.

[53] Thomas R. Gruber, A translational approach to portable ontologies, Knowledge Acquisition 5 (2) (1993) 199–229.

[54] Ontotext, Annotation is the process, to find, interpret and reuse, https://www.ontotext.com/knowledgehub/fundamentals/semantic-annotation/#:~:text=Semantic, 2021, Ontotext.

[55] Amelie Gyrard, Martin Serrano, Connected smart cities: interoperability with seg 3.0 for the internet of things, in: 30th International Conference on Advanced Information Networking and Applications Workshops (WAINA), IEEE, 2016, pp. 796–802.

[56] Jean-Paul Calbimonte, Oscar Corcho, Alasdair J.G. Gray, Enabling ontology-based access to streaming data sources, in: International Semantic Web Conference, Springer, 2010, pp. 96–111.

[57] Jean-Paul Calbimonte, Ontology-based access to sensor data streams, Informatica (2013).

Role of IoT and semantics in e-Health

2

Abinaya Inbamani[a], A. Siva Sakthi[b], R.R. Rubia Gandhi[a], M. Preethi[c], R. Rajalakshmi[c],
Veerapandi Veerasamy[d], and Thirumeni Mariammal[e]

[a]*Department of EEE, Sri Ramakrishna Engineering College, Coimbatore, India*
[b]*Department of BME, Sri Ramakrishna Engineering College, Coimbatore, India*
[c]*Department of IT, Sri Ramakrishna Engineering College, Coimbatore, India*
[d]*Dept. of Electrical Engineering, University of Putra Malaysia, Serdang, Selangor, Malaysia*
[e]*Rajalakshmi Engineering College, Chennai, India*

Contents

2.1 Introduction

The Internet of Things or IoT may be a system of interconnected measuring equipment, mechanical gadgets, or things with a unique identifier (UIDs). It has the ability to deliver information over a network without any help from human beings. In modern generations, the IoT has become a steaming telecommunication device that is achieving a reputation with scientists and specialties. Because of the interdisciplinary methods, the IoT has been extraordinary in changing several aspects of the ancient, standard healthcare. Internet of Things is a bridging platform, which combines the physical world and the cyberspaces [1]. At the same time, in deep analytics intention, the internetworks and design of the IoT for economical massive deployment have been needed to fulfill spaces of satisfactory service of quality, efficient fulfillment, and operations [2]. A lot of wireless technologies inventions lead to the production of more IoT devices. This integration will lead to diagnosis and treatment of illness. The IoT revolution has led to the drastic change in the healthcare industry. The middleware architectures play a major role in the integration of the heterogenous nodes. For a ubiquitous type of communication, the middleware is needed. Its helps in maintaining the communication between the various elements in the network. A detailed overview of MOM and SOA type of architecture is explained in this chapter. Section 2.1 gives a brief overview on the connectivity of IoT objects. Section 2.2 explains the various middleware architectures for IoT in e-Health. Section 2.3 details more on the interoperable environment with respect to the Internet of Things in the healthcare industry.

2.2 Internet of Things in the healthcare industry

The Internet of Things (IoT) is gaining fame and achievement in many fields like agriculture, oil corporations, defense, healthcare, and transportation. In particular, healthcare is one of the quickest developing fields within recent years. It affects the complete sphere of society in the healthcare industry. Moreover, new industries have been concerned with numerous e-Health utilizations and they are attempting to exploit it [3]. Sharp alertness concerning health and fitness is the most important issue that drives the healthcare industry. The user having the Internet of Things gadgets can discard the clinic visits and high-rated specialists avoiding long queues in hospitals. There is an outstanding development within the usage of Internet of Things gadgets like wearables and implantable appliances.

The customers have their data in their mobile phones, and the major challenge lies in the interpretation of the data. The Internet of Things in e-Health has hugely expanded in various fields to manage important instrumental devices, patient-care, medical assets, track devices usage, etc. The IoT sensor network should be scalable, interoperable, and heterogeneous. Fig. 2.1 gives an overview of the basic connection between IoT data sources and IoT applications via middleware.

2.2.1 Characteristics of IoT ontology and challenges in e-Health

The IoT is machine-to-machine interaction between objects, devices, and people. Nowadays, most of the communications and information processing is performed by IoT systems. Data processing has changed with the introduction of IoT. It is difficult to operate between different types of data sources and formats due to the size of the data, change in velocity, different variety of data sources, and data formats. To describe concepts and relationships between different entities in various domains, the semantics community has developed ontologies. With IoT, it is difficult to process the real time data due to the increase in the number of sensors and multiple parameters. Lightweight models with a minimum number of concepts and relationships are used to improve the processing time of IoT data.

The healthcare sector consists of many parameters to process and analyze, which in turn creates a huge amount of data that is difficult to process and maintain a database. IoT plays a crucial role in the medical sector to process such information efficiently. Before the invention of the IoT, patient's interactions with doctors were limited to hospital visits. The doctor could not monitor the patient continuously. IoT enables devices, makes remote monitoring of patients possible, and thereby has transformed the healthcare sector in a huge way. The beneficiaries of IoT in the healthcare sector are patients, physicians, hospitals, families, and insurance companies. This reduces the hospital stay and prevents

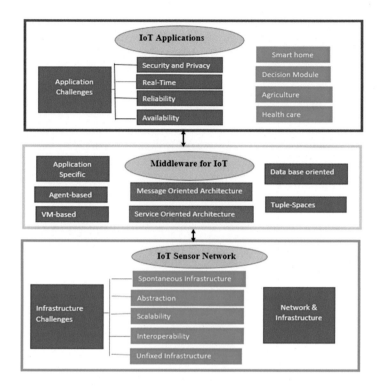

FIGURE 2.1

Connection between IoT data sources and IoT applications via middleware.

readmission as they are monitored remotely on a continuous basis. The stages of IoT solutions in the healthcare industry include data acquisition, data collection and pre-processing, data storage, and data analysis. The various stages of the IoT in the healthcare sector is shown in Fig. 2.2.

2.2.2 Characteristics of IoT in the healthcare industry

The main goal of IoT is to provide an independent system, which is capable of sharing information between uniquely identifiable real world objects using RFID tags. The common characteristics of IoT in the healthcare industry is shown in Fig. 2.3.

Device connection

With everything going on with technology, there needs to be a connection to be established between sensors and other electronics and connected hardware and control systems. This is an important characteristic of the IoT in the healthcare sector and it is the foremost process to establish good connectivity before data acquisition.

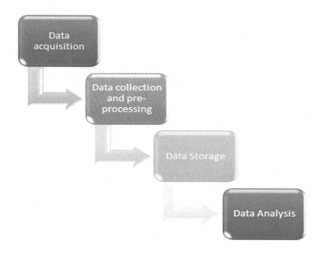

FIGURE 2.2

Stages of IoT in the healthcare sector.

FIGURE 2.3

IoT and its characteristics.

Data sensing

In the healthcare industry, there is a huge amount of data being generated day to day, and this data needs to be sensed and managed effectively for future use. Data sensing can be done using various types of sensors like the mechanical sensor, electrical sensor, chemical sensor, biological sensor, etc. Each sensor acquires different physiological parameters, which are essential to identify the patient's condition. The basic data collected from the patient includes general information like the name, date of birth, mobile number, age, gender, family history, previous history, etc. and other physiological data such as blood pressure, temperature, pulse rate, oxygen saturation, ECG, etc. to know about the patient's health condition.

Data collection and communication

Data is the important component of the IoT, and it is the first step toward development of any system. The data generated during various procedures can be collected and stored for further processing and communication. The communication between the device and data can be established either over short-distance or long-distance communication pathways. The technology used for the communication of data from the nodal center to the user-end is through WIFI, LoRa, NBIoT, etc. These communication pathways should be secured, so that information loss will not occur.

Data Analysis

Various tools are available in the market for performing data analysis. These tools perform predictive analysis of patient information, and predict patient illness accurately. They can extract clinical parameters such as age, gender, and other physiological information, define data elements from the database, and create statistical models to respond to various outcomes and hypotheses for treatment, diagnosis, and research. Data analysis involves the following steps such as descriptive analysis, diagnostic analysis, predictive analysis, and prescriptive analysis. By performing data analysis, one can track the individual practitioner's performance, and can track the details of healthy people and be able to identify the people at the risk of chronic disorders at the earliest possible.

Data value

The data gathered from the big data analytics and the sensing capabilities in IoT devices plays a major role in identifying the data value. Data value is the consequence of intelligence in the medical field to diagnose and treat the disease appropriately. This can be either a manual action or the action based upon the debates regarding phenomena and automation, and is often the most important piece of the IoT in the healthcare industry.

Human value

Human value is what other technologies, communities, and goals give the place of the IoT. The Internet of Everything dimension, the platform dimension, and the need for solid partnerships is to be considered while incorporating the IoT in the medical field.

2.2.3 Challenges of the IoT in e-Health

Application of the IoT in e-Health is tremendous, as it helps to connect patients, physicians, and hospitals at a wide range irrespective of its geographical location. There are however many challenges and research issues also that need to be more carefully addressed before implanting them in real time. The major challenges of the IoT in e-Health is depicted in Fig. 2.4.

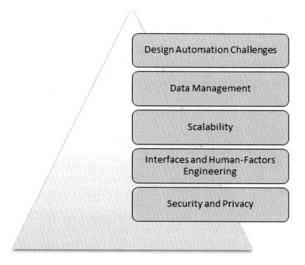

FIGURE 2.4

Challenges of the IoT in e-Health.

2.3 Middleware architectures for the IoT in e-Health

To overcome all of the challenges of the IoT in e-Health middleware is essential. The middleware solution offers a technological framework that allows two or more systems to communicate with one another. Its historical function has been mainly done to secure the transmission of a message from one subsystem to another with varying degrees of coupling. Various middleware architectures are projected to solve the application challenges and the infrastructure difficulties in the IoT, which are considered critical issues and are needed to be addressed before providing a final solution in healthcare [4]. This chapter gives an overview of various techniques stressing their features and problems.

2.3.1 An overview of IoT middleware solutions

The overview of the IoT middleware solutions is necessary to understand its impact when implemented in real time. Razzaque et al. [5] offer a specific study among a lot of others. The authors of this study examined many existing solutions in depth and provided a current overview of the various forms of middleware architecture. Heinzelman et al. [6] and Espeholt et al. [7] provide another categorization of IoT middleware methods.

Application-specific

The application specific method focuses on resource management and it is based on the demands of a specific application. As a result, there is a tight connection between application and data sources, resulting in the need for specialized middleware. This middleware technique is shown by MidFusion. On behalf of applications, the technique finds and picks the optimal set of sensors or sensor agents. The best set of sensors are picked using the algorithms in MidFusion techniques. The sensors utilizing the Bayesian methods are used to choose the optimal network ensuring that Quality of Service is the main drawback in the MidFusion technique. The MiLAN type of middleware technique offers a policy to integrate the data and the sensor network. The heterogeneity characteristics are not specified in the MiLAN architecture.

Agent-based

The agent or modular-based method divides applications into modular programs to facilitate network distribution thereby utilizing mobile agents. They may be emphasized by offering decentralized systems and are capable of addressing middleware's reliability, availability, and resource management needs. The complexity of a middleware architecture can reduce methods like Agila, Impala, and smarty messages. However, it has certain drawbacks, such as the inability to conduct code management duties and the unpredictability of system agents during runtime.

Tuple-spaces

It is an information repository that may be accessed by several people at the same time. Each device from the physical layer is represented as a tuple space in the middleware. On the gateway, all of the tuple spaces combine to produce a federated tuple space. This technique is ideal for the IoT devices because it allows them to communicate data in real time while staying within gateway connectivity restrictions. Intended for sensor networks and mobile ad hoc networks, TinyLime and TeenyLime are tuple-space middleware solutions. They solve frequent disconnections and asynchronous communications issues yet has a flexible design that allows middleware to be utilized in a variety of contexts. They do, however, have drawbacks in terms of resource management, scalability, security, and privacy. Virtual machines (VMs), mobile agents, and interpreters are all part of adaptable strategy. The middleware is made up of two layers based on this technique. In the first layer of the middleware, every physical device is deployed as a virtual machine. A generic VM understands the modules in the second layer and gives data to the application that expresses its demands through a query. This method supports self-management transparency in distributed heterogeneous IoT infrastructures while addressing architectural needs such as high-level programming abstractions and adaptivity. However, the methodology introduced by the swapped instructions makes this technique ineffective.

Database-oriented

The entire sensor network is viewed as a distributed and virtual database in this middleware approach. It collects target data across the network using SQL-like queries. While this technique provides strong programming abstraction and data management support, other IoT needs such as scalability, real-time, and spontaneous interactions are not addressed. Furthermore, its centralized design makes it difficult to manage the IoT network's dynamic and heterogeneous features. The designs listed above have been utilized in a variety of IoT applications and research areas. However, in the last decade, there has been a significant increase in IoT projects. The types of middleware architectures are explained in Table 2.1.

Service-Oriented Architecture

Two major IoT trends have arisen in recent years: first, hardware is becoming smaller, more powerful, and second, the software industry is shifting toward service-oriented integration solutions. It is a technique of thinking about and building information systems that have been utilized in corporate IT systems for a long time. Smart sensors are shown as services for consumer applications in SOA-based

Table 2.1 Types of middleware architectures

Types of middleware	Techniques	Advantages	Drawbacks
Application-specific	MidFusion [5]	Best of sensors are selected from the group	Quality of service is provided only for limited algorithms
	MiLAN [6]	Specific Policy is used for the integration technique	Heterogeneity characteristics of the nodes are not addressed
Agent-Based	Impla (Importance Weighted Actor-Learner Architecture) [7]	Scalable to thousands of machines	Updates in the policy can lag behind the rule
	SmartyCat [8]	Navigation techniques are supported	Control over the sensors is provided by agencies
	Agilla [9]	Self-adaptive	Suitable for low power consumed devices
Tuple-spaces	TinyLime [10]	Flexible architecture	Resource management is constrained
	TeenyLime [11]	Flexible architecture and self-adaptive	Scalability is less
Database-oriented	GSN (Global Sensor Network) [12]	Data management support is provided	Less scalable and not suitable for heterogenous type of communication.
Service oriented architecture	SStreaMWare, USEME, SensorWeb	Management of heterogeneous devices is assured	Suitable for low power consumed devices
	SenseWrap [13]	Sensors are virtualized and supports resource discovery	Actuators cannot be virtualized
	TinySOA [14]	Supports resource management	Complex algorithms are not supported
	SensorsMW [15]	Strict Quality of service rules is followed	Reconfiguration of sensor nodes are difficult
	MOSDEN [16]	User friendly architecture	Service discovery is difficult
Message Oriented Middleware	Event based architecture [17]	Components and applications interact with each other through events	Decoupling mechanism is absent
	Publish subscribe model [18]	Decoupling mechanism is assured	Broker needs to be synchronized with the publisher and subscriber

IoT techniques. The service interface is separate from the implementation and is crucial. Providers characterize their services (sensor attributes) and place them on the market. Providers explain their services (sensor properties) and make them available to customers. The standard for describing services is Web Services Description Language (WSDL).

In [8], the state of the art of SOA-related middleware for wireless sensor networks is declared and defined; it includes SStreaMWare, USEME, SensorWeb 2.0, OASiS, and others. Middleware proposals include B-VIS, MiSense, SOMDM (SI), and SOA-MM. Wireless sensor networks have been proposed for finding the solution. Their primary features include real-time monitoring, heterogeneous device

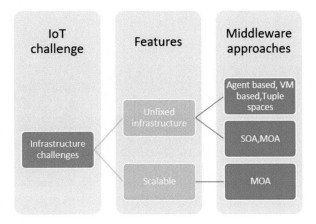

FIGURE 2.5

Middleware approaches for the infrastructure challenge.

management, data collecting, and filtering. However, the assessment reveals that none of these systems address all of the needs for sensor network administration in intelligent settings.

Message Oriented Middleware

MoA has long been used in network communications, notably in industrial networks such as integrated production systems. It has an event-based design with a publish/subscribe communication model. Components, programs, and all other applications interact through events in event-based architecture. The publish/subscribe paradigm is used to send events from the sending-to-the-receiving application components. The publish/subscribe model is a form of interaction that involves both publishers and subscribers. The various middleware approaches with the infrastructure challenge and application challenge is shown in Fig. 2.5 and Fig. 2.6.

2.3.2 Middleware solutions in healthcare

An IoT middleware must confirm the variety of data sources and services in order to guarantee the availability of services and statistics sources; hence the data should be available to the end user. As the healthcare applications are sensitive, in the event of a failure of data, an IoT middleware should continue to function. To ensure high-level system dependability, each component of the middleware must be validated. For many IoT applications in healthcare, instantaneous information distribution is important. In those situations, delayed data may be both worthless and hazardous. A delayed notice of a fall monitoring application can result in a person's death.

Many gadgets communicate with one another, and a lot of statistics are shared. These statistics might be private, personal, or even information about one's everyday activities. As a result, trust, security, and privacy are critical concerns that must be addressed in any IoT middleware solution. All of these middleware development projects have the same goal: to provide a framework that can enable a plug-and-play adaption layer. Ambient Assisted Living (AAL) and medical applications are two types of

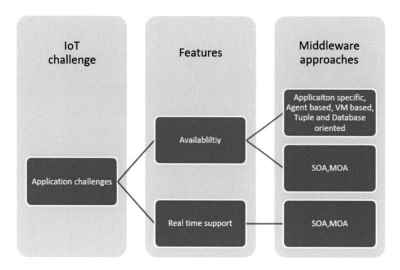

FIGURE 2.6

Middleware approaches for the application challenge.

IoT applications that are currently being researched. Medical applications are mostly used in retirement homes and hospitals to detect, prevent, and even monitor patient health.

AAL apps are often placed in a patient home to track and monitor his or her health condition and activity. Middleware solutions have been developed in this area to meet the needs of the IoT and healthcare fields. The Sphere project, which uses a clustered-sensor method, is included in this strand [19]. Its goal is to provide a general platform that fuses sensor data to create rich data sets that may be used to diagnose and manage a variety of health problems. Radhika and Malarvizhi [20] propose another monitoring method in which body sensors are connected to a smartphone through a Bluetooth to gain the data such as heart rate and body temperature. Its purpose is to combine data from various wireless sensors into a network of body sensors mediated by a smartphone.

A case study is considered wherein the blood glucose sensor, heartbeat sensor values are measured and sent to the Intel Galileo processor board. The choice of board varies depending on the user needs. With the common connectivity framework, the board is connected to the broker model. The message queuing telemetry transport is the publish subscriber model used in this case study. The broker model preferred is muzzley. The model is then connected to the mobile application thereby heling in remote monitoring. The MQTT models have various message types as shown in Fig. 2.7.

The Intel Galileo boards are connected to the device gateway. The device registry, authorization, and authentication of the nodes are the initial steps undertaken before connecting it to the gateway. The rules engine helps in transforming the messages to AWS services, and the device shadow enhances the persistent connection between the things (Fig. 2.8).

MESSAGE TYPE	DESCRIPTION
CONNECT	Client request to connect to Server for connection
CONNACK	Connection Acknowledgement sent to the client
PUBLISH	Message which represents a separate publish
PUBACK	QoS 1 Response to a PUBLISH message
PUBREC	First part of QoS 2 message flow
PUBREL	Second part of QoS 2 message flow
PUBCOMP	Last part of the QoS 2 message flow
SUBSCRIBE	Message used by clients to subscribe to specific topics
SUBACK	Acknowledgement for a SUBSCRIBE message
UNSUBSCRIBE	A message used by clients to unsubscribe from specific topics
UNSUBACK	Acknowledgement of an UNSUBSCRIBE message
PINGREQ	Heartbeat message
PINGRESP	Heartbeat message acknowledgement
DISCONNECT	Graceful disconnect message sent by clients before disconnecting.

FIGURE 2.7

Message types of MQTT protocol.

2.4 Semantic interoperability of objects in e-Health

The IoT device connectivity with the middleware is assured thereby enhancing the semantic interoperability with the nodes. Heterogeneous devices need to be interacted in the e-Heath industry to ensure interoperability among the devices. This interaction allows semantics between the objects ensuring connectivity and continuity of services. In the internet era, a lot of devices remain connected to the internet. Ubiquitous access to data is mandatory as the devices are dependent on each other. Depending on the value obtained from one device, the other device reacts. This event driven mechanism poses a major challenge on the design of the devices as they are heterogeneous and interoperable among each other [21]. The heterogeneous nature of the devices is due to the variety in the operating system, hardware architecture, and logic behind the system. The Intelligent Information System will in turn address these

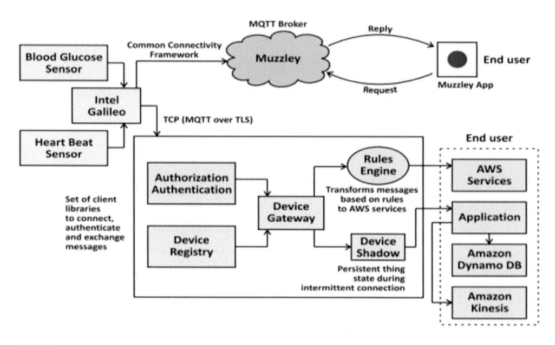

FIGURE 2.8

IoT device connectivity with the middleware.

issues and bring about a seamless integration. e-Health is providing such a paradigm to the users enhancing and improving the data communication and integration between the end user and the doctor. The emerging services is shown in Fig. 2.9.

Hence, semanticity in e-Health support telemedicine, and consumer health informatics align with the mobile devices supporting the healthcare system. The two computers on either end interact with each other and interpretation should be better on the receiving end. Presently, the challenge is more on the document management and its interpretation along with the security. The semantic models proposed by Ogden and Richards stand ahead when compared to the other models. The models use three important components in the semantic triangle, namely symbol, concept, and referent. For the concept to be understood in a better way, let us consider the semantic triangle as shown in Fig. 2.10 with an example.

Doctor Chandran asks Doctor Raghul whether he remembers the pill prescribed for the patient Mano for the strange fever. The pill indicates the symbol. The pill can be either generic or exclusive for a particular disease. The concept of understanding about the pill by Doctor Chandran will lead to a suggestion of the medicine. This conceptual understanding will lead to analysis of the symbol giving an answer. Doctor Chandran has to attach the referent along with the symbol to enhance the correctness of the answer. He can send a message to Doctor Raghul inculcating the symbol and the referent. He could send a message such as, "Can you please send me the pill in which you prescribed to the patient Mano for the strange fever last month"? The doctor will interpret the message correctly and send the pill as required. This way of semanticity is needed as the application domain is critical as it may endanger the

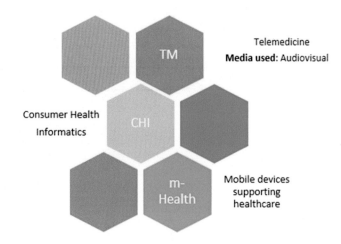

FIGURE 2.9

Emerging services.

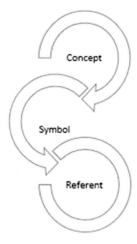

FIGURE 2.10

Semantic model.

life of a human being. There can be two patterns incorporated in the semanticity. The patterns can be either point-to-point or web semantic as shown in Fig. 2.11.

The modeling of messages is the first step in the model prescribed. The messages may be a text, number, structured data, or unstructured data. The message should hold an entity and a fact to understand the reference of the model. Once the referent has good clarity, the interpretation of the results is also correct. UFR and LFR are the various facts and references present with respect to universal and local. To ensure proper communication of the messages, the following steps in Fig. 2.12 have to be

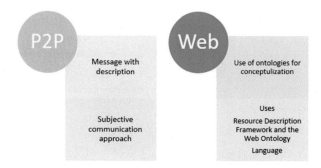

FIGURE 2.11

Types of patterns.

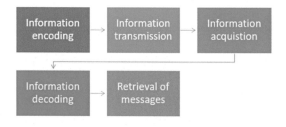

FIGURE 2.12

Steps to transfer the message from one end to another end.

carried out. The figure gives a good understanding of the concepts along with its tag and facts. The messages are validated and then understood by the end-user [22].

2.5 Interoperability in healthcare

Interoperability drives the enhancement medium of the operation, which is going to work with the data and information in the e-Health environment. Looking into the complexity in the e-Health sector, the major outcome is the patient needs to benefit in a faster way without facing the hurdles, which he faces in the real time scenario. That needs to be made effectively possible by integrating the interoperability tasks as shown in Fig. 2.13.

The standardization of the healthcare information can be made by interfacing the integration with interoperability. The major concern in the application entity is to perform the task [23]. Healthcare is a complex domain with many vertical organizations in which many horizontal processes are to be completed to have virtual growth. This makes most healthcare companies investigate semantic interoperability and invest in it.

FIGURE 2.13

Flow of interoperability in healthcare.

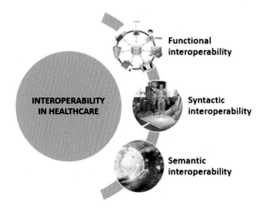

FIGURE 2.14

Levels in interoperability.

2.5.1 Interoperability levels

The three main levels in the interoperability are:

- Functional interoperability
- Syntactic interoperability
- Semantic interoperability

Functional and Syntactic Interoperability

End user is the one who finally judges the operability. The capacity for the treatment can be improved due to the shortened distance. This also increases the e-data entry in clinical records rather than focusing for the manual entry of the patients and other relevant data. With the functional interoperability, the cost of care, which includes reductant testing on the patients can be reduced. This also avoids the development testing and integration costs through the usage of commonly developed standards in implementing it as shown in Fig. 2.14.

Table 2.2 Increasing capability of interoperation Turnista model hierarchy.

Category	Levels	Interoperability	Performance in the content delivery
Interoperable	LEVEL 5	DYNAMIC	Dynamic content understood
	LEVEL 4	PRAGMATIC	Context Understood
Integrable	LEVEL 3	SEMANTIC	Meaning Understood
	LEVEL 2	SYNTACTIC	Common format
Interfacable	LEVEL 1	TECHNICAL	Common and Physical transport layers
	LEVEL 0	NONE	Standalone

Semantic operability

This semantics make the syntactic distinct by getting the information to be exchanged as computer processable so there is no end-user community involved in the system. All the shared and retrieved information are a pre-processed one, as it can be directly fetched and used [24].

2.5.2 Turnista's model

In the healthcare sector, the increasing capability of interoperation can be achieved by adapting different models of IEEE standard. The model discussed here is the Turnista model. In the healthcare domain like this model, there are many models that can be adapted for the efficient increase in interoperability. From this model, it can be understood that the semantic and syntactic levels need to be integrated, which benefits the end-user. For these syntactic levels, HL7 CDA (Clinical Document Architecture), IEEE 11073, DICOM are some examples can be used for enriching the levels in the best way. For this semantic level enhancement, it can be enriched using Snomed -CT, IHD-PCD RIM (Reference Information Model), IEEE 11073, LOINC, etc. (Table 2.2).

Regarding the interaction between two or more computers when exchanging information, the corrections are automatically interpreted by receiving systems, which needs to be assured by interoperability. By interfacing the computational facility with the interconnected units, the functional interoperability has become a major challenge [25].

2.5.3 Semantic web

The overall semantic interoperability can be efficiently improved with the advent of the semantic web. Knowledge extraction from the shared ontologies on the semantic annotations helps in metadata prediction. Machine-to-machine level transition for automatic interpretation of information can be made as machine interpretable markup by software agents. As was discussed earlier, for the clinical documents, CDA and RIM are some reference models for making the documents machine readable and can be processed and constructed electronically. From there, the user can take use of the document. For the clinical documents, the exchange model HL7 (Health Level 7) CDA can be used for exchanging the information. Using the HL7 RIM (Reference Information Model) with the CDA, the information is machine coded, readable, and pooled in the semantic web. This information can be accessed by all the users. For the message received by the end-user, the local semantic ontology vendors need to re-

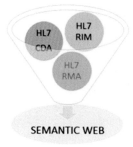

FIGURE 2.15

Semantic web ontology.

construct the semantic ontology using RMA (Referent Mapping Activity) [28]. Anyhow, the message needs to be sent with the appropriate meaning. For this condition to be satisfied, overall subjectivity of the information shared and processed has to be maintained by centralized agents, which need to periodically update the web. The overall system can be made user friendly. Anyhow, the semantic ontology needs maintenance and the scalability in the semantic web should be protected and prevailed by the vendors as shown in Fig. 2.15.

2.5.4 OpenEHR archetype

Some of the older proposed interoperability in healthcare is EHRs which is the Electronic Healthcare Records. EHRs are defined as "digitally stored healthcare information about an individual's lifetime with the purpose of supporting continuity of care, education and research, and ensuring confidentiality at all times" [26]. Simply put, it can be defined as collection of clinical related data about an individual's lifetime in the form of document structure. The same HL7 CDA, EHRcom and openEHR are some standards that can be followed for healthcare interoperability. OpenEHR is a developing archetype, which ensures semantic interoperability. They are developed by domain knowledge governance tools like CKM (Clinical Knowledge Manager). According to the statistics given by CKM, discussing openEHR foundations, there are nearly 227 numbers of archetypes for ensuring semantic interoperability [27].

The openEHR is creative in the field for holding semantic interoperable EHRs. It has presented the execution of the task. It plans to give another plan of action for the electronic clinical records. The most recent release of Microsoft's Connected Health Framework incorporates openEHR (ISO 13606-2) models as a component of its do-primary information design. The openEHR Reference Model depends on ISO and CEN EHR norms and is interoperable with HL7 (Health Level Seven) and EDIFACT (Electronic information trade for organization, business, and transport) message guidelines [28]. This empowers openEHR-based programming to be coordinated with other programming and frameworks.

2.6 Conclusion

The IoT in e-Health is a cross-domain as it needs the knowledge of biomedical engineering, electronics, data management, and network design [29]. Therefore, it is necessary to have a multidisciplinary knowledge. The data that is collected from the healthcare sector is heterogenous and it is difficult to process it [30]. The IoT in e-Health is highly related to human beings; therefore, it is important to design a system that has a role of the human and machine interfacing system. In healthcare sector, IoT faces a major challenge in maintaining the data collected from each and every sensor, which is a huge amount of data every minute. These data can be stored in different formats for easy accessing in the database. Maintaining such a huge amount of data needs standardization, such that the data can be stored and accessed with high resolution. To integrate the healthcare sector with the IoT, all users need to have direct access to medical services through their handheld mobile phones. These services need their own sensors and integrating systems to function efficiently and to have a secure browser to process the user request. This arrangement can be scaled up to a small level, i.e., at the hospital level or at the bigger level, i.e., accessibility of data throughout the city. The user must be trained to use those devices effectively. Expert involvement needs to be minimized through use of patient-friendly interfaces. To make the device more comfortable, one possible approach is to include the stakeholders in the feedback process. This has a significant concern especially in the field of healthcare as data corruption may lead to a serious threat to the patient. [31]. Security has to be ensured through the data fed into the database and viewed by means of queries [32].

References

[1] Gorrepati, Rajani Reddy, Do-Hyeun Kim, Sitaramanjaneya Reddy Guntur, Ubiquitous Internet of Medical Things on Interoperable Ssn Ontology Platforms for e-Health Monitoring of Connected Objects, 2021.

[2] T. Poongodi, Balamurugan Balusamy, P. Sanjeevikumar, Jens Bo Holm-Nielsen, Internet of Things (IoT) and e-Healthcare system – a short review on challenges, IEEE India Info 14 (2) (2019).

[3] Jayashree R. Prasad, Shailesh P. Bendale, Rajesh S. Prasad, Semantic Internet of Things (IoT) interoperability using Software Defined Network (SDN) and Network Function Virtualization (NFV), in: Semantic IoT: Theory and Applications, in: Studies in Computational Intelligence, vol. 941, 2021.

[4] Soma Bandyopadhyay, Munmun Sengupta, Souvik Maiti, Subhajit Dutta, Role of middleware for internet of things: a study, International Journal of Computer Science and Engineering Survey 2 (3) (2011) 94–105.

[5] Binbin Xu, et al., Mid-fusion: octree-based object-level multi-instance dynamic slam, in: 2019 International Conference on Robotics and Automation (ICRA), IEEE, 2019.

[6] W.B. Heinzelman, A.L. Murphy, H.S. Carvalho, M.A. Perillo, Middleware to support sensor network applications, IEEE Network 18 (1) (2004) 6–14.

[7] Lasse Espeholt, et al., Impala: scalable distributed deep-RL with importance weighted actor-learner architectures, in: International Conference on Machine Learning, PMLR, 2018.

[8] P. Kang, C. Borcea, G. Xu, A. Saxena, U. Kremer, L. Iftode, Smart messages: a distributed computing platform for networks of embedded systems, ACM Transcations on Embedded Computing Systems 47 (4) (2004) 475–494.

[9] Walter Tiberti, et al., A model-based approach for adaptable middleware evolution in WSN platforms, Journal of Sensor and Actuator Networks 10 (1) (2021) 20.

[10] Lalita Mishra, Shirshu Varma, Middleware technologies for smart wireless sensor networks towards Internet of Things: a comparative review, Wireless Personal Communications 116 (3) (2021) 1539–1574.

[11] P. Costa, L. Mottola, A.L. Murphy, G.P. Picco, TeenyLIME: transiently shared tuple space middleware for wireless sensor networks, in: Proceedings of the International Workshop on Middleware for Sensor Networks, ACM, 2006, pp. 43–48.

[12] Kanu Patel, Hardik Modi, Analysis and review on service-oriented-based IoT middleware, in: Emerging Technologies in Data Mining and Information Security: Proceedings of IEMIS 2020, Vol. 3, Springer, Singapore, 2021.

[13] P. Evensen, H. Meling, Sensewrap: a service oriented middleware with sensor virtualization and self-configuration, in: 2009 5th International Conference on Intelligent Sensors, Sensor Networks and Information Processing (ISSNIP), IEEE, 2009, pp. 261–266.

[14] Christos Theologou, Design and implementation of a distributed Web IoT application using microservice architecture for device control and measuring, PhD diss., University of Thessaly, 2021.

[15] Mohammadreza Parvizimosaed, et al., A containerized integrated fast IoT platform for low energy power management, in: 2021 7th International Conference on Web Research (ICWR), IEEE, 2021.

[16] Daniyal Alghazzawi, Ghadah Aldabbagh, Abdullah Saad AL-Malaise AL-Ghamdi, ScaleUp: middleware for intelligent environments, PeerJ Computer Science 7 (2021) e545.

[17] Nagender Aneja, Sapna Gambhir, Recent advances in ad-hoc social networking: key techniques and future research directions, Wireless Personal Communications 117 (3) (2021) 1735–1753.

[18] N. Brouwers, K. Langendoen, Pogo, a middleware for mobile phone sensing, in: P. Narasimhan, P. Triantafillou (Eds.), Middleware 2012, in: LNCS, vol. 7662, Springer, Heidelberg, 2012, pp. 21–40.

[19] Rabia Afzaal, Muhammad Shoaib, Data recoverability and estimation for perception layer in semantic web of things, PLoS ONE 16 (2) (2021) e0245847.

[20] J. Radhika, S. Malarvizhi, Middleware approaches for wireless sensor networks: an overview, International Journal of Computer Science Issues (IJCSI) 9 (3) (2012) 224–229.

[21] Nidhi Pathak, et al., HeDI: healthcare device interoperability for IoT-based e-Health platforms, IEEE Internet of Things Journal (2021).

[22] Hitha Alex, Mohan Kumar, Behrooz Shirazi, Midfusion: an adaptive middleware for information fusion in sensor network applications, Information Fusion 9 (3) (2008) 332–343.

[23] Wenyuan Yang, et al., A secure heuristic semantic searching scheme with blockchain-based verification, Information Processing & Management 58 (4) (2021) 102548.

[24] Bander A. Alzahrani, Secure and efficient cloud-based IoT authenticated key agreement scheme for e-Health wireless sensor networks, Arabian Journal for Science and Engineering 46 (4) (2021) 3017–3032.

[25] João Moreira, et al., SAREF4health: towards IoT standard-based ontology-driven cardiac e-Health systems, Applied Ontology 15 (3) (2020) 385–410.

[26] Nader Mohamed, Jameela Al-Jaroodi, Service-oriented middleware approaches for wireless sensor networks, in: 2011 44th Hawaii International Conference on System Sciences (HICSS), IEEE, 2011, pp. 1–9.

[27] Asuman Dogac, Tuncay Namli, Alper Okcan, Gokce Laleci, Yildiray Kabak, Marco Eichelber Escherweg, Key Issues of Technical Interoperability Solutions in eHealth and the RIDE Project, Software R&D Center, Dept. of Computer Eng., Middle East Technical University 06531.

[28] Maria Koutli, et al., Secure IoT e-Health applications using VICINITY framework and GDPR guidelines, in: 2019 15th International Conference on Distributed Computing in Sensor Systems (DCOSS), IEEE, 2019.

[29] Tarek Elsaleh, et al., IoT-stream: a lightweight ontology for internet of things data streams and its use with data analytics and event detection services, Sensors (Basel) 20 (4) (2020 Feb) 953.

[30] R. Pandey, M. Paprzycki, N. Srivastava, et al. (Eds.), Semantic IoT: Theory and Applications, Springer Nature, Switzerland, 2021.

[31] Farshad Firouzi, Mohamed Ibrahim, From EDA to IoT ehealth: promise, challenges and solutions, IEEE Transactions on Computer-Aided Design of Integrated Circuits and Systems (February 2018).

[32] I. Abinaya, D. Manivannan, Remote monitoring using wireless sensor node with IoT, Research Journal of Pharmaceutical Biological and Chemical Sciences 6 (3) (2015 May 1) 1480–1484.

Evaluation and visualization of healthcare semantic models

Anastasija Nikiforova[a,b]**, Vita Rovite**[b]**, Sanju Tiwari**[c]**, Janis Klovins**[b]**, and Normunds Kante**[b]

[a]*University of Tartu, Tartu, Estonia*
[b]*Latvian Biomedical Research and Study Centre, Riga, Latvia*
[c]*Autonomous University of Tamaulipas, Victoria, Tamaulipas, Mexico*

Contents

3.1 Introduction and motivation

Today, data is the most common and extensive resource, which can differ not only in terms of content, but also in terms of different structure and data sources from which data come. Data can be both structured, unstructured, or semistructured, as well as may vary in terms of the language, dictionary/vocabulary used (if any), units of measurements, while sources may vary from very traditional relational databases to the most striking and sometimes unexpected examples of the Internet of Things (IoT). Even our iron, refrigerator, and kettle can collect and exchange with the data. In the medical sector,

Semantic Models in IoT and eHealth Applications. https://doi.org/10.1016/B978-0-32-391773-5.00009-1

this is linked to a huge number of systems that communicate with each other and continuously increase the number of sensing devices such as those used in smart healthcare systems. These data should be collected not only by one device or system, but also processed and integrated with others, forming a network for their further use. At present, there is little clarity on the universal way of managing heterogeneous data effectively. On the other hand, it should be acknowledged that attempts to reach this objective are becoming more popular at both the universal and domain-specific levels, where semantic approaches come as a rescue to this interoperability problem incurred due to heterogeneous data, which can be both health history, sensing data, patient's data, device, especially sensor data from IoT device, and other types of data [1].

This problem is current and crucial in the context of (bio)medicine and (bio)medical studies, as data come from a variety of national and international sources and in different forms. As example, in the case of the Latvian Biomedical Research and Study Centre, these are: (1) (semi)internal data of the Latvian biobank, including data of Latvian Genome Centre, Latvian Genome Database, Centre of Disease Prevention and Control (SPKC), and cofunded enterprises, where the digital data are collected and processed, (2) doctorates and their surveys completed by hand, (3) external national and international systems such as registries or medical institutions coming in the form of both structured and unstructured data. These data are often difficult to integrate into the internal system, taking into account the different types of notation for different objects largely dependent on the system, person, or process that acquire and process the original data in the original source. These problems have a practical nature with regard to the Latvian Biomedical Research and Study Centre, which face challenges when the data need to be processed and entered into their system (data types violation, data formats, different nomenclatures, incorrect designs of systems, or databases used, etc.) for their analysis of patients and diagnosis, for data-based research that they carry out and their subsequent reuse by stakeholders involved. At the moment, a number of ad hoc approaches, which vary from one case to another, are utilized. Although it is difficult to complain about the current effectiveness of the research center concerned, which is sufficiently competitive and collaborates actively with many third parties, we argue that its efficiency could be greatly improved if more modern data management solutions were introduced.

This issue is also of theoretical nature and important to the scientific community, as there is no one universal solution or "silver bullet" at the moment. According to [35,36], the healthcare domain needs advances in current big data management approaches since there is a lack of fully Findable, Accessible, Interoperable, and Reusable, thereby complying with FAIR, data resources. This is a matter of great importance, particularly for institutions related to biobanks, and also because the European Union signed a joint declaration on cross-border access to the 1 million human genomes by the end of 2022 [54]. But, as [55] stress, handling data on a large, transnational scale does not go without challenges. The authors emphasize that researchers and clinicians will need remote access to human data outside the country's borders to assemble and manage very large cohorts or identify people with a rare phenotype. They stress that at present, each European country sets its own regulatory/legal framework for the processing of health and genetic data. Even more, genetic and associated data collected of that generated through healthcare are not shared widely compared to research data.

In practice, there are sometimes quite primitive and manual ad hoc data integration mechanisms. However, the most appropriate option for these trends, especially for larger data sets and larger organizations and medical-related practices, is the use of ontology and semantic models, which are based on them. They are particularly important for widely-studied topics, since ontologies standardize terminology and determine relationships between the elements under consideration, thereby facilitating researchers and combating unresolved issues.

A natural question to be asked is *whether there is a difference between typical widely-used ontology and those for the (bio)medical sector?* According to [29] unlike ontology developed for Artificial Intelligence (AI), which highlights formalism and first-order logic, domain ontology such as ontology developed for biomedical informatics prioritize data sharing and interoperability. Thus, the objective of these ontologies is to disseminate structured, controlled vocabularies and to support data annotation where experimental data are associated with terms in ontology [37]. This task-oriented approach to the development of ontologies is a distinguishing feature of healthcare and biomedical ontologies [38]. For example, the Human Phenotype Ontology (HPO), which is one of the most widely used ontologies for phenotypes, standardizes and promotes the consistent use of phenotype terminology, and determines the relationships between phenotypes at both semantic, logical, hierarchical levels [29,31]. Their suitability and increasing popularity can easily be explained by the emphasis on concepts of the modeling and hierarchical relationships between them, which is unquestionably the main feature of ontology. It plays a key role in data integration, knowledge representation, and decision support systems (DSS) [2].

Virtual reality (VR) is another emerging technological development in the healthcare domain [6]. Virtual reality applications provide the visualization tasks of treatment, disease analysis, diagnosis, and prevention. Semantic portals also play an important role to visualize the ontologies as a graphical representations and help to make the user-interface dashboards [64].

While the concepts and benefits of both semantic models and medicine-related ontologies have been covered in the previous chapters, this chapter takes a step further and looks at the visualization of ontologies. Although ontology can be seen as an important step toward more efficient management of heterogeneous data, where heterogeneity can appear both in the data and in the data sources, visualization is often seen as a facilitator and a tool of more efficient use of ontology. This section therefore poses the following research questions:

RQ1:   *What a visualization term stands for in the context of ontology and/or semantic models? What types/methods of visualization exist? And when and how could they be beneficial?*

RQ2:   *What specific aspects should the institution or particular project take into account to choose and implement healthcare ontology visualization? What specific features (if any) should affect choice? Does a silver bullet exist?*

RQ3:   *What is evaluation in the context of ontology/semantic models? What methods of this evaluation exist? What specific aspects should affect choice?*

In order to meet the aims, the chapter is organized as follows: Section 3.2 provides an overview of the concept of visualization, both in very general, ICT- and semantic models-related areas, as well as it classifies existing approaches and emphasizes the key cornerstones. Section 3.3 is devoted to evaluation of both an ontology itself and ontology visualization. Section 3.4 sets out initial results and discussion, and the paper concludes in Section 3.5 highlighting future directions for research and clarifying the primary contributions of this chapter.

3.2 Role of visualization: background
3.2.1 Visualization concept and its cognitive capacity

Let us define the overall role of visualization with a special interest in computer-aided artifacts. It is not a secret that visual information is easier and faster to capture, perceive, modify, or transform compared to textual information. This is related to our cognitive processes. Visualization externalizes knowledge and helps to share with them. This can lead to more successful solutions at very different levels, allowing users to get information that might otherwise be difficult to capture [3,7,9]. Even more, visual artifacts help to relieve the human working memory [5], thereby improving thinking capacity. This, however, can emerge new ideas [8,10]. Visual artifacts group relevant information by clarifying how and whether concepts are related [5], which reduces the effort that would otherwise be needed to seek elements to make problem-solving inferences [4]. In other words, according to previous studies [12,17,16], visualization can increase cognitive memory and processing resources, reduce the search for information both in terms of efforts needed from the human and the overall time for process to be completed. In addition, it is capable to improve the detection of patterns (if any), enable perceptual inference operations and attention mechanisms for monitoring, encoding information in a manipulable medium. It allows the processing of large volumes of information, keep an overview of the entire picture pursuing details, keep track of things, and perform an abstract representation of the situation using the omission and recoding of information.

3.2.2 Visualization in ICT, healthcare, and biomedical sectors

The above mentioned benefits are applicable to all the information we deal with in the real world and are particularly representative for human-computer interaction (HCI), where visualization is the main factor. This also applies to data and knowledge management, particularly in domain-specific fields other than ICT, such as the medical sector. And the larger the object of interest, such as a single database or system, the more valuable visualization becomes. It makes it easier to understand each particular element and its relationships with another element (if any) and to manipulate them more easily.

In the context of ICT, visualization was the subject of debate, even for classical databases, which we have been forced to use decades ago. More specifically, Jog and Shneiderman [57] highlighted that the exploring of large multiattribute databases is greatly facilitated by visual representation of information as it enables users to dynamically query the database. There are also known later examples, such as migration to graph databases [14], which has been achieved with the support of Neo4j and GraphQL. The built-in graph representation of data sets ensured better manageability and sustainability, not only in terms of access and use of data, but also in a compelling overview of data quality and data accuracy. Today it is even more so. More and more organizations cooperate and geographical borders are no longer a restriction even in times of the COVID-19 pandemic, where these borders can affect our daily activities. This also applies to the cooperation of sectors, with collaboration becoming increasingly interdisciplinary. This leads to a significant increase in the volume and diversity of data storage facilities, which as a result also increases the need to visualize data and relationships between multiple entities, thereby making them more manageable, actionable, easier to perceive, traceable, and reusable. In fact, it boosts the FAIRness of the data, more precisely their Findability, Accessibility, Interoperability, and Reusability [13], which become at least partly supported.

Given that each information system has the data they are dealing with, it is clear that the solutions allowing the parties involved to process these data, i.e., collect, process, etc., should not respect limitation on people skills and knowledge—they should be as user-friendly as possible and facilitate users' interaction with data storage in all possible ways. This also applies to semantic models that were considered suitable only for IT people some time ago. However, these days it is observed that they become a daily phenomenon, and more and more non-IT people use them because of the wide range of possibilities they provide, where in addition to more coherent and simpler navigation by moving users from one concept to another in the ontology structure. Another valuable feature is that, by establishing the relationships between the concepts, it enables automated reasoning of data. Such reasoning is easy to implement in a semantic graph database where ontology is used as semantic schemata [58]. In addition, ontology is easy to extend because relationships and concept matching are easy to add to existing ontology. As a result, this model develops with data growth without affecting dependent processes and systems if something goes wrong or needs to be changed.

However, although the list of benefits that could be gained from the use of ontology is impressive, sometimes people with no domain knowledge face challenges [2], so the facilitating techniques should help to tackle them. Some of these issues can be resolved by means of visualization aimed at improving the understanding of the overall state of the art, i.e., data and their interrelation, both within and outside the organization, their accuracy, correctness, and complexity as well as an existence of patterns. This problem is even more complicated when healthcare systems are under question (also in line with [19]).

The benefits of the visualization of data and information were seen in the (bio)medicine and (health)care-related sectors even before the concepts of semantic models and domain-specific ontologies have been involved. According to [40], information visualization techniques have great promises to improve the quality of medical care. Rind et al. [40] emphasize that the benefits extracted from data visualization are observed in (a) analysis of patient medical history, including monitoring the overall condition of the single patient, (b) analyzing a patient's response to a treatment, and (c) aggregating multiple patient records in the cohort as a baseline to compare a development of individual's symptoms, etc. Visualization tools, even those, developed years ago, i.e., different types of charts and plots, were useful decision-making aids for clinicians when prescribing treatment for acute or chronic patient problems. These visualizations were often focused on complete Electronic health records (EHR), which could be very heterogeneous and a combination of both structured, semistructured, and unstructured data. However, while they were widely used and recognized as sufficiently efficient for general cases, there was little focus on more specific healthcare domain-related visualization and a lack of more advanced solutions has been noticed.

According to Iakovidis and Smailis [59], the diversity of data and information sources and the increasing amounts of data produced by different sources continuously, pose significant challenges in data mining. Knowledge gathered by manual annotation or extracted by unsupervised data mining from one or more "feature spaces" should be modeled using generalized high-quality spatial semantics. Semantic models based on such ontology have been proposed for a variety of medical applications, including computer-aided reporting, medical decision making, and data mining.

But does the above mentioned benefits of visualization still apply to ontology, i.e., whether a simple visualization of artifacts–elements, properties and relationships (or just a few of them), bring all those benefits? And why, given all these benefits, this is not a de facto standard for all models proposed? These questions are sufficiently complex, particularly in view of the fact that the ontology visualization is a relatively broad concept where the visualization term can be divided into different methods depending on the purpose to be achieved by their use. The next section therefore refers to the visualization of semantic models and diversity of definition and application of this concept.

3.2.3 Basics of visualization

Visualization as the same as modeling is understood as a tool for communication. However, in order to make them valuable and appropriate for the establishment and maintenance of "communication," some prerequisites need be taken into account. Let us first refer to most general principles that are applicable to all types of visualization. All types of models are generally understood to be a set of affirmations, premises, constraints, or rules to gain a higher abstraction of the problem [21,20]. The term abstraction is generally defined as the ability to ignore information that is not of interest in a particular context [21], with three basic forms: reduction, generalization, and classification. The meaning of reduction and generalization implies from their naming, however, by the classification Mellor means: "grouping important information based on common properties," which is met by default, considering the nature of semantic models. However, depending on the interpretation of this concept and the level to which it is applied, it can be considered as only partly resolved. As a counterexample, Dudáš et al. [16] have stressed that the grouping can be used in spatial encoding by emphasizing similar concepts that should be placed closer to each other, thereby favoring users' comprehension and understanding, allowing them to be easily interpreted as similar concepts at the cognitive level.

In short, the main consensus that should be achieved in creating visualization and how to make a comprehensive and whole-grained overview of the entire picture easy to understand, while not overfilling it with too many details. But, depending on the application domain, other prerequisites may appear. These prerequisites for the healthcare domain will be emphasized in the following subsections.

3.2.4 Visualization of semantic models

First, it should be understood that visualization of the semantic model is not a visualization in its traditional sense when a textual artifact is replaced by an image. Visualization rather complements the textual representation of the artifact. This is found to be more effective for human cognition, where it can be understood as an intuitive way to understand, discover, detect, and interpret data [14,11], and as a facilitator for more effective data exchange across multiple sites.

While the visualization of semantic models is often understood as "by default," the number of studies covering the topic of visualization is significantly lower compared with those covering semantic models. More precisely, our digital library analysis, which query them as respective keywords, shows that only 3% of semantic models-related studies mention visualization. What is more, the understanding of the "visualization" term varies in these studies. While some study "visualization," they understand the graphical representation of elements and their relationships—graphical ontology editors such as WebVOWL [45], BioOntoVis [46], OWLgrEd [33], Welkin [60] to name only a few, other focus on the graphical construction of queries (such as SPARQL queries) over a predefined data schema (for instance, OptiqueVQS [41], OZONE [42], VisiNav [43], and ViziQuer [44]). In addition, as we will discuss in the following sections, the visualization can also be used as an ontology evaluation technique. The aim of the visualization tool can also vary, i.e., it can be used not only to create a visual representation of ontology that will effectively display all the information, but also to allow the user to easily perform various operations on the ontology [19], where the nature of operations can vary. Here, Katifori et al. [19] stress that the visualization of ontology is used and can be useful at three levels, i.e., ontology design, management, and browsing levels. Another classification is provided by Dudas et al. [26]. It slightly differs from the previous one and distinguishes at least four different high-level use-cases of ontology visualization methods:

- *learning* is used to explore and learn ontology for its further use. Suitability for a specific purpose is particularly important, given that many existing tools are tailored to a specific task or use special types of diagrams that need to be learned first to understand the visualization (also in line with [25]);
- *editing* is used to manipulate visualized ontology, for instance, by drawing lines between classes to add new triples that link classes, and may be less demanding for users who are little familiar with the textual OWL syntax (only a few of tools support visual editing);
- *inspecting* is used to check the already defined ontology to find errors (if any), missing or inaccurate elements, or links between elements. It is considered as well suited for checking ontology for model adequacy, i.e., how well the ontology covers its domain;
- *sharing* is used if the creation of figures based on the ontology developed is needed, i.e., if a specific part of or all ontology should be provided to an external third-party or article, book, blog, or website of the project under development or in other form of documentation and dissemination that is not directly related to the actual use of ontology.

The "editing" method is highly valuable and beneficial for users with preliminary or basic knowledge of the textual OWL syntax, where graphical representation should be much easier to understand and manipulate with. However, "learning" and "inspecting" methods are more beneficial for users with advanced knowledge, who can benefit from the visualizations ability to give an overview of the ontology and from corresponding filtering capabilities that allow us to focus on specific parts. To make a more clear understanding, in Table 3.1, we summarize these methods by providing (a) a purpose of use, which we divide into (a1) development or (a2) use, (b) the required level of details, where we distinguish it as (b1) high, where the overall overview, such as class hierarchy, is sufficient or (b2) low-detailed view, including axioms associated with classes, (c) some use-cases examples, which we have derived from [16], with minor changes regarding purpose of use. Here, in contrast to Dudas et al. [16], we assume that the editing method is suited for both not only development, but also use, i.e., editing could be beneficial for daily tasks. In addition, for each of the methods covered, we provide a list of the tools to be used and considered to be the most appropriate, where our conclusion is based on a review of relevant studies [16,19,25], etc., which have undergone testing of these tools.

    This chapter does not suppose testing existing visualization tools. First, this was done several times by other researchers (including in recent years), and second, we found that the choice of the most appropriate tool is highly dependent on tasks. Third, the ranking is often becoming outdated, which is based on both the development and evolution of tools that are modified and improved continuously, and the fact that some of them are developed in a specific project, and then when the project is completed, the support of these tools is no longer provided. This was the case for many tools that we have selected for further inspection in this study (including GoSurfer).

    Although this classification seems popular and widely used, we would rather classify visualization methods depending on the level of knowledge and skills of its users, i.e., learning, using, inspecting, where the methods can overlap:

- *learning* method gives users and medical staff in particular greater benefit to understand whether they will be able to adapt to it and its use will facilitate their work and its productivity;

Table 3.1 Summary of ontology visualization methods.

Method	Purpose	Level of detail	Use-cases	Best tools	Main characteristics and features
learning	usage	high (overview)	deciding about the suitability for a specific purpose—analyzing an ontology to create or annotate data with it	KC-Viz	allows to automatically select "important" concepts; displays an overview of large ontologies
editing	development/usage	low	building a new ontology; adapting an existing ontology, i.e., by adding entities or customizing it, to fit a specific purpose	Jambalaya, OWLGrEd	allows text search for entities to be edited and easy access to properties of an entity
Inspecting	development	low	building a new ontology; detecting structural errors; adapting an existing ontology; checking for model adequacy; analyzing an ontology to map it to another ontology	Ontodia, Jambalaya and KC-Viz	allows focusing on the desired part to be inspected
sharing	usage	high (overview)	creating illustrations of selected parts of an ontology (or its overall content and structure)	KC-Viz, Ontodia, Jambalaya	allows limitation to a selected part of the ontology; enables the user to manually layout the visualized entities

- *inspecting* is mostly beneficial for data operators and managers who maintain the system, although compared to traditional data storages such as SQL and NoSQL databases, there is a greater likelihood that medical stuff will be involved here because of the strong focus on domain-specific elements, their properties, and relationships;
- *using*, however, is what Dudas et al. [16] call editing.

This diversity has risen the question *what kind of visualization methods can be defined in addition to them? What else can we meet named as "visualization," but with another semantic meaning,* which we have briefly addressed in this section and present in Fig. 3.1. In this chapter, we mainly treat the "visualization" concept as an opportunity to use and edit ontology in a graphical way. Although this division was domain-agnostic, in the next section we will cover the concept of ontology visualization in health(care) and (bio)medical context. This will include answering the following questions: *What types / methods of visualization exist? And when and how could they be beneficial? What specific aspects should the institution or particular project take into account to choose and implement in healthcare ontology visualization?* To determine this, we will refer to both our experience and scientific literature on this issue and identify at least several categories on the basis of which visualization techniques can be classified. Thus, in the following sections, we will cover these categories, which should form a knowledge base on this subject and allow us to identify relevant key aspects to be taken into account when selecting the ontology visualization tool.

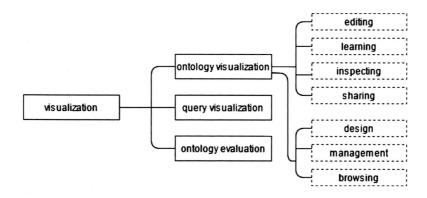

FIGURE 3.1

Ontology visualization methods.

3.2.5 Visualization of ontology and semantic model in healthcare and (bio)medical sectors

Even though some time ago there was the concept of "ontology," which data scientists were forced to adapt to the specificalities of the system, today there is a great variety of ontologies tailored to the specific domains. As an example, as a representative of Latvian National Biobank (BBMRI Latvian Node), let us mention the ontology inherent to our area—OBIB—an ontology built for the annotation and modeling of the biobank repository and biobanking administration. The reason and benefits derived from such domain-specific ontology are based on the fact that they are adapted to a specific area of application, following the common agreed guidelines or preferably standards adopted in the given field. When taking a step back to OBIB (based on OWL language), it is based on MIABIS (MIABIS Core 2.0), i.e., Minimum Information About Biobank data Sharing, which standardizes data elements used to describe biobanks, research on samples and related data, according to general attributes to describe biobanks, and sample collections and studies at an aggregated/metadata level [56]. Even more, although it could be assumed that the biobank-specific ontology should primarily support and facilitate highly medical-oriented data such as sample-related data, they are well suited for the acquisition and processing person-related facts, more precisely patient-related, such as "biological family relationship" and "relationship by marriage" labels being the SubClassOf "family relationship." This significantly simplifies the work of data scientists and data managers, compared to the building of relevant data schemes in relational databases. In some cases, these ontologies are also adapted to specific countries. As an example, BRO-Chinese—Biomedical Resource Ontology with Simplified Chinese annotations, OCMR—Ontology of Chinese Medicine for Rheumatism, ADHER_INTCARE_SP—Adherence and Integrated Care in Spanish, MDRFRE—MedDRA French, Medical Dictionary for Regulatory Activities Terminology (MedDRA), French Edition.

On the one hand, this could naturally raise the question of their compatibility with those from other countries. However, it is not a problem today, and among very traditional solutions, some although intuitive but previously lacking, have been proposed. One of the examples is MLGrafViz (plug-in for Protégé), which allows the user to translate and visualize the ontology into 135 languages by means of integration with the open-source Google translate API proposed by Florrence [32] (we will cover it in a bit more detail in further sections). Unfortunately, none of these ontologies cannot be considered as universal for the whole medical sector or at least its particular subdomain.

Today, with introducing and increasing popularity of ontologies, including domain-specific ontologies, the number of visualization techniques and tools developed to study patient EHRs has increased significantly. However, a "silver bullet" still has not been found. A natural question arises—why is this the case?

First, the vast majority of medical organizations and systems deal with data falling within the category of large ontology. Although a uniform and well-defined definition of the "large ontology" concept does not exist, it can be characterized by a large number of elements assigned to sometimes complex properties and multiple relationships between different elements. The number and diversity of the various ontology available today is huge. As an example, BioPortal—the world's most comprehensive repository of biomedical ontologies has 905 ontologies with a total of 13,495,119 classes, 36,286 properties and 55,648,584 mappings. Another example is OLS, i.e., Ontology Lookup Service—biomedical ontology repository that provides 264 biomedical ontologies supporting 6,504,429 terms and 32,832 properties. Just a few examples are: Alzheimer Disease Relevance Ontology by Process (AD-DROP), Cancer Research and Management ACGT Master Ontology (ACGT-MO), Biomedical Image Ontology (BIM), Biomedical Informatics Research Network Project Lexicon (BIRNLEX), etc. The popular Gene Ontology project's sequence database, which is a part of a larger project known as Open Biomedical Ontologies (OBO), deals with huge amount of data that, when processing using cluster analysis results, presented researchers with a directed graph (Aleksakhin, 2012) that has 10,042 vertices and 24,155 edges and 3918 are unconnected components. It has 1 root, 2729 nodes and 7312 leafs (terminal nodes). This makes its visualization by means of traditional and widely-known ontology visualization tools, such as yEd (graph editing tool using hierarchical layout) inappropriate because the visualization becomes too "messy" due to so many artifacts.

Thus, in view of the likely need to use large ontology and the prerequisite for user-friendly tools to support and facilitate their use by users who may not have IT knowledge, a list of more stricter and more advanced requirements for visualization techniques implies. More precisely, the need for more advanced techniques is more expressive for the (bio)medical sector. The so-called "concept clustering" [16], where the selection and/or aggregation of key and/or the most relevant concepts of ontology and the limitation of visualization to those called "smart" filtering, becoming almost "must have" for them. This should make it easier for users to further customize visualization and navigate it, especially when the (large) ontology is not hierarchically organized. But, obviously, it is not the only feature to be present in "suitable" visualization tool.

Second, the needs imposed for a tool depend to a large extent on the task carried out by the data user and vary significantly from one case to another. Thus, efficient management of data should either use a very advanced tool with a list of predefined functionality, including the ability to switch between visualization techniques in addition to a broad range of general features, or a set of different tools should be used. The later option is not the best, so ad hoc solutions are sometimes developed to meet the needs and expectations of data users. But this also requires significant knowledge based on what should be taken into account.

3.2.5.1 Task-dependency of visualization

Let us first refer to the question of *how the task can affect the visualization tool and the requirements posed for it*. This question has been thoroughly addressed in [47]. The authors revealed that depending on field of bioinformatics, i.e., molecular structure, genome and sequence annotation, sequence analysis, molecular pathway, ontology and taxonomy, phylogeny, and the list of crucial features to be in-built in the tool differs significantly. As an example, when inspecting the molecular structure, the expected features of the tool are zoom, filter, and detail-on-demand, but the expression profile analysis (microarray) does not consider them important, where an overview is more important. The later case, however, sets out an additional requirement for geometric techniques that must be in place and support a colored mosaic and scatter plot, as well as hierarchical techniques such as a graphical tree, expandable tree, and treemap. At the same time, the genome and sequence annotation, which mainly deal with 1-D linear and circular data, benefits from interaction met by visualization technique, where concepts such as semantic zoom, magic lens, and brushing play a critical role in efficient data management. This makes it difficult to propose a single "one-fits-all" solution because the requirements are determined by the specificity of the task, which could be well known to the domain expert but not always is well known to the developer of the tool. Therefore, even meeting the needs of several medical-sector related subdomains, there could be those overlooked for which these tools will not be suitable.

There are, however, some attempts to create a slightly more comprehensive solution suitable to more than one application area. This was presented in [29], who has tried to avoid the use of different tools—one per task—by developing a more general tool suitable for a wider range of phenotype-related tasks, although only one subdomain is covered. To achieve this goal, he has proposed an Ontology-Based Analytical Abstraction for visualization. This was obtained by developing first the three case-specific tools called PhenoStacks, PhenoLines, and PhenoBlocks, where: (1) PhenoBlocks—supports visual comparison of phenotypes between patients or between the characteristics of patient and disorders, (2) PhenoStacks, which is a visual analysis tool that supports the inspection of phenotype variation in cross-sectional patient cohorts, (3) PhenoLines—an analysis tool to interpret the subtypes of disease derived from the application of topic models to clinical data—this allows a comparison of the prevalence of the phenotype within and between disease subtype topics, thereby supporting the characterization of the subtype, a task that includes the determination of the dominant phenotypes of the proposed subtype, exposure age, and clinical validity. Each tool has been both designed, developed, and validated involving people carrying out the task in consideration. This allowed the author to formulate the so-called pillars on which further visualization tools related to the medical sector should be based, divided into (1) topological simplification, (2) visual comparison, and (3) generalization. Topological simplification should cover problems such as unified representation, elimination of redundancy, and control of the level of detail. Visual comparison, however, applies to pairwise comparison, many-to-many comparison, and trend comparison. In this way, according to the authors' findings, at least a slightly more general tool can be developed. However, as in the previous case, the involvement of the domain expert and the acknowledgment of all related tasks should be taken into account.

Another type of classification is related to the visualization technique underlying the ontology visualization, i.e., a graphical representation, in which it is represented. This type of classification also includes a debate on the best choice, where the consensus has also not been reached. The next section is devoted to visualization techniques underlying visualization of the semantic model.

3.2.5.2 Visualization of semantic models: classification by the visualization technique

Generally, ontology and semantic model visualizations depending on the visualization technique, are divided into indented lists and graphs [23,16,25]. Ontology visualization as a graph is considered to be a very natural way to depict the concepts and relationships structure in a domain of knowledge (in line with [25]). However, although the graph is a very traditional and intuitive way of representing knowledge, there are studies that prove that indented lists, such as Protégé Class Browser, are sometimes more efficient compared to graphs [23,30]. Yang [30] explains this in simplicity and for users familiar with this visualization technique, which is very similar to classical file system navigation. Perhaps this is the reason why they are the main tools used for ontology editing.

Another study, which attempts to determine which type is better, is an eye-tracking user study [23] aimed at evaluating and comparing the two above-mentioned techniques, when they are used by biomedical staff who are most likely not familiar with them. However, results of around 500 MB of eye movement data containing around 30 million rows of gaze data generated by a Tobii eye tracker did not allow such a conclusion to be reached; however, they found some evidence suggesting that indented lists could be a little more efficient to support the search for information, while graphs are more efficient in supporting the processing of information. Glueck [29] stresses, however, that in the biomedical domain, the preference is usually given to tree structures obtained by converting graphs into hierarchies, duplicating nodes with multiple parents, so that large data sets are more easily understood compared to hierarchical visualizations. In the context of this chapter, however, it would be beneficial to note that this experiment involved biomedical staff that, on the one hand, allows to bring some results suited for the purpose of this study, and on the other hand, demonstrates to representatives of (bio)medical institutions and organizations that ontologies are well suited for this sector.

However, in addition to these traditional types, a few more visualization techniques may be distinguished. A study aimed at presenting techniques of ontology representation and categorize their features to help select the most appropriate, which is often considered to be one of the most comprehensive reviews on the topic, is [19]. Katifori et al. [19] divide them into: (1) *indented list, (2) node–link and tree, (3) zoom'able, (4) space-filling, (5) focus+context or distortion*, where the overlapping of methods is allowed. Moreover, depending on the space dimension used by the relevant tools, the methods can be divided into 2D and 3D.

2D use the screen space as a plane and do not use any notion of depth. The authors acknowledge the so-called 2.5D, which is applied to 2D visualizations that use a perspective view to create a 3D sense without permitting movement or manipulation in the 3rd dimension. However, most studies still consider them to be 2D. 3D visualization method; however, it has been found to be less suitable for most users, particularly those not familiar with them previously and so-called novice users (also in line with [19]). In addition, it requires more advanced system resources to ensure both representation and navigation, which is rarely available. Therefore, the 3D methods will not be considered in this chapter, the less likely to be suitable for users representing the domain. Florrence [32] derives a few more techniques, where the above list is supplemented by the so-called *(1) Squartified graph, (2) Grid-Alphabetical* and some other more exotic types. Dudas et al. [16] in addition to the most classical types, divide them into *(1) label-based* and *(2) layout-based*. As their name implies, label-based techniques give more attention to labeling and their naming, thereby contributing to a better level of detail. However, they are divided into (1a) *UML-inspired*, where the node label consists not only of the name of the entity it represents, but also of other information, such as a list of data properties associated with the class represented by the node, and (1b) *name-label-only*, where each node is labeled by name only, and does not (or only very few) textual information. The first group is more popular and is presented by such widely-known examples as Ontodia, OWLGrEd, TopBraid Composer, and VOWL. They are considered relatively easy in use, but not for all users, because they require some knowledge of UML class diagrams. However, although they are not considered well suitable for very specific domains such as (bio)medicine, they are still developed and adapted to them and sometimes turn out to be very efficient in use.

A layout-based category is far more richer, represented by: (2a) *force-directed layout/spring embedded*—simulates a physical system where the edges act like springs, while nodes repel each other and drag forces that ensure the nodes covered. By correctly setting the magnitudes of the forces, visualization reaches a balance where nodes with the most connections to other nodes are arranged in the visualization center and the least connected nodes are at the outer edge of the acquired graph. Although it tends to avoid crossings of edges and overlaps of nodes, it is characterized by the inherent randomness of the most force-directed algorithms, which result in a different arrangement of node each time they are executed. This group is represented by tools such as Jambalaya, NavigOWL, Neon Toolkit ontology visualizer, OLSVis, Ontodia, SOVA, TGViz, WebVOWL, Cytoscape, VOWL 2, and GLOW; (2b) *tree layout/treemap* (already covered above) representing the ontology class hierarchy, where the entity at the top of the ontology hierarchy is visualized at the top of the display area. Its children are placed at one level below the root node, according to the hierarchy, while their children are placed at the second level below them, etc. While compared to the previous category, the layout remains unchanged in each run of the algorithm, and if the inheritance structure remains unchanged, this category is typically characterized by a risk that nonhierarchical relationship links will cross other links and nodes due to the level-based node arrangement, and a lack of consideration of other property relations (mainly classes). This group is represented by tools such as Jambalaya, OWLViz, GLOW, and OWL-VisMod; (2c) *radial layout*, where the root entity is placed in the center of the display area, and its children are placed in an orbit around it, which children are placed in the next orbit, one step further from the center, etc. It may be more space-efficient than the classical tree layout, because the number of nodes usually grows at each level of hierarchy. This group is represented by tools such as OntoStudio, SOVA, and GLOW; (2d) *circle layout* in which entities are arranged so that they form one large circle. This allows to visualize a large number of relationships between entities, especially if there is no need to display individual labels of the relationships. Probably the most expressive example of it is NavigOWL; (2e) *inverted radial tree layout* in which nodes are located in a circle that consists of several concentric rings if there is a hierarchical relationship between entities represented as nodes. Ring parts represent nodes, where nodes appear at the top of the hierarchy as parts of the outer ring, where their children are in the ring one level closer to the center, etc. Links start and end in the nodes in the inner ring to specify other nonhierarchical relationships, as is in NavigOWL.

Other researchers have observed that although graphs rendered in force-directed or hierarchical layouts often result in appealing visualizations, most of them, focus only on some aspects of ontology. For instance, OWLViz, OntoTrack [61], and KC-Viz [62] represent only the ontology class hierarchy. OntoGraf and FlexViz [63] represent different types of property relations, but do not show data-type properties and property characteristics required to fully understand ontology. While there are some examples attempting to provide a more comprehensive overview, they often fail because the visualization becomes messy, where the various elements are often difficult to distinguish, particularly if additional features such as zoom and filters are not supported. Some steps to resolve this were taken by TGViz [66] and NavigOWL [67], which, while using very simple graph visualizations, where all nodes and links look the same, have used different colors to allow the user to distinguish between them. Lohman [25] found the attempts made by GrOWL and SOVA [65] more successful, because they define more elaborated notation using different symbols, colors, and node shapes. But this negatively affects the readability of visualization, which sometimes becomes characterized by a large number of crossing

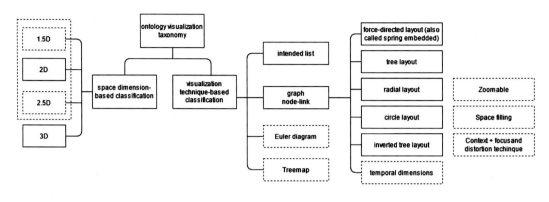

FIGURE 3.2

Ontology visualization taxonomy.

edges. Although some tools deal with this issue by making additional efforts in this regard, this is done in very rare cases. In addition, they become inappropriate for normal/typical users, since the notation of both rely on symbols from description logic and abbreviations.

Another nontypical but today very crucial observation is made by [32], who finds that there are very few tools supporting *natural languages* (NL), where one example is the OWLGrEd tool of Latvian origin [33], which has added a verbalization to their graphical ontology editors by means of a controlled natural language (CNL). So, the author proposes a multilingual ontology visualization plug-in for Protégé, called MLGrafViz, developed as an extension of OntoGraf, which, in contrast, allows the user to translate and visualize the ontology into 135 languages. This is achieved by means of integration with the open-source Google translate API. The authors have also ensured an opportunity to switch the layout between the Radial, Spring, Horizontal Tree, and Vertical Tree and uses Zoomable techniques, which is considered a "good practice." It makes the tool more universal, when depending on the task, which can require another visualization technique, i.e., there is no need in change of the tool. In addition, it facilitates sharing the defined and visualized ontology, which makes this new proposal highly competitive and prospective, supporting the elimination of differences in geographical location and language among users of ontology.

In view of this diversity, we present an ontology visualization tool taxonomy, while Table 3.2 summarizes the most widely discussed graphical representation techniques, giving them a brief description, pros and cons, as well as some examples found as the most competitive.

To sum up, visualizations of ontologies may be classified at least in (a) *space dimension-based visualizations* and (b) *technique-based visualization* in accordance with Fig. 3.2. Dotted lines in refer to less popular and commonly accepted categories which, however, are useful to take into account. They serve as a knowledge base to further derive the features to be taken into account when selecting a ontology visualization for a given case. We will not go into more detail, because this has already been a matter for research. Instead, we aim to establish understanding of benefits and differences between different techniques, which should facilitate the choice. In addition, we intend to focus on more medical-related issues, to which this knowledge will form a knowledge base.

Table 3.2 Summary of ontology visualization methods.

Visualization technique	Description	Advantages	Disadvantages	Examples
Indented List	a list of entities, where entities are categorized as a tree view	easy implementation and representation; quick browsing and effective information retrieval; systematic exploration of the whole ontology; retraction and expansion for focusing on specific parts of the hierarchy;	represents a tree (not a graph); multiple inheritance cases are not apparent; nodes at the same level are not immediately recognized as siblings if subhierarchies are expanded; not well suited for the overall ontology structure, determining the depth of the hierarchy or finding nodes that have many children, deep hierarchies	Protégé Entity Browser
tree-like node link	a taxonomy with top–down or left-to-right layout	effective for an overview of the hierarchy for small trees; various levels and features (depth / width) are easily distinguishable;	inefficient use of screen space, leaving the root side empty; a hundred nodes trees often require multiple screens or scrolling to be fully displayed;	OWLViz, Ontodia, OntoGraf, Onto Studio
Zoom'able	allows the user to zoom-in to child nodes to enlarge them making them the current level of viewing	effective to find and locate to specific node – comprehensive view of the hierarchy level where the user is zoomed in; successful for browsing to locate specific nodes	do not offer an effective overview of the hierarchical structure; do not support the user in forming a mental image of the hierarchy;	Jamba laya
Space filling	use the entire screen space by allocating the available space for the node between its children.	visualizing trees that have property values at the instance level; effectively if the user cares mainly leaf nodes and their properties but does not need to focus on the topology of the tree or the topology is trivial	not effective for structure-related tasks; not effective if user is interested in the topology of the tree of the topology; not effective for non-trivial topology;	Tree Maps
Focus + Context and Distortion	distorts the view of the graph to combine context and focus	effective in presentation and displaying multiple nodes at once; quick browsing; each node can be moved to the center of the tree to display while retaining the context of the focal nodes related	does not maintain a constant positioning of nodes; does not offer an intuitive representation of the structure of the hierarchy;	Touch Graph

Thus, the conclusion implying from these findings (including Fu et al. [23]) is in line with Dudáš et al. [16] and Florrence [32], who suggest that better results could be achieved if the tool supports more than one visualization technique, allowing the user to switch between them. However, this creates a lot of resource-consuming tasks for developers. In addition, given that many very intuitive and simple implementation features remain an issue for many visualization techniques, it is difficult to imagine that such tools will become available in the coming years. As we have mentioned above, we understand and further consider "visualization" as an opportunity to edit ontology in a graphical way. This choice seems to be not only the most correct in general case, but even more correct in case of health(care)-related domain, where such an opportunity is particularly relevant and beneficial for healthcare-domain experts who may not have an in-depth IT-knowledge to build queries over the data schema in languages like SPARQL. Although it is quite intuitive and even relatively similar to classical SQL, it brings significant challenges for medical personnel, when more complicated queries need to be developed. It is also in line with [16], according to which even though the user-friendly OWL syntax, such as Turtle or the Manchester syntax, visualization is able to provide simpler ontology inspection and editing. The authors pose graphical visualization as an important enabler in many practical tasks. However, although there is a trend of domain-specific ontology that is highly adapted to a specific use in a given domain that is becoming more granular, i.e., not only health(care)-related ontology, but also more specific, such as biobank- or phenotype-related ontology, visualization tends to be more universal. In fact, there is both the demand and the supply of a universal ontology visualization framework that implements a core set of visual and interactive features that can be extended and customized to the use cases [16].

However, despite the universalism of most visualization frameworks, the current needs of the semantic web community are not satisfied, i.e., there are no visualization frameworks that could be considered as standard and recommended for their use. Therefore, the choice remains in the hands of users. This, however, means that there should be provided guidelines for this choice so that the community can do so based on their needs. Dudas et al. [16] explain this as follows: (1) there is no "one-size-fits-all" solution, and various tasks and use-cases may require different methods of ontology visualization (also in line with [19,47]); (2) there is a little improvement on the state of the art in a field, where new methods and tools for ontology visualization are often developed from scratch, learning from the experience of previous visualization methods.

Even more, visualization of an already existing ontology can make it easier to take a decision on its reuse to annotate new data by inspecting the suitability of classes and properties of ontology and whether they meet new needs [18]. This is particularly beneficial for the health(care) sector, where such assessment can be carried out at both levels, at the level of one research group, different research groups of one organization, and different institutions and exchange of their experience with relevant artifacts and even different countries. As an example, different biobanks representing different countries could do so to reach the highest possible level of data integration when using the same ontology, thereby ensuring the consistency of the data acquired and used at all levels.

3.3 Requirement of evaluation

Typically, deployed semantic models are ready for use in large and complex systems. This leads to an intuitive assumption that developers of the models concerned have accurately and cordially developed each modeling language artifact of the semantic model, i.e., both elements, relationships, attributes, visual representations, and model types that form them. However, the practice demonstrates (also in line with [15]) that in many cases the developer refers to formerly existing, similar artifacts and the integration of best practice, thereby trying to make the proposed solution as easy to use as possible, making it similar to what the user is most likely to be dealing with. In the case of modeling languages, for instance, the typical choice is to refer to the UML and BPMN (Business Process Modeling Notation) specifications. Although, on the one hand, it gives a lot of advantages and attracts public, it would be logical to ask—*whether the derivation of components from an external and differently oriented solution has not negatively affected the features of the proposed solution and allowed the identification and presentation of all the necessary objects, giving them all intended features (which were not representative to the existing ones).* So, when the idea of the need for ontology comes to organization or institution, it is crucial not only to determine and clearly understand all needs and expectations, as well as data and all the relationships between them, but also the appropriateness of the identified candidate ontology. The question to be asked is whether it is really appropriate in the sense of containing the classes and properties needed to represent the data.

Ontology evaluation is one of the key phases in both ontology development and its further use and reuse. The quality of ontology affects the accuracy of results provided to a user [49]. It is therefore an emerging area with different approaches, taking into account various set of aspects or criteria (in line with [1,48]). It involves both the adaptation of older approaches and introducing new ones. This subsection forms an overview of them.

3.3.1 Taxonomy of the evaluation

While most studies focus on the ontology evaluation tools and their classification, we will provide a very brief insight on them, and will focus on the less discussed but at the same time prospective approaches, mainly referring to evaluation of ontology visualization. This should provide more added value, since knowledge on this subject is very limited, but very crucial, especially for the (health)care and (bio)medical sectors, which are less likely ready to try a number of approaches, if one of the selected has failed to meet expectations.

3.3.1.1 Recommenders

This is even more the case in the light of the current development of (bio)medical- and (health)care-related ontologies, which in addition to very specific ontologies and their detailed description, provides recommenders as does BiOSS [70], content-based recommender proposed in [71], text mining-based proposed in [72], the flexible biomedical ontology selection tool [73], and Bioportal (repository of biomedical ontologies developed under National Center for Biomedical Ontology) considered to be the most comprehensive biomedical ontology recommendation system [69,68]. Bioportal provides web-based NCBO Ontology Recommender 2.0 (initially developed in 2010), which provides the user with suggestions for the most relevant ontologies based on a biomedical text corpus or a list of keywords. This suggestion is based on four criteria taken into account by recommender, namely: (1) *coverage*—the extent to which the ontology covers the input data; (2) *acceptance*—the acceptance of the ontology in the biomedical community; (3) *detail*—the level of detail of the ontology classes that cover the input data; and (4) *the specialization of the ontology to the domain of the input data.* The developers of this recommender emphasize that these criteria are the most relevant and widely-used criteria for ontology recommendation, although the definition of these criteria may differ.

Another example is OntoKeeper—a web-based application that allows to assess the quality of the ontology [52] based on semiotic measures devised by [50].

And the last but not least example that can be considered interesting and useful for the decision making is called Ontology Visualization Tools Recommender. It can be used for both, finding the most appropriate ontology visualization tool for the specific use-case and ontology in use. This recommendation is based on four questions, which the user is expected to answer. The first relates to the assessment of the OWL features and their significance for the given purpose, and the second—preference of ontology development environment with the opportunity of choosing between Protégé 3, Protégé 4, and the NeOn Toolkit. The third question concerns the intended use, i.e., make screenshots, check domain inconsistencies, structural inconsistencies, reuse ontology, adapt an existing ontology to a new use case, develop a new ontology, analyze mapping on the ontology, and the last, but of course not least, the size of ontology. Each question with its subquestions is assigned with weight (within the interval <-3;3>) representing uncertainty of the importance of the respective attribute/feature.

3.3.1.2 Ontology evaluation tool

Another solution is the web tool for evaluating and ranking ontologies proposed by Jimborean et al. [53]. The authors propose a solution that evaluates the quality of ontologies in two different dimensions, namely *structural* and *semantic*. With the structural dimension, the authors mean metrics that indicate how well ontologies were developed in terms of schema size, depth, width, density, richness, and inheritance. The semantic dimension examines how instances were placed in ontology and the usage of the real-world knowledge representation. The idea under recommender systems is beneficial for those with little knowledge on the topic in question, thereby greatly facilitating the choice by (semi)automatic assessment. However, in many cases this is not sufficient because the set of criteria may not meet the needs of the user.

Therefore, this issue is less expressed for the domain under question, as is also pointed in [69], while the selection of the visualization tool and evaluation of its applicability remains open. The choice should be more justified, providing the most appropriate and correct option for the task. Even more, as we have found previously, the practice and current reality shows that in some cases new task-specific solutions are developed. Therefore, the further content should be valuable for them, pointing on the aspects to be taken into account while developing a new visualization tool with its consequent evaluation.

The ontology evaluation tool is probably the most popular and broad category, which includes additional subclassifications. Their overview has been provided in [51]. As regards ontology evaluation tools, one of the most recent studies on this matter, [48] proposes a framework for evaluation of ontologies called OnE (Ontology Evaluation), which considers aspects such as accuracy, interrelatedness, consistency, exhaustiveness, and reusability. It represents a set of criteria to be taken into account by developers when developing an ontology. Another set of criteria was discussed by Mishra et al. [1]. It includes correctness or accuracy, adaptability, clarity, completeness or competency, computational efficiency, conciseness, consistency or coherence, and organizational fitness. But perhaps one of the most popular classifications is the one proposed by Burton-Jones et al. [50], which metric suite includes syntactic, semantic, pragmatic, and social quality, which are further supplied with attributes such as (a) lawfulness—correctness of syntax, (b) richness—breadth of syntax used, (c) interpretability—meaningfulness of terms, (d) consistency—consistency of meaning of terms, (e) clarity—average number of word senses, (f) comprehensiveness—number of classes and properties, (g) accuracy—accuracy of information. It also formed the background for OntoKeeper that we have mentioned above.

3.3.1.3 Ontology visualization

Most visualizations focus on classes, their hierarchical relations, and properties. However, OWL has more constructs, i.e., restrictions that determine which instances can belong to a class, relationships between classes expressing disjointness, or other constraints [16]. For this purpose, one of the most obvious choices, particularly in view of the previous subsection on visualizations, is ontology visualization. Ontology visualization tools use the OWL code of the ontology as input and typically display it as a node-link visualization, where nodes represent classes and links represent properties—based on domain/range definitions [19].

Although they are not intended to be focused on an evaluation, they are very beneficial in practice since they allow easily detection of deficiencies. In addition, they are useful in cases when the existing ontology is considered for reuse for annotating new data, when it should be checked that the ontology contains classes and properties suitable for describing the data. Ontology visualization itself in some available tools, such as LODSight, WebVOWL, IsaViz, BioOntoVis, OWLVix, FrOWL, OWLgrEd, Welkin, LODSight, OWL Viz, KC-Viz, and WebVOWL, to support such a decision, is an option. We also argue that they could be the most efficient when rule-based or logical approach is applied. Here, the idea posed in Pak et al. [79] seems appropriate, i.e., to design anomalies. This means that a set of rules is defined to check how the ontology behaves and whether any conflicts are identified such as disjoint classes, missing or incorrect relationships, etc.

3.3.1.4 Briefly on other approaches

Another one widely-known approach is the model coverage assessment defined as the *completeness* of ontology, more precisely availability of appropriate classes and properties that can be used to describe relationships and entities in a given data model. It typically allows model tests to be validated by assessing how thoroughly the model objects are tested, i.e., by calculating the extent to which a model test case performs simulation pathways through a model. In the case of ontology visualization, the model coverage can be associated with a visualization test on the test data instances to find out whether it covers all ontology concepts to find errors in the ontology (if any) as has been proposed in [34]. However, the authors point out that the "appropriateness" here is very relative, where its precise definition depends both on the use-case and on users' understanding of the concept. For example, in the case of medicine, a class *Patient* can be used as a type of an instance representing a *Patient of a Medical Institution*, although a class like *Entity* would be "too distant" and inappropriate.

An approach derived from the model coverage assessment is ontology usage visualization (OUV), proposed by Dudas et al. [18]. OUV is a visualization of a data set schema that supports merging schemes into a single visualization, filtering the visualization to focus on selected ontology, and showing all of the classes and properties in the data set, i.e., not only the most commonly used classes since there is a need to see all "capabilities" of ontology. In other words, the aim of OUV, which partly stems from its title, is to show how and whether ontology entities have already been used in existing data sets, thereby ensuring that actual usage and needs are identified. This makes it possible to learn by example and to define the requirements to be set for the particular situation by identifying cases not included in the list of requirements (if any). According to Dudas et al. [18], it can be used to (1) clarify whether ontology is suitable for modeling the given problem, (2) learn how to use ontology to annotate data, and (3) detect errors in usage. It uses one or more RDF data sets as an input, while the output displays all combinations of classes and properties that are used in data sets. More specifically, classes whose instances are associated data set properties are displayed and linked with the properties in the ontology usage graph/node-link visualization. The result is similar to ontology visualization covered above, but it does not depend on domain/range relationships—it depends on the actual data of the data set only. The visualization of data set schema refers to a single data set and aims to help users find out what the data set contains and how to query it. The authors believe it is suitable for evaluating the model coverage, yet it can be used in addition to common ontology visualization and textual documentation. It depends heavily on the use-case in which it is tested, on its complexity, number and comprehensiveness, i.e., whether it supposes all cases in which ontology will be used. In an experiment conducted by the authors, it was done in an artificial environment and with very trivial examples, so very accurate and comprehensive conclusions about the suitability of OUV to the task in consideration are not possible, although the authors did some.

One of the very significant challenges encountered by the authors was the inability to find for suitable data sets with correct use of the ontology, which is an essential prerequisite for OUV. Another point to be taken into account in the context of this book is that the authors found that OUV is more appropriate in cases when ontology lacks a domain relationship. Therefore, in the case of the health(care) domain, given that most tasks are very domain-specific, OUV is not appropriate. This approach will therefore not be discussed in more detail.

3.3.1.5 Feature-based evaluation of ontology visualization tools

A way to evaluate the ontology visualization tool and method or to compare multiple tools is to refer to some specific features. We refer to this category as a *feature-based evaluation* of ontology visualization tools. This, in view of the nature of this task and above discussed (specificity of tasks posed in the (bio)medical sector), could be one of the most appropriate, if the needs of the users are known. Let us try to identify the main features to be taken into account.

One very specific example that aims to help in their determination is proposed by Schaaf et al. [22]. It evaluates the performance, maintenance, usability, functions, and topicality of ontology visualization tools. More precisely, the authors have developed an ontology of information management at hospitals to support knowledge transfer. They have also faced problems with identification methods that can support the visualization ontology and its use, as well as posed awareness that there will be similar problems with bioinformatics. In order to bring added value for efficient data and information management in the biomedical sector, they have carried out an analysis of eight tools for visualization of large ontologies to assess their fitness in the medical informatics field. They have focused solely on the assessment of tools capable of dealing with large ontologies, which are considered to be one of the most important prerequisites for the (bio)medical sector, which deals with large and complex data sets that come not only from different institutions but also from different devices. Unlike other studies mainly concerning visualization and use, the authors share our awareness that the query of these large ontologies is very important and needs to be addressed. Because only the knowledge representation without their efficient use (for decision-making, reasoning, pattern retrieval, etc.) is useless, especially in the medical sector, where querying of ontologies for nondomain experts is a challenge. However, this challenge is addressed relatively rarely. It is thus beneficial for us to take a look on their findings in this regard.

The authors believe the tool to be a performant, when ontology visualization and navigation can be performed along with the querying without noticeable skips or dropped frames, which is a problem for many visualization tools. With maintenance, the authors mean the support of both .csv and .owl files import and export and ontology manipulation. Usability refers to general tool handling capacity, documentation, plug-in management, while functionality refers to support of concept search, concept property overview, subgraphs creation and querying, implementation of graph algorithms, including "shortest path," "concept neighborhood," static views, hierarchy overview, and display filters. The last criteria they put forward are the topicality or up-to-date'ness of the tool, and its further development. The authors have chosen a list of 20 visualization tools (their selection was based on [24]) for further inspection, but only eight of them—Ontograf, OWLViz, GraphViz, IsaViz (all are Protégé plug-ins), Tulip, Gephi, Cytoscape, Neo4j, were thoroughly assessed because most of the tools were not suitable for large ontology. More precisely, either their import was unsuccessful or the functionality was not suited for their efficient management. Their analysis shows that Gephi, Tulip, and Cytoscape are promising tools capable of visualizing large ontology, yet they still lack some important functions, such as edge detection and filtering, that play an extremely important role in a given domain. However, they stress that the extendibility and the possibility of using a number of plug-ins that could at least partially resolve this. Thus, an important point to be taken into account when selecting the most appropriate tool is to make sure that the features that are not implemented in the tool can be added by extending it.

Because meeting users' requirements is a very complicated issue, Lohmann et al. [25] attempted to propose a user-oriented ontology representation called VOWL 2, which is an improved version of their previously proposed VOWL. VOWL 2 is based on a set of well-defined graphical primitives and an abstract color scheme. Reference to a color scheme seems to be taken into account rarely, although it is effective as a communication tool, thereby supporting the main objective of visualization. It was developed to be easily understood by typical users with only a little training, by following Visual Information Seeking Mantra of "overview first, zoom and filter, then details-on-demand" [28] thereby giving users with an overview of the complete ontology first, and then allowing them to subsequently explore parts of it in deep. This is not the most typical choice, since many tools choose the opposite path— start with the root and then allow the graph to be expanded. The benefits of this study rely also in the "think-aloud" method that they use, while testing their tool and comparing it to two more (SOVA and GrOWL). This allows to gain insight on typical (the authors call them "casual") users' experience and expectations, i.e., what does the usual/typical user expect from the "good" ontology visualization tool? Their study has revealed that visualization tools allow users to process different tasks even if they have not previously dealt with ontology. More specifically, 84% of the tasks were successfully completed, with 33% of participants who have gave up while completing them in GrOWL. The negative result, shown by GrOWL, is due to the fact that participants were not able to find elements under question due to a lack of a feature to search for a specific element, i.e., participants were forced to process the graph by visually scanning.

The authors therefore defined a set of six criteria for evaluating visualization tools. It includes clarity—perceived visualization clarity, learnability—perceived ease of learning how to use the visualization, findability—ease of finding specific elements, mappings—perceived understanding and comprehensibility of the mapping between elements of ontology and visualization, colors—helpfulness of colors for comprehension of the visualization, shapes–helpfulness of shapes for understanding of the visualization. Users acknowledged that clarity was the most challenging for all visualization tools under question. Participants have appreciated the presence of zooming and the search opportunity to quickly find an element by its name. They also found it helpful to make a clear visual distinction of ontology elements when they are distinguished by means of different shapes and sizes compared. And also the high level of details is sometimes very useful; it was stressed that this results in too many tangles and edges, especially those crossing each other. The authors found out that explicit labeling of all elements has a positive effect to make visualization more self-explanatory, but it may cause distraction, so it would be advisable to give users the choice of whether such labels are displayed. Since testing was conducted using ontology of a relatively moderate size, i.e., no more than 53 classes, the authors have expressed awareness that graph visualization would not be suitable, i.e., the overview of larger ontology could be lost and the graph would no longer be usable anymore. Thus, other graphical techniques could be more appropriate, which means that switching between different graphical methods would be very important (also in line with [19]). However, other studies suggest that there are additional mechanisms that contribute to the interaction and readability of graphs modeled for large ontology.

Dudas et al. [16] define a set of criteria, including the implementation of visualization methods, the support of interaction techniques, the OWL constructs visualized that they call OWL coverage, i.e., a list of 15 OWL constructs, retinal properties such as color, shape, size, saturation, and texture. This list is applied to 37 ontology visualization tools using an expert examination, i.e., users are not involved. Out of the 37 selected tools, the authors found that only 17 works, including those, which are only partly usable and very limited in terms of functionality. The authors have defined and evaluated a very broad range of interaction techniques—*radar view, graphical zoom, entity focus, history, pop-up window, incremental exploration, search and highlight, filter parts, filter entity types, fish eye distortion, edge bundling, 3D navigation, panning, drag and drop, clustering, textual editing,* and *visual editing*. The most popular interaction techniques implemented by the most concerned tools were *graphical zoom, search and highlight, filter entity types, panning, drag and drop*. The most exotic techniques, however, are: *3D navigation* (OntoSELF and OntoSphere), *clustering* or representing in a dendrogram plot (KC-Viz and OntoTrix), *edge bundling* (Knoocks, OntoTrix, and OWLGrEd), *visual editing* (CmapTools OntologyEditor, GrOWL, OntoTrack, OWLGrEd, Triple20), *fish eye distortion* (Jambalaya, OntoRama, Multiview ontology visualization, Ontoviewer, TGViz), *history* (KC-Viz, Ontodia, Ontology visualizer, OntoStudio, OWLGrEd, Triple20).

For retinal properties, typically understood as a simple but a sufficiently efficient feature of supporting and facilitating interaction between model and its user, the most popular is a *color property*, followed by *shape*. However, properties such as *size* (4), *texture* (2), and *saturation* (1) are used by only several tools. In their study, the most feature-rich tools were KC-Viz, Jambalaya, and Ontodia. Dudáš et al. [16] have concluded that many tools do not provide all well-known and sometimes easy-to-implement features. This is because most of them have been developed as experimental prototypes for research projects—there is not enough time and other resources to implement extensive coverage. We share this conclusion and this is what we have faced when we examined some of these solutions, i.e., some of them have not been found because they are no longer supported.

Surprisingly, but the most frequently and widespread problem is *performance*. More precisely, just 5 of 37 tools (Ontodia, OntoGraf, Entity Browser, TGVizand TopBraid) were able to load a large ontology without unreasonable delay. This finding is also consistent with other studies, such as [22].

Summary of the results from the study led the authors to conclude that the choice of the most suitable and appropriate tool can be a challenge because even if the tool can be characterized by a high performance and rich list of features implemented, as is for Ontodia, OntoGraf, and TopBraid, they may be still use-case specific. As an example, Ontodia although meeting above criteria, displays only part of the ontology manually selected by the user. Thus, the choice of the most appropriate tool should be a task-specific choice, which should assess both the functionality and features of the tool, but most importantly, the requirements posed to it by a specific use-case should be acknowledged. We therefore make a stronger focus on points that should be taken into account when the choice should be made without a repetition or conducting a new comparative study on existing tools, which usefulness can be very limited, particularly in the use-case of domain-specific-medical semantic models.

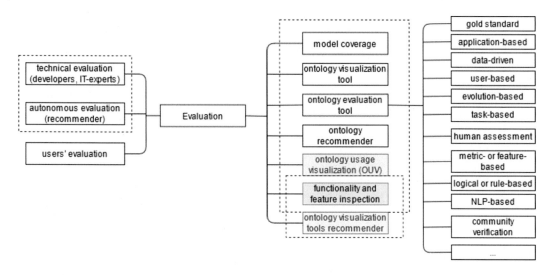

FIGURE 3.3

Ontology evaluation taxonomy.

3.3.1.6 Taxonomy of evaluation techniques

As a summary of this section, we propose a taxonomy of evaluation techniques (Fig. 3.3) that we obtained from the literature review of the evaluation techniques and approaches. It allows us to distinguish six primary evaluation categories: (1) *model coverage*, (2) *ontology evaluation tools*, (3) *ontology visualization tools*, (4) *ontology recommender*, (5) *ontology usage visualization*, (6) *ontology visualization tools recommender*, and (7) *functionality and feature inspection* for both ontology or semantic model and visualization evaluation. In addition, they can be classified in accordance with the stakeholder involved, i.e., technical evaluations that suppose the involvement of the developer, IT-expert etc., users' evaluation or autonomous, where the evaluation is carried out in a background, although it is closely linked with the technical evaluation since both refers mainly to the technical side, while the users' evaluation supposes the overall acceptability level of the method/tool under question.

3.3.2 How to make a choice of a visualization tool

Based on the knowledge we have gained from the literature and those we have from our own experience, we define the list of features that are critical or at least very important to make a choice of the tool. This list consists of (1) feature found to be important, (2) area to which it is applicable, (3) its brief description, (4) level of importance, where we distinguish (4a) critical, (4b) high, (4c) task-dependent, and (5) list of literature supporting our opinion (Table 3.3). To sum up, most important points to be considered are related to (1) *performance* to allow the user working with it, (2) *extendability* to allow to adapt the tool to the needs if the feature of interest has not been implemented, (3) *interaction* to allow the user to use it as a communication tool, thereby facilitating daily activities. For the later point, the consensus between an attempt to provide a comprehensive overview, and sufficiently appropriate

Table 3.3 Features to consider for ontology visualization.

Feature	Area	Definition	Importance	Ref.
Performance	Tool	allows to load a large ontology without unreasonable delay	Critical	[27,19,22,16]
Expandability	Tool	allows to extend the tool with the features that are not implemented in it	High/critical	[77,22]
Multiple visualization techniques	Tool	allows the representation in different models	High/critical	[78,19,32,16,23]
Search	Visualization	allows to search for a specific elements by its name	High	[28], [16], [25], [29]
Overview	Visualization	allows to gain an overview of the entire collection	High	[19], [47], [16], [27], [28]
Zooming and Panning	Visualization	allows to zoom in and out items of interest, while retaining global context	High	[47], [19], [29], [32], [28], [16], [27]
Semantic Zooming	Visualization	allows to control the level hierarchy in the visualization by zooming	Task-dependant	[16], [27], [19], [78], [47]
Concept clustering	Visualization	the selection and aggregation of key or the most relevant concepts of ontology and the limitation of visualization	High	[16], [76], [75]
Filter	Visualization	filters out uninteresting items to avoid or at least minimize clutter	Task-dependent/ high	[47], [29], [19]
Details on demand	Visualization	allows to select an item or group and get details when needed	Task-dependent/ high	[47], [19], [28], [16]
History	Visualization	allows to keep a history of actions to support undo, replay, and progressive refinement	Task-dependent/ high	[19], [16], [62]
Colors	Visualization	allows the user to distinguish between ontology elements	High	[25], [16]
Shapes	Visualization	allows the user to distinguish between ontology elements	Task-dependent/ high	[25], [16], [65]
Natural language	Tool	allows the user to translate and visualize the ontology in the user's language	High	[32]

distinguishing of different elements and ability to navigate between them and search for element or area of interest should be reached.

In addition, as we have discussed above, depending on the task and user intents, additional requirements can be posed as potentially facilitating interaction and use of the ontology and its visualization. This includes features such as *brushing, magic lens, symbols* (in addition or instead of *color and shapes* for different elements of ontology), *saturation and texture, 3D navigation*, etc. From popular features, *labeling* has not been deliberately mentioned here, although in most cases it has a positive effect on visualization making it more self-explanatory, but it may also cause distraction. Therefore, for this as well as for many other features that are not considered to be "must have," it would be advisable to give users the opportunity to choose whether and when the feature under question is used, i.e., dynamic interaction with the tool features should be ensured. Otherwise, extendability becomes "must have."

3.4 **Discussion**

Health(care) ontology and their terminologies play a key role in the representation of knowledge and integration of health data. In healthcare systems, the Internet of Technologies (IoT) or more precisely Internet of Medical Things (IoMT), provides a data exchange between different entities and ontologies that provide a formal description to present the knowledge of healthcare domains. It is recommended that these ontologies ensure the quality of their adoption and applicability in the real world. However, despite the popularity of medical sector-related ontologies and diverse supporting tools, it is still not clear *how do the system holders, who worked with relational databases over the decades should migrate to the ontologies and semantic models.* And even in case of a positive answer, although there is a wide variety of different approaches and domain-specific ontologies and their visualizations, the issue of more efficient management of all data in (bio)medical organization remains open.

This is even more the case when the question on what is understood as (bio)medical data is asked, i.e., whether these are only data collected in and exchanged between medical organizations or they include scientific literature published on these concerns. As a result, there are may ad hoc approaches to this issue. One such attempt was Sarkans [39] focusing on more efficient and sustainable scientific report-related data management through the BioStudies public database that organizes data from biological studies. Another more recent attempt is PharmSci [36], which is an ontology for modeling pharmaceutical research data aimed at making it easier to access, reuse, curate, and integrate data from documented research toward providing services, thereby unveiling hidden knowledge. What makes this solution different compared to others is that it mainly refers to scientific outputs already produced by medics and their effective reuse—it combines scholarly metadata with domain-specific metadata. Both of these studies emphasize the significance of scientific outputs, their easier access and further reuse, which opinion we share. This question has not been discussed in this chapter, but could be seen as one of prospective directions to be explored in the future.

Another point to be taken into account is posed by [54] and [55], according to which, given that healthcare is a national competence that is covered to national law, it is often challenging to export health data from one country outside regional or national jurisdictions. They therefore stress that the transformation of the (European) field of life science and health data will only be possible only by coordinating national and international initiatives, connecting developments between projects and countries in long-term standards-based infrastructure operating on a continental scale, and by providing a procedural framework that will guarantee the rights of patients' while ensuring controlled access to data across borders. So, apart of the intuitive questions asked and answered in this section, there is still a long list of open questions, including general ones related to legal issues that should also be taken into account when dealing with the data.

3.5 Conclusion

The aim of ontology is to achieve a common and shared knowledge that can be transmitted between people and application systems [74]. Thus, ontology plays an important role in ensuring interoperability between organizations and on a semantic web, as they aim to acquire knowledge of areas, and are tasked with creating semantics explicitly in a generic way, providing a basis for reaching an agreement in an area. Thus, ontology has become a popular research topic in many communities, including the medical sector. Semantic models provide crucial machine-interpretable information to the knowledge discovery process [36].

In this section, we have provided an overview of both the concept of ontology visualization and ontology visualization evaluation, as well as have transformed this knowledge in a taxonomy that seems to be lacking and, therefore, misunderstanding about the concept has been observed in both practice and literature. This allowed us to point on the solutions that are not appropriate for medical and (health)care sectors, as well as identify areas that although are challenging for most other areas, are almost resolved in this particular domain. The proposed taxonomies provide an overview of the diversity of tools and approaches, their objectives and, consequently, requirements posed to them, where there is still no "silver bullet" and "one-fits-all," although there are many prospective solutions for a wide range of purposes and users, starting with those with very limited knowledge, where recommenders are provided for the selection of both ontology and ontology visualization. They provide users with suggestions that most likely will meet their needs, and continue with those suitable for both advanced and experienced users or even developers. We found that the evaluation of ontology in the medical sector was almost resolved, where a continuously growing number of very specific medical ontologies forced the development and continuous improvement of ontology selection supporting tools. However, the same question posed to a visualization technique or tool remains open, where the correctness of choice depends to a large extent on the specific nature of task. This leads to cases when organizations and research teams are forced to use either a set of tools, or to develop their own visualization tools, or to pass and refuse the task. The list of requirements posed for the tool depends on the subdomain and may vary from one task to another, even in scope of one domain. This list should therefore be determined on the basis of the requirements imposed from the task.

In this section, we have provided a list of different types of features that could facilitate the choice of a visualization tool by determining a list of the key features required for the task. This should allow researchers to define a whole-grained set of requirements for visualization technique or a tool that is less likely to change afterwards. At the same time, it could be beneficial for developers that aim to develop more universal tools.

Acknowledgments

This research was funded by the European Regional Development Fund project "DECIDE – Development of a dynamic informed consent system for biobank and citizen science data management, quality control, and integration," grant 1.1.1.1/20/A/047.

References

[1] Sanju Mishra, Sarika Jain, Ontologies as a semantic model in IoT, International Journal of Computers and Applications 42 (3) (2020) 233–243.

[2] Sanju Tiwari, Abraham Ajith, Semantic assessment of smart healthcare ontology, International Journal of Web Information Systems 16 (4) (2020) 475–491.

[3] Katie Li, Ashutosh Tiwari, Jeffrey Alcock, Pablo Bermell-Garcia, Categorisation of visualisation methods to support the design of Human–Computer Interaction Systems, Applied Ergonomics 55 (2016) 85–107.

[4] Jill H. Larkin, Simon Herbert, Why a diagram is (sometimes) worth ten thousand words, Cognitive Science 11 (1) (1987) 65–100.

[5] Barbara Tversky, Suwa Masaki, Thinking with Sketches, Oxford University Press, 2009, pp. 75–84.

[6] Mohamad Ummul Hanan, Mohammad Nazir Ahmad, Youcef Benferdia, Azrulhizam Shapi'i, Mohd Yazid Bajuri, An overview of ontologies in virtual reality-based training for healthcare domain, Frontiers in Medicine 8 (2021).

[7] Jooyoung Jang, Schunn D. Christian, Physical design tools support and hinder innovative engineering design, in: Proceedings of the Human Factors and Ergonomics Society Annual Meeting, Sage CA, Los Angeles, SAGE Publications, CA, 2012, pp. 1279–1283, 55(1).

[8] Daniel Fallman, Design-oriented human–computer interaction, in: Proceedings of the SIGCHI Conference on Human Factors in Computing Systems, 2003, pp. 225–232.

[9] Elizabeth Sanders, Colin William, Harnessing people's creativity: ideation and expression through visual communication, in: Focus Groups, CRC Press, 2002, pp. 147–158.

[10] Anne Romer, Martin Pache, Guido Weißhahn, Udo Lindemann, Winfried Hackera, Effort-saving product representations in design—results of a questionnaire survey, Design Studies 22 (6) (2001) 473–491.

[11] Yue Guan, Wei Qiang, Chen Guoqing, Deep learning based personalized recommendation with multi-view information integration, Decision Support Systems 118 (2019) 58–69.

[12] S.K. Card, Information visualization, in: Julie A. Jacko, Andrew Sears (Eds.), Human–Computer Interaction Handbook, Lawrence Erlbaum Associates, 2002, pp. 544–582.

[13] Barend Mons, Erik Schultes, Fenghong Liu, Annika Jacobsen, Deep learning based personalized recommendation with multi-view information integration, in: Data Intelligence, MIT Press, 2020, pp. 1–9, 2(1–2).

[14] Hannes Ulrich, Ann-Kristin Kock-Schoppenhauer, Petra Duhm-Harbeck, Josef Ingenerf, Using graph tools on metadata repositories, German Medical Data Sciences: A Learning Healthcare System (2018) 55–59.

[15] Hans-Georg Fill, On the conceptualization of a modeling language for semantic model annotations, in: International Conference on Advanced Information Systems Engineering, Springer, Berlin, Heidelberg, 2011, pp. 134–148.

[16] Marek Dudas, Steffen Lohmann, Vojtech Svatek, Dmitry Pavlov, Ontology visualization methods and tools: a survey of the state of the art, Knowledge Engineering Review 33 (10) (2018) 1–39.

[17] Marek Dudas, Steffen Lohmann, Vojtech Svatek, Dmitry Pavlov, Starting ontology development by visually modeling an example situation—a user study, in: Proceedings of the 2nd International Workshop on Visualization and Interaction for Ontologies and Linked Data, 2016, pp. 114–119.

[18] Marek Dudas, Vojtech Svatek, Ontology reuse decision support: visualize the ontology or its usage?, in: Proceedings of the Third International Workshop on Visualization and Interaction for Ontologies and Linked Data, 2017, pp. 13–19.

[19] Akrivi Katifori, Constantin Halatsis, George Lepouras, Costas Vassilakis, Eugenian Giannopoulou, Ontology visualization methods—a survey, ACM Computing Surveys (CSUR) 39 (4) (2007) 10-es.

[20] Jose Barranquero Tolosa, Oscar Sanjuan-Martinez, Vicente Garcia-Diaz, B. Cristina Pelayo G-Bustelo, Juan Manuel Cueva Lovelle, Towards the systematic measurement of ATL transformation models, Software, Practice & Experience 41 (7) (2011) 789–815.

[21] J. Stephen, Mellor MDA Distilled: Principles of Model-Driven Architecture, Addison–Wesley Professional, 2004, pp. 10–134.

[22] Michael Schaaf, Franziska Jahn, Kais Tahar, Christian Kücherer, Alfred Winter, Barbara Paech, Visualization of large ontologies in university education from a tool point of view, in: Exploring Complexity in Health: An Interdisciplinary Systems Approach, IOS Press, 2016, pp. 349–353.

[23] Bo Fu, Natalya F. Noy, Margaret-Anne Storey, Eye tracking the user experience – an evaluation of ontology visualization techniques, Semantic Web 8 (1) (2017) 23–41.

[24] M.K. Bergman, Listing of 185 ontology building tools, AI3: Adaptive information, [Online], available: http://www.mkbergman.com/904/listing-of-185-ontology-building-tools/, 2010.

[25] Steffen Lohmann, Stefan Negru, Florian Haag, Thomas Ertl, VOWL 2: user-oriented visualization of ontologies, in: International Conference on Knowledge Engineering and Knowledge Management, 2014, pp. 266–281.

[26] Marek Dudass, Ondrej Zamazal, Vojtech Svatek, Roadmapping and navigating in the ontology visualization landscape, in: International Conference on Knowledge Engineering and Knowledge Management, Springer, Cham, 2014, pp. 137–152.

[27] Steffen Lohmann, Stefan Negru, Florian Haag, Thomas Ertl, Visualization of gene ontology and cluster analysis results, Linnaeus University, 2012, pp. 1–64.

[28] Ben Shneiderman, The eyes have it: a task by data type taxonomy for information visualizations, in: The Craft of Information Visualization, Morgan Kaufmann, 2003, pp. 364–371.

[29] Michael Glueck, Ontology-based Context in Visualizations to Facilitate Sensemaking: Case Studies of Phenotype Comparisons, Doctoral dissertation, University of Toronto, 2018.

[30] Bin Yang, Memory Island: Visualizing Hierarchical Knowledge as Insightful Islands, Doctoral dissertation, Universite Pierre et Marie Curie, Paris, 2015, pp. 1–200.

[31] Sebastian Kohler, et al., The human phenotype ontology in 2017, Nucleic Acids Research 45 (1) (2017) 865–876.

[32] Merlin Florrence, MLGrafViz: multilingual ontology visualization plug-in for Protégé, Computer Science and Information Technologies 2 (1) (2021) 43–48.

[33] Uldis Bojars, Renars Liepins, Normunds Gruzitis, Karlis Cerans, Edgars Celms, Extending OWL ontology visualizations with interactive contextual verbalization, in: VOILA@ ISWC, 2016, pp. 5–16.

[34] Viktoria Pammer, Peter Scheir, Stefanie Lindstaedt, Ontology coverage check: support for evaluation in ontology engineering, in: The 2nd Workshop: Formal Ontologies Meet Industry, 2006.

[35] Xiaoming Wang, Carolyn Williams, Zhen Hua Liu, Joe Croghan, Big data management challenges in health research—a literature review, Briefings in Bioinformatics 20 (1) (2019) 156–167.

[36] Zeynep Say, Said Fathalla, Sahar Vahdati, Jens Lehmann, Sören Auer, Ontology design for pharmaceutical research outcomes, in: International Conference on Theory and Practice of Digital Libraries, Springer, Cham, 2020, pp. 119–132.

[37] Janet Kelso, Robert Hoehndorf, Kay Prufer, Ontologies in biology, in: Theory and Applications of Ontology: Computer Application, Springer, Dordrecht, 2010, pp. 347–371.

[38] Robert Stevens, Phillip Lord, Carole Goble, Application of ontologies in bioinformatics, in: Handbook on Ontologies, Springer, 2009, pp. 735–756.

[39] Ugis Sarkans, Mikhail Gostev, Awais Athar, Ehsan Behrangi, Olga Melnichuk, Ahmed Ali, Jasmine Minguet, Juan Camillo Rada, Catherine Snow, Andrew Tikhonov, Alvis Brazma, Johanna McEntyre, The BioStudies database—one stop shop for all data supporting a life sciences study, Nucleic Acids Research 46 (1) (2018) 1266–1270.

[40] Alexander Rind, Wolfgang Aigner, Silvia Miksch, Sylvia Wiltner, Margit Pohl, Thomas Turic, Felix Drexler, Visual exploration of time-oriented patient data for chronic diseases: design study and evaluation, in: Symposium of the Austrian HCI and Usability Engineering Group, Springer, Berlin, Heidelberg, 2011, pp. 301–320.

[41] Ahmet Soylu, Martin Giese, Ernesto Jimenez-Ruiz, Evgeny Kharlamov, Dmitry Zheleznyakov, Ian Horrocks, OptiqueVQS: towards an ontology-based visual query system for big data, in: Proceedings of the 5th International Conference on Management of Emergent Digital EcoSystems, 2013, pp. 119–126.

[42] Bongwon Suh, Benjamin B. Bederson, OZONE: a zoomable interface for navigating ontology information, in: Proceedings of the Working Conference on Advanced Visual Interfaces, 2002, pp. 139–143.

[43] Andreas Harth, VisiNav: a system for visual search and navigation on web data, Journal of Web Semantics 8 (4) (2010) 348–354.

[44] Karlis Cerans, Uldis Bojars, Juris Barzdins, Julija Ovcinnikova, Lelde Lace, Mikus Grasmanis, Arturs Sprogis, ViziQuer: a visual notation for RDF data analysis queries, in: Research Conference on Metadata and Semantics Research, Springer, Cham, 2018, pp. 50–62.

[45] Vitalis Wiens, Steffen Lohmann, Soren Auer, WebVOWL editor: device-independent visual ontology modeling, in: International Semantic Web Conference (P&D Industry BlueSky), 2018.

[46] Nassira Achich, Alsayed Algergawy, Bassem Bouaziz, Birgitta Konig-Ries, BioOntoVis: an ontology visualization tool, in: EKAW, 2018, pp. 49–52.

[47] Ying Tao, Yang Liu, Carol Friedman, Yves A. Lussier, Information visualization techniques in bioinformatics during the postgenomic era, Drug Discovery Today: BIOSILICO 2 (6) (2004) 237–245.

[48] Ying Tao, Yang Liu, Carol Friedman, Yves A. Lussier, OnE: an ontology evaluation framework, Knowledge Organization 47 (4) (2020) 283–299.

[49] Ravi Lourdusamy, Antony John, A review on metrics for ontology evaluation, in: 2018 2nd International Conference on Inventive Systems and Control (ICISC), IEEE, 2018, pp. 1415–1421.

[50] Andrew Burton-Jones, Veda C. Storey, Vijayan Sugumaran, Punit Ahluwalia, A semiotic metrics suite for assessing the quality of ontologies Data &, Knowledge Engineering 55 (1) (2005) 84–102.

[51] Muhammad Amith, Zhe He, Jiang Bian, Juan Antonio Lossio-Ventura, Cui Tao, Assessing the practice of biomedical ontology evaluation: gaps and opportunities, Journal of Biomedical Informatics 80 (2018) 1–13.

[52] Muhammad Amith, Frank Manion, Chen Liang, Marcelline Harris, Dennis Wang, Architecture and usability of OntoKeeper, an ontology evaluation tool, BMC Medical Informatics and Decision Making 19 (4) (2019) 1–18.

[53] Ioana Jimborean, Adrian Groza, Ranking ontologies in the ontology building competition boc 2014, in: 2014 IEEE 10th International Conference on Intelligent Computer Communication and Processing (ICCP), IEEE, 2014, pp. 75–82.

[54] Declaration of Cooperation, Towards access to at least 1 million sequenced genomes in the European Union by 2022 European Commission, http://ec.europa.eu/newsroom/dae/document.cfm?doc_id=50964. (Accessed 15 May 2019).

[55] Gary Saunders, et al., Leveraging European infrastructures to access 1 million human genomes by 2022, Nature Reviews Genetics 20 (11) (2019) 693–701.

[56] Roxana Merino-Martinez, Loreana Norlin, David van Enckevort, Gabriele Anton, Simone Schuffenhauer, Kaisa Silander, Linda Mook, Petr Holub, Raffael Bild, Morris Swertz, Jan-Eric Litton, Toward global biobank integration by implementation of the minimum information about biobank data sharing (MIABIS 2.0 Core), Biopreservation and Biobanking 14 (4) (2016) 298–306.

[57] Ninad Jog, Ben Shneiderman, Starfield visualization with interactive smooth zooming, in: Working Conference on Visual Database Systems, Springer, Boston, MA, 1995, pp. 3–14.

[58] M.R. Umamaheswari, A role of ontology in banking, in: International Conference on Emerging Trends in Banking, Insurance and International Trade, 2019, pp. 481–485.

[59] Dimitris Iakovidis, Christos Smailis, A semantic model for multimodal data mining in healthcare information systems, in: Quality of Life Through Quality of Information, IOS Press, 2012, pp. 574–578.

[60] Enrico Motta, Paul Mulholland, Silvio Peroni, Mathieu d'Aquin, Jose Manuel Gomez-Perez, Victor Mendez, Fouad Zablith, A novel approach to visualizing and navigating ontologies, in: International Semantic Web Conference, Springer, Berlin, Heidelberg, 2011, pp. 470–486.

[61] Thorsten Liebig, Olaf Noppens, OntoTrack: a semantic approach for ontology authoring, Journal of Web Semantics 3 (2–3) (2005) 116–131.

[62] Enrico Motta, Silvio Peroni, Ning Li, Mathieu d'Aquin, KC-Viz: a novel approach to visualizing and Navigating ontologies, in: The 17th International Conference on Knowledge Engineering and Knowledge Management (EKAW 2010), 11–15 Oct 2010, Lisbon, Portugal, 2010, p. 674.

[63] Sean M. Falconer, Chris Callendar, Margaret-Anne Storey, FLEXVIZ: visualizing biomedical ontologies on the web, in: International Conference on Biomedical Ontology, Software Demonstration, 2009.

[64] Ronak Panchal, Priya Swaminarayan, Sanju Tiwari, Fernando Ortiz-Rodriguez, AISHE-onto: a semantic model for public higher education universities, in: DG. O2021: The 22nd Annual International Conference on Digital Government Research, 2021, pp. 545–547.

[65] Piotr Kunowski, Tomasz Boiński, SOVA—simple ontology visualization API, Protege, online, https://protegewiki.stanford.edu/wiki/SOVA, 2009.

[66] Alani Harith, TGVizTab: an ontology visualisation extension for Protégé Knowledge Capture (K-Cap'03), in: Workshop on Visualization Information in Knowledge Engineering, Sanibel Island, Florida, United States.

[67] Ajaz Hussain, Khalid Latif, Aimal Tariq Rextin, Amir Hayat, Masoon Alam, Scalable visualization of semantic nets using power-law graphs, Applied Mathematics & Information Sciences 8 (1) (2014) 355.

[68] Clement Jonquet, Mark A. Musen, Nigham H. Shah, Building a biomedical ontology recommender web service, Journal of Biomedical Semantics, BioMed Central 1 (1) (2010) 1–18.

[69] Marcos Martinez-Romero, Clement Jonquet, Martin J. O'Connor, John Graybeal, Alejandro Pazos, Mark A. Musen, NCBO Ontology Recommender 2.0: an enhanced approach for biomedical ontology recommendation, Journal of Biomedical Semantics 8 (1) (2017) 1–22.

[70] Marcos Martinez-Romero, Jose M. Vazquez-Naya, Javier Pereira, Alejandro Pazos, BiOSS: a system for biomedical ontology selection, Computer Methods and Programs in Biomedicine 114 (1) (2014) 125–140.

[71] Harith Alani, Natalya F. Moy, Nigam Shah, Nigel Shadbolt, Mark A. Musen, Searching ontologies based on content: experiments in the biomedical domain, in: Proceedings of the 4th International Conference on Knowledge Capture, 2007, pp. 55–62.

[72] Tan He, Patrick Lambrix, Selecting an ontology for biomedical text mining, in: Proceedings of the BioNLP 2009 Workshop, 2009, pp. 55–62.

[73] Maiga Gilbert, Ddembe Williams, A flexible biomedical ontology selection tool, International Journal of Computing and ICT Research, Special Issue 3 (1) (2009) 53–66.

[74] Mohammad Mustafa Taye, Understanding semantic web and ontologies: theory and applications, Journal of Computing 2 (6) (2010) 182–192.

[75] Chang-Shing Lee, Yuan-Fang Kao, Yau-Hwang Kuo, Mei-Hui Wang, Automated ontology construction for unstructured text documents, Data & Knowledge Engineering 60 (3) (2007) 547–566.

[76] Pat Hayes, Thomas C. Eskridge, Raul Saavedra, Thomas Reichherzer, Mala Mehrotra, Dmitri Bobrovnikoff, Collaborative knowledge capture in ontologies, in: Proceedings of the 3rd International Conference on Knowledge Capture, 2005, pp. 99–106.

[77] Nils Malzahn, S. Weinbrenner, P. Husken, J. Ziegler, H.U. Hoppe, Collaborative ontology development—distributed architecture and visualization, 2007, pp. 5–10.

[78] Julia Dmitrieva, Fons J. Verbeek, Multi-view ontology visualization, in: The 11th International Protege Conference, 2009, pp. 1–4.

[79] Jinie Pak, Lina Zhou, A framework for ontology evaluation, in: Workshop on E-Business, Springer, Berlin, Heidelberg, 2009, pp. 10–18.

Role of connected objects in healthcare semantic models

4

Gustavo de Assis Costa[a]**, Inaldo Capistrano Costa**[b]**, and Ayush Goyal**[c]

[a]*Department of Computer Science, Federal Institute of Education, Science, and Technology of Goiás, Jataí, GO, Brazil*
[b]*Division of Computer Science, Aeronautics Institute of Technology (ITA), São José dos Campos, SP, Brazil*
[c]*Department of Electrical Engineering and Computer Science, Texas A&M University, Kingsville, TX, United States*

Contents

4.1 Introduction

As a government policy, the healthcare of citizens of any country must be a priority. Thousands of initiatives have been taken worldwide to solve problems and optimize patient care by physicians, nurses, hospitals, and clinics. In this sense, one of the great allies for the advances obtained so far was the constant advancement of information technology.

Several information systems for healthcare were developed and implemented in different parts of the world, giving rise to different solutions, and consequently, different standards such as business rules, data models, protocols, and several other aspects inherent to the software development process.

One of the results of this computerization process was the implementation of Electronic Health Records (EHRs). According to Gunter and Terry [1], EHR is a systematized collection of patient health, clinical, and demographic information stored in digital format and shared in different environments, such as information systems in hospitals, health plans, clinics, physicians, etc.

Even with the advances already obtained in developing and implementing standards in managing and using health and medical information, the interoperability between the most different types of actors and systems involved with EHR remains one of the biggest and existing challenges. In addition to managing data concerned to healthcare such as medical appointments and drug prescriptions, the possibility of including various other types of information obtained in the treatment of patients, such as imaging exams and videos, for example, increases the complexity of systems.

Semantic interoperability has been used as a critical solution to enable data-centricity. According to Veltman [2], semantic interoperability is "the ability of information systems to exchange information based on shared, preestablished, and negotiated meanings of terms and expressions." In EHR software, an archetype model is an approach that enables semantic interoperability, decreasing software maintenance needs since all knowledge will reside in software-independent archetypes, and changes can be made independent of software or databases [3].

Data integration between different clinical health records is still difficult to resolve but essential for the efficient use of EHR for personalized and quality patient care. Different information systems used by the various healthcare providers must be able to interoperate [4] and, in this sense, one system must understand the context and meaning of the information provided by another system [3]. This contextual and semantic synergy is possible through the use of essential but challenging semantic interoperability.

Remote health monitoring activities are gaining momentum with the advent of the Internet of Things (IoT), one of healthcare's most popular technological trends [5]. Each industrial segment has specific demands, and the application of IoT-based technologies needs to be adapted to meet the specific requirements of each one [6]. In the healthcare domain, the application of IoT technology is considered with a specific definition. The use of remote monitoring devices to analyze vital signs and other measures relating to the patient's environment allows continuous observation and monitoring [7].

Visioning the future of the fully automated industry, IoT is one important technology that is boosting great advances, as in Industry 4.0 [8]. Regarding the e-Health industry, Manogaran et al. [9] defined the term Internet of Medical Things (IoMT) (also known as healthcare IoT). With this kind of infrastructure, it is possible to check, directly over the internet, signals such as temperature, ECG, EEG, blood pressure, blood oxygenation, among others. Consequently, the idea of patient-centered service provision becomes more plausible, allowing for greater precision and increased quality of life for patients.

According to Jabbar et al. [10], issues like standards, scalability, devices heterogeneity, scalability, standards in addition to many others are still open challenges and topics of recent research. In line with EHR systems, interoperability is a key issue to allow effective communication and knowledge sharing between the most remote monitoring devices and other systems employed in industries. Also, in line with the interoperability solution strategies from e-Health systems, Ganzha et al. [11] state that devices interoperability, like IoT-based, can be solved if semantically interoperable.

The solution space for the interoperability problem in the focus areas of this text is directly associated with adopting semantic technologies. Recent work has demonstrated this trend, promising results [12–18]. Nevertheless, some concerns must be taken into account so that it is possible to realize the expected benefit not only for patients but also for other users of all the systems. To make the most of the environment made up of different connected objects and healthcare system software to achieve full interoperability, it is necessary to think of semantic models that consider both remote health monitoring devices (IoT) and the e-Health systems.

An increasing number of IoT ontologies were proposed in the literature to overcome the interoperability issue between connected devices in the most different kinds of industries. Among the most commonly discussed, the W3C Semantic Sensor Network (SSN/SOSA) and the ETSI SAREF are the most promising [7,19,20], which in turn could be aligned [21].

On the other hand, there are some standards in e-Health systems as the openEHR information model, defined by the openEHR Foundation as "a set of models, software, and open specifications used to create standards and build information and interoperability solutions for healthcare" [22,23]. HL7 is another attempt from the health community, defined by the Health Level Seven International (HL7) organization as "a framework and standards for exchanging, integrating, sharing, and retrieving e-Health information. Features like package format, peer communication, structure, and data types are some examples of standards defined for integration systems" [24]. There is not yet a consensus in standards specification to e-Health systems. Nevertheless, openEHR is one of the most widely used standards and significantly influences developing other international standards like HL7 itself, CEN (European Committee for Standardization), and ISO [3].

To align these standards and technologies, one must still consider another important aspect of these ecosystems, like communication protocols, security and privacy concerns, cloud services, APIs, and data analysis capability. Combining these technologies demands a huge effort and requires the effective participation of industry, academia, standardization bodies, and government.

Despite the great advances, there is much to be done. In the figure below (Fig. 4.1), it is possible to depict a general landscape view of the context in question. Without sticking to the details, one can state that a semantic layer will be a key component of any proposed architecture. Semantic annotations and ontologies will be essential tools to overcome the interoperability issues.

To better illustrate the problem, we will consider a case study where a patient with chronic heart disease has some of his vital signs, like blood pressure, body temperature, electrocardiograms (ECG), and oxygen saturation, constantly monitored. In addition, sensors and devices collect environmental parameters that directly influence the patient health conditions. He also has a health record stored in the EHR system of his city.

First, the ideal situation would be to consider that all data could be shared between all EHR systems in his state and country. Under this panorama, we have to regard the interoperability between different standards adopted between the various systems. Each standard adopts different attributes and relationships regarding data. Each of the systems must understand what the other does. This understanding is precisely what semantic interoperability offers, a set of technologies that will allow systems to share the meaning of data through previously defined models.

The next sections will present an overview of the role of sensor devices and semantic technologies in the e-Health domain through descriptions of technologies and patterns, existing works, current trends, and future insights. Although not exhaustive, this chapter aims to present and consolidate the main ideas developed or under development to seek maximum quality in the health services offered to society.

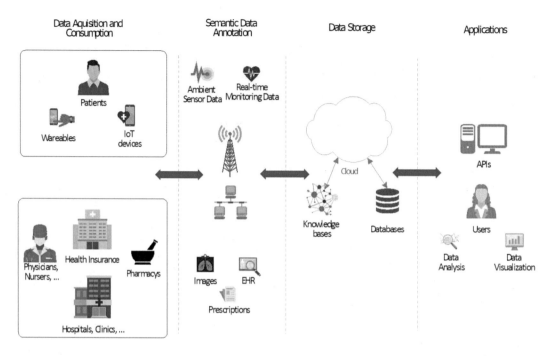

FIGURE 4.1

HealthCare system landscape under semantic interoperability perspective.

4.2 The EHR ecosystem

Hovenga and Garde [25] present a set of functions that can benefit from information extracted and managed by e-Health systems. This set includes decision support, managed care, resource management, public health, health policy analysis, and more. Beyond the common sense of electronic transactions employment, these systems were designed to extrapolate the doctor-patient relationship from both doctors and their patients, making way for new opportunities to create better medical services, health policies improvement, control and transparency increase, worldwide system integration.

Clinical data registry, also known as the clinical registry, is the atomic element of information for e-Heath systems, and it has been widely used around the world. The National Institutes of Health describe a registry as "a collection of personal information, usually focused on a specific diagnosis or condition" [26]. National Center for Biotechnology Information (NCBI) defines clinical registry as "a real-world view of clinical practice, patient outcomes, safety, and comparative effectiveness [27]."

In turn, International Organization for Standardization (ISO) [28] defines the "Integrated Care Electronic Health Record" (EHR) as "a data set with information about the specific healthcare in computer processable form" and, it has a commonly agreed logical information model that specifies the structures and relationships between information but is independent of any particular technology or implementation environment."

Claiming frustration with promised IT automation to the health systems and processes and the lack of support for continuity of care, healthcare professionals and stakeholders supported the openEHR community, which proposed the openEHR specification aiming to improve the quality of technology in medicine, consequently improving patient care and the quality of the entire health system.

Some healthcare standards were proposed to provide interoperability between different healthcare systems and for different purposes. These standards relate to both syntactic and semantic interoperability, defining the meaning of the communication. The most widely used patterns are Health Level Seven (HL7) International like HL7, HL7 v2, and HL7 v3 [24], which are related to message interchange; Systematised Nomenclature of Medicine – Clinical Terminology (SNOMED-CT [29]), which are related to terminologies and openEHR [22] and HL7 Clinical Document Architecture-CDA, which are related to clinical information and patients records [30].

Interoperability with EHR systems is still a problem to be solved and many opportunities and challenges to overcome. Adel et al. [31] list some examples of limitations that persist, among these:

- the need for information adaptation to the standards;
- complex understanding, and consequently, the complex implementation of some standards;
- adaptability to future uses;
- some standards, such as HL7, do not have full support to semantic interoperability;
- some standards can be ambiguous due to inaccurate information, causing interpretation problems by physicians;
- the great number of EHR standards and systems would not contribute to the seamless exchanging of clinical data.

In the subsections below, we give an overview of some EHR technologies that support most modern systems.

4.2.1 OpenEHR

According to openEHR Foundation [22] openEHR is a set of open specifications, templates, and software used to create standards and build healthcare information and interoperability solutions for the healthcare area.

Different worldwide projects influenced the openEHR architecture, leading to a model-driven approach with a multilevel modeling framework separating data representation from domain content. According to [22], this multilevel approach provides three levels of models [23].

Reference model (RM): the first level of modeling is a stable reference information model and the only implemented in software. The reference model only involves concepts relating to the administrative and services context. Reusable content element definitions: archetypes as artefacts with definitions of clinical information. The different kinds of care events and subjects of care are defined in archetypes and templates; formal definitions for data created from a set of elements obtained from archetypes in openEHR models. These data are used for forms, documents, messages, etc.

Archetypes and templates are "well-defined semantic gateway to terminologies, classifications and computerized clinical guidelines" [23]. A data schema, usually based on an archetype model that considers data types definition from RM, is used to define the structure of the data persistence model in EHR systems [7].

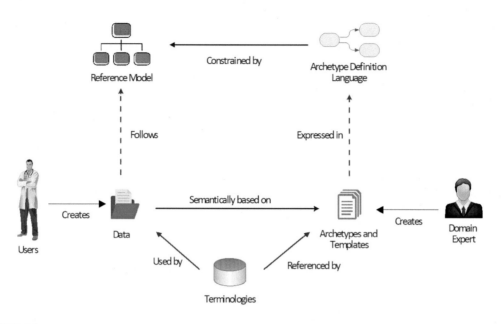

FIGURE 4.2

openEHR Archetype architecture overview.

   Domains experts can create or modify archetypes and templates, maintaining compatibility to the information system. Terms within an archetype can have relationships with external terminologies like SNOMED-CT. Archetypes are constrained by RM and Archetype Definition Language (ADL), specifically by the archetype type, data structure, and data type defined in RM. One can consider there are three elements in an ontological way in the openEHR environment: the reference model, the archetypes, and terminology, both internal and external. However, various archetypes may be developed for the same clinical concept, hindering semantic interoperability [17] (Fig. 4.2).

4.2.2 Health level seven international (HL7)

HL7 provided a set of standards and a framework for exchanging, integrating, sharing, and retrieving EHR data, defining the process of information packaging and communication, and setting essential elements for integration between systems as language, structure, and data types required [24]. In order to improve existing standards, HL7 FHIR [32] was created to facilitate the exchange of healthcare information between organizations. This new framework is composed of a content model in the form of resources and a specification for real-time resource sharing through RESTful interfaces and messaging, and documents.

   According to Peng and Goswami [33], the FHIR working group is also working on the FHIR linked data module, which is based on semantic web technologies. Standards such as Resource Description Framework (RDF) and Web Ontology Language (OWL) are some of the resources used.

4.2.3 ISO 13606

"The ISO/EN 13606 standard specifies the communication part or all of the EHR of one or more identified subjects of care between EHR systems, or between EHR systems and a centralized EHR data repository" [34].

It was developed by CEN/TC251, a European technical committee responsible for setting standards in Health Information and Communications Technology. The standard provides a model for representing the information that can be included in an EHR and how EHRs can exchange data between them. However, it does not define the architecture details and data storage. ISO/EN 13606 is based on a reference model and an Archetype Object Model (AOM), where knowledge is represented by the concepts of the clinical domain by means of archetypes [35].

4.2.4 Semantic models in healthcare

Although all the efforts, there is no standard de facto to overcome the interoperability problem concerned to EHR systems. We have not yet a solution, but standards are maturing independently. An ideal semantic model should minimally "understand" the different standard health formats like HL7, OpenEHR, ISO 13606, and others.

The semantic model is a high-level conceptual data model description and structuring formalism that includes semantic information, permitting the interpretation of the meaning from the instances. Semantic web technologies are used for annotating and sharing data using web protocols. Data integration and reuse in a machine-readable format is a crucial aspect when we use these tools. The Semantic Web Stack, standardized by W3C [36], among several other elements, defines the Ontology Web Language (OWL), a specification of an ontology language based on a description logic, RDF (Resource Description Framework), a specification that describes resources in the form of triples, and SPARQL, a query language based on triple patterns modeled with RDF.

Ontology as a data model can formally represent concepts from a specific domain and their relationships. Thus, humans and machines may understand the meaning of the exchanged data. This is the essence of interoperability, and nothing is more natural than applying ontologies to resolve interoperability issues in EHR systems. Because of this, most recent works are supported by the use of semantic web technologies.

Approaches like in Boscá et al. [37] propose to use archetype methodology to perform data transformation between FHIR and other specifications such as HL7 CDA, EN ISO 13606, and openEHR. Nevertheless, there are some drawbacks, like separating the narrative definition and the formal definition of FHIR resources when dealing with FHIR archetypes.

Papež et al. [38] states the urgent need to develop and evaluate a machine-readable standard to facilitate the systematic creation, sharing and reuse of EHR derived phenotyping algorithms. They proposed the use of OWL and RDF for enabling computable representations of EHR-driven phenotyping algorithms.

Another approach proposed to integrate EHRs from heterogeneous resources [39]. They build semantic data virtualization layers on top of data sources and then use SPARQL templates and a reasoner to complete the transformation to a source-independent RDF model based on a canonical ontology. According to them, the canonical ontology model enables to export into formats accepted by various other systems.

In [40], the authors proposed an open-source tool based on a two-step approach to perform RDF data transformation by applying semantically rich models to data sets. The goal is to standardize concise RDF data sets, originally transformed from relational sources or even other schemas, with a semantically rich model that is schema-independent. An important aspect of the tool is the definition of a set of constraints (target model) on the application ontologies.

In their first work, Adel et al. [31] considered different formats as ADL (Archetype Definition Language) archetype from OpenEHR, relational databases, spreadsheets (CSV, ...), and XML documents. This proposal was also based on a two-step approach where in the first step, they convert each input source to an OWL ontology. After that, they integrated all the ontologies into a canonical ontology. In [41], they proposed a framework containing an expanded version of the canonical ontology into a fuzzy ontology aiming to better deal with uncertainty.

Regarding our cardiac patient in the case study, it is known that he is an executive who has a hectic life and who frequently travels all over the world to solve his company's business. With his lifestyle and health condition, it would be expected that he would need medical attention sometimes. EHR semantic models could then offer full conditions to access patient records, last clinical attendances, exams, and other relevant information to any health system participant of any country.

Suppose a local physician needs to prescribe medication to the cardiac patient during his travels. In that case, he will not do so without first knowing all the patient's condition and history and the physician accompanying the patient also knowing about this. The EHR systems available to physicians will access all the information previously modeled by ontologies and stored in the cloud.

What about if this patient would need to have his vital signs monitored during his travel? What if the physician who accompanies the patient in his city would need to follow up of his ECG and environmental conditions during his travel?

In the next section, we will discuss the technical features of connected objects, some standards and the importance of also applying semantic interoperability in this context.

4.3 Connected objects in healthcare

Healthcare has always been a concern of any citizen, but even so, health problems have never ceased to impact people's lives. However, the lifestyle of modern society has increasingly influenced the neglect of signs that generally indicate that something is wrong. With the advancement of electronic technology and especially with the advent of pervasive computing, or ubiquitous computing, thousands of possibilities emerged to allow some limitations when monitoring a person's clinical conditions and vital signs to cease to exist.

With the daily advancement of wireless access device technology and electronic gadgets for personal use, many applications and functionalities are created and made available to consumers. This scenario is no different in applications for monitoring vital signs and other information concerned to health and environmental conditions present in smartphones, smartwatches, and wearable devices.

The development of standards that allow the articulation between remote monitoring objects and EHR systems by academia and industry has contributed a lot to the evolution of this ecosystem. However, along with this evolution, an environment of heterogeneous technologies and systems has emerged. This problem increases even more when comparing initiatives from different countries and continents.

Nevertheless, the path taken to solve this problem through semantic interoperability is the most promising. A large part of the related problems lies in the definition and proper treatment of data models, i.e., how the data will be extracted, transmitted, and manipulated. Next, some of the main semantic interoperability technologies and standards that have been addressed in the most recent works are briefly discussed.

Before discussing semantic models with connected objects, we will present an overview of some technologies and standards that support communication between different monitoring devices currently used in healthcare.

4.3.1 Internet of things

With the advent of IoT technologies, the pervasive e-Health market could be further boosted. Several applications could be planned and developed to monitor real-time information from different patients, greatly helping people with more severe comorbidities and the elderly, which require greater and continuous attention. Consequently, there could be reductions in hospitalizations, a cost decrease for all stakeholders, including own patients, and mainly the quality of life improvement.

An important aspect of being considered is the integration of IoT technologies with existing EHR systems, which, at first, may seem simply advantageous but is a necessity. The combination of information from EHR systems, with patients records from hospitals, clinics, pharmacies, caregivers, etc., and data gathered from remote devices with environmental and personal data obtained all the time, can improve the healthcare system with higher fidelity information. Recording this information opens the opportunity for a series of treatments that can be applied to the data, such as the search for patterns and the increasingly precise refinement to support decision-making. Technologies and solutions based on big data management and data analytics (machine learning, data mining) can be clearly adopted to deal with the huge amount of data.

From a scenario with different manufacturers, different standardization bodies, thousands of customized solutions, demands, and particular needs on the part of stakeholders, both EHR systems and IoT-based systems suffer from the same problem, interoperability. The most commonly adopted strategy has been the approach through the use of semantic technologies.

Furthermore, in the specific case of IoT, there are some other challenges in terms of complexity because of the nature of technology, which includes solutions to low-level aspects like the device itself, networking, middleware, application services, and data models. Each one of these levels must be considered when looking for an integrated solution. In this sense, several existing and emerging ontologies were designed for this domain. For instance, standardized solutions as W3C SSN (Semantic Sensor Network) ontology [42], SensorML (Sensor Model Language Encoding Standard) [43], IoT-Lite [44], IoT-A information model and IoT.est ontologies [45].

4.3.2 Semantic sensor network

According to Alamri [46], SSN has been one of the most widely adopted standards to describe sensors and IoT-based devices, describing the notions of sensor and physical devices, actuators, observations, and other concepts. With the exception of the SensorML, the rest of the standards above are in some way based on SSN. SensorML provides a framework to define sensors and sensor systems and the measurement processes, with descriptions of geometric, dynamic, and observational characteristics.

THE W3C SSN Incubator group states that "SSN uses high-level specifications to allow the efficient management, organization, understanding, and querying of data resultant from the network and its sensors. Ontologies can be a great tool to describe sensors, allowing classification and reasoning on the capabilities and measurements of sensors. They can also track the provenance of measurements and allow reasoning about individual sensors and the connection of several sensors as a macro instrument."

From a movement on rethinking SSN, Janowicz et al. [47] proposed SOSA (Sensor, Observation, Sample, and Actuator) ontology, one of the most general frameworks of semantic sensor models, provides representation for the elements, like relations and entities, involved in sensing, sampling, and actuation.

In a comprehensive survey of semantic sensor technologies in IoT [13], Honti and Abonyi provide an exciting overview and guidelines for any stage of semantic-based projects development. Calvillo-Arbizu [48] published another survey that analyzes the data life cycle;: trust, security, and privacy, and human-related issues. In this article, the authors reviewed the literature and concluded that most existing approaches that consider IoT application in healthcare do not regard the specific requirements of this domain.

4.3.3 M2M

ETSI (European Telecommunications Standard Institute) is one of the leading partners of oneM2M [49] global standards initiative. It involves some specifications and requirements, architecture, APIs, security solutions, and interoperability.

Lucic et al. [50] discussed the standardization of M2M as the direct communication between devices through a mobile or fixed network. The Machine-to-Machine (M2M) principle allows communication between two or more entities without direct human intervention, thus automating communication processes. As with IoT technology, M2M can be used by a wide range of ubiquitous applications, like e-Health.

A fundamental difference between M2M and IoT is the latter's dependence on IP-based networks so that monitored data can be sent from devices, usually on a cloud platform. In the case of M2M, devices are generally embedded in equipment and communicate through Wi-Fi networks, cellular networks, and can communicate with other objects by using peer-to-peer architecture. The underlying network structure must be transparent to the M2M system, including any network addressing mechanism [51].

To bridge the gap between M2M and health standards, the manager and the server entities do this role with openEHR and HL7 used in communication [52]. M2M devices are similar to IoMT devices. After sensing patient and environmental parameters, data is forwarded to the final applications through gateways, enabling integrating sensors and the platform and interchanging data through the semantic web paradigm [7].

4.3.4 oneM2M

oneM2M specification proposes to bring together all components in the IoT solution stack, with an architecture based on standard middleware technology in a horizontal layer, which interlinks devices, communications networks, cloud infrastructure, and IoT applications.

Recently, some articles [7,53–58] have proposed and discussed new architectures of IoT-based e-Health systems, also known as IoMT. For example, based on M2M specifications, these architectures are commonly based on multilayering abstraction where network components are involved in the overall specification, allowing tasks like device discovery, device identification, and binding between device and applications.

4.3.5 Smart appliances reference ontology

The ETSI Technical Committee for Smart Machine-to-Machine Communications (SmartTM2M) [59] has been promoting oneM2M Base Ontology with extensions in many IoT domains with SAREF. [60, 61] (Smart Appliances REFerence ontology), an open ontology for sharing and understanding semantic data across IoT devices and servers using different technologies among several domains.

Moreira et al. [62] presented SAREF4Health, an extension of SAREF, where authors proposed to address the verbosity problem of SAREF messages in IoT scenarios of real-time electrocardiography (ECG), where data needs to be represented as frequency-based time series of measurements observed by sensors.

Although several studies have been carried out or in progress that addresses semantic interoperability in IoT-enabled e-Health systems, it remains an open challenge in this field. Thus, in addition to characterizing a scenario similar to that of EHR systems, this situation demonstrates the great effort required to overcome the problem of semantic interoperability under these two different perspectives.

These two environments must complement each other to obtain efficient semantic health models, and thus build consistent systems that can benefit from all available data. In the current scientific literature, few solutions rely on the use of semantic models for EHR systems and integrate with models for monitoring and remote communication systems.

4.4 Semantic-based connected objects in e-Health

With the introduction of sensor-based systems in healthcare, new challenges concerned to the management of a huge amount of data generated by sensors and monitoring devices became a reality. Big data technologies would easily address this issue if it were not to interlink potential health data sources to view patient conditions comprehensively.

One of the recent advances in the healthcare industry involves continuously monitoring patients. Physicians can monitor patients' conditions through an information system integrated with data from various records such as these patients' history and their real-time health conditions and environmental parameters that could interfere with the patient's treatment. We can highlight a wide variety of applications in healthcare that are based on remote monitoring devices like monitoring glucose level, electrocardiogram, blood pressure, body temperature, oxygen saturation, rehabilitation system, medication management, in addition to many others.

Thus, linking device and sensor data with EHR data would be an expected requisite to any e-Health solution. The interoperability issue then would become even more latent because of the complexity and heterogeneity of existing devices. The communication between these different objects is not trivial, mainly because manufacturers create equipment that can maintain different proprietary protocols and systems.

These two distinct contexts must be considered together aiming to solve the interoperability problem and achieve the maximum use of available data. As discussed in the text, the semantic models approach is a consensus in solutions for the issue. A data-driven resolution must so be the starting point

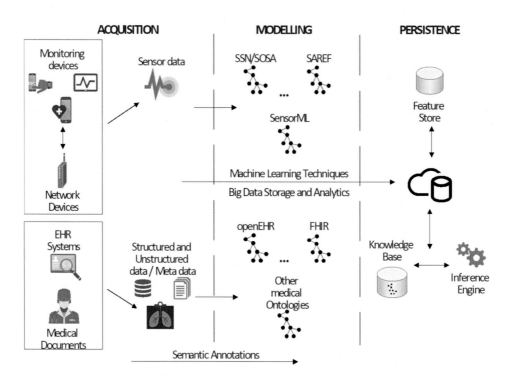

FIGURE 4.3

An overview of the data-driven e-Health life cycle.

for systems and architectures. Following this approach, some requirements are essential. As shown in Fig. 4.3, the data flow, from acquisition to persistence, can involve some important aspects that will influence solutions.

Semantic-based connected objects are essential for increasing the quality and efficiency of e-Health solutions, reducing costs, increasing the quality of life, and enriching the user's experience [63]. We can use semantic web technologies to define and model data collected from devices and sensors and then normalize structures and relationships from EHR systems in a fully-integrated and fully-interoperable perspective.

Back again to our case study, we can now glimpse an integrated system where sensors data will be part of the EHR, allowing all users to get comprehensive patient data. The accuracy of diagnoses and treatments will be further increased, increasing the patients' quality of life. In the next few years, connected objects like smartphones, weareables, and medical devices will be an integral part of EHR systems worldwide.

4.4.1 Semantic integration of IoT and EHR systems

Just a few works in the literature address the interoperability issue between EHRs and IoT/IoMT, proposing solutions composed of frameworks and architectures (Table 4.1). Although the ontology alignment problem is still an open research problem [64], some of these works made progress in this area considering the health domain. Standardization efforts will also be decisive factors for this reality to materialize.

In [7], authors proposed a solution based on the alignment of SSN ontology and an open EHR-based ontology for IoT platforms. In addition, they also presented an extension of the M2M ETSI standard based on semantic technologies to provide interoperability between IoMT and EHR. Following semantic web principles and considering the use of both standards, they intend to provide an interoperable environment with a standardized query mechanism. Despite the progress in semantic interoperability between EHR and IoT platform, this work only considers the openEHR and SSN standards.

Peng and Goswami [33] presented a method to integrate health services, implemented by RESTful Web APIs or published as Linked Data, and SSN/SOSA ontology-described WoT devices that adopted HL7 FHIR. Data is modeled with the Linked Health Resources (LHR) ontology to aggregate health and environment data, which are integrated as a data resource graph. In the same way, as discussed in the previous work, this method is based on one standard, FHIR.

A conceptual semantic-based healthcare collaboration framework was presented in [30]. The framework is based on an IoT infrastructure to allow knowledge exchange between different healthcare systems. In a fog layer at the edge of the network, there is a semantic gateway that offers a restful API to give access to healthcare information of each system for collaboration. The framework performs a mapping of different data sets to local ontologies for each healthcare system and after that to a global ontology. Nevertheless, the authors do not detail the ontology mapping process and nor how they consider the IoT infrastructure in the semantic model.

Alamri [46] proposed a middleware to support semantic interoperability for IoT data and EHR systems by using ontology mapping. The middleware is composed of three components, which are based on triple stores where data is stored and queried. In this work, the author proposed to model EHR data with SNOMED-CT ontology and IoT data with SSN. A significant drawback is that the author does not detail the ontology mapping process, and there is no definition of EHR systems.

In [62], authors proposed a discussion about the verbosity problem of SAREF when used for real-time electrocardiography signal sensing with time-series data. To overcome this problem, they proposed SAREF4Health, an extension of SAREF that addresses this problem, combining ontology-driven conceptual modeling and RDF implementation of stream data. The model is based on a reference ontology enhanced by a standardization procedure that considers the RDF serialization of the HL7 FHIR standard. There is no reference to any EHR system, and that is a drawback of this solution.

Dridi et al. [65] proposed a framework for IoT and healthcare data called SF4FI-EHR. It integrates three categories of health-related data, EHR unstructured and structured data and IoT/medical device data. Unstructured data are treated by Natural Language Processing techniques and the Metamap tool, which map medical text to UMLS terminologies and after that transmitted to a RESTful FHIR server, which is also used to deal with structured data. IoT data is annotated with a proposed ontology that is an extension of SSN ontology. The integration process is supported by mapping ontologies to FHIR.

A platform for common data interchange format and semantic interoperability was proposed in [52]. Authors considered M2M devices for communications between devices and services. A smartphone acts as a Gateway employing the HL7 message format in the communication with openEHR-based storage. They do not address interoperability problems at the sensor level and consider a few different devices with sensing capability.

Table 4.1 Semantic models concerning Connected Objects and EHR systems.

Work	Communication	IoT	EHR	Restful
Rubí and Gondim [7]	M2M	SSN	openEHR	No
Peng and Goswami [33]	–	SSN/SOSA	HL7 FHIR	Yes
Sigwele et al. [30]	–	–	Custom	Yes
Alamri [46]	–	SSN	*SNOMED-CT	No
Moreira et al. [62]	–	SAREF	HL7 FHIR	No
Dridi et al. [65]	–	SSN	HL7 FHIR	Yes
Pereira et al. [52]	M2M	–	openEHR	No
Villanueva-Miranda et al. [66]	–		–	No
Abinaya et al. [67]	–	Custom	–	No

Villanueva-Miranda et al. [66] have just analyzed the enhancement of semantic interoperability in IoMT platforms with two main approaches of ontology alignment. Nevertheless, there is no standardized EHR definition. Abinaya et al. [67] developed a centralized ontology-based system for decision making in the Internet of Things (IoT) environment. The integration with IoMT semantic devices is not explicit, and there is no standardized EHR definition for the ontology.

Some proposals [30,46] implemented their storage systems based on EHRs, but they do not address the standards to allow interoperability in the exchange of clinical data between EHR systems. Regarding functional architecture, proposals like [7,52] adopted the ETSI M2M standard approach as a way to give more control over communication platform, and consequently, over the overall solution. Regarding the provision of a Restful API, there are three proposals that offer it someway, [30,33,65]. Two of these works [66,67] have not considered standards in the interoperability in IoT and EHR systems.

4.4.2 Data-centric e-Health perspective

Although the relevance of devices, infrastructure, protocols, software systems, etc., data is the first-class citizen in the context of semantic models for e-Health systems and IoT-based systems. Thus, data management must receive special treatment in all aspects, including processes of acquisition, modeling, and persistence.

Taking into account the complexity and heterogeneity issues of the healthcare sector by itself is a challenge. Besides many others, it includes patients with different health and social conditions, government regulations, different stakeholders, specialized clinical knowledge bases such as specific medical therapies, prescriptions, and treatments.

Several advances were made in e-Health with EHR systems, for example. However, as discussed in this text, the great challenge is to ensure interoperability between systems due to their integration need. The advantage of using semantic models, in this case, is indisputable and, in addition, it brings another perspective to the solutions because of data treatment.

Semantic web technologies are one of the most powerful tools to enable data-centric solutions [68]. The data model precedes the implementation of any given application and will be around and valid long term. In this sense, after the model definition, a significant part of the effort should focus on carrying out data extraction, preparation, and persistence procedures.

4.4.2.1 Modeling

The data modeling step will dictate the integration capability of systems. Healthcare semantic models composed of ontologies from IoT-based and EHR systems are a trend. According to [7] and [48] and as far as we know until now, the literature lacks proposals for interoperability between EHRs and IoT/IoMT.

In e-Health and IoT/M2M systems, standardization bodies do not agree on semantic interoperability proposals to support healthcare systems. Thus, this is the first great challenge that must be overcome by both industry and academy. New solution proposals can include any e-Health and IoT/M2M ontology as long as they ensure their horizontal (same domain) or vertical (different domain) alignment to integrate these two contexts effectively.

In the data-centric paradigm, the modeling step must be the starting point to solutions proposals. As ontology construction is a complex task, it demands the use of methodologies to give adequate support. Methontology, a framework presented by Fernández-López et al. [69] was conceived to facilitate the construction of ontologies. It defines a development process and life cycle, which consists of the following steps: specification, conceptualization, formalization, implementation, and maintenance.

Neon methodology [70] is supported by a reengineering perspective to integrate ontologies by their reuse from a public repository in addition to a set of known ontology design patterns. A methodology proposed by Noy and McGuinness [71] suggests guidelines as the definition of domain and scope of the ontologies, ontologies reuse, enumerating relevant terms, classes and their properties, class hierarchy definitions, in addition to others.

Whereas that ontology assessment is part of ontology construction and maintenance, a methodology was proposed in [72] to assess the semantic model of SHCO [73]. SHCO is a smart healthcare ontology that has been designed to deal with healthcare data and IoT devices in a way that data collected from devices are transferred to a knowledge base and vice versa. Authors claim that any existing healthcare ontology can align with this ontology and can interact with IoT devices.

Moreover, graph-based healthcare data can be subjected to machine learning algorithms looking for hidden patterns from data, opening opportunities for applying both existing and new algorithms.

4.4.2.2 Preprocessing

Most solutions are based on cloud where data are stored, processed, and consumed by applications. Data can be collected from sensors/devices, relational databases, file systems, and knowledge bases. Cloud storage allows systems data access interchangeably in addition to the other advantages offered by this environment.

As with other domains, healthcare IoT-based systems impose several challenges related to the big data five v's, namely velocity, volume, value, variety, and veracity. Due to the most different acquisition methods (streaming, queries, file system access), representation forms (e.g., numeric values, texts, images), and the most different data sources, it is important to establish the importance of data modeling in this case. Healthcare IoT-driven solutions should consider data generation, processing, storing, and exploration [74].

Data must be preprocessed to overcome unwanted problems like those supported by data cleaning or imputation, for example. Methods like this could bring data trusted to different processes like data storage or when applying machine learning algorithms. As important as these aspects but elementary to the data-centric treatment, one should consider the semantics annotation to process data from the two contexts in integrated.

The integration of huge data collections from patient records and vital signs obtained from IoT devices could improve the performance of many different machine learning algorithms. However, it would be essential to consider an architecture that involves tools to manage and exploit data to generate cross-domain knowledge, which is difficult if we consider each data source separately [48].

4.4.2.3 Persistence

When using semantic models, it is necessary to specify the persistence mechanism to facilitate access to data and the possibility of carrying out the reasoning using an inference engine. As Triple Stores, RDF, and SPARQL, semantic web standards have been used in this case.

Due to the heterogeneity and the volume of gathered data, NoSQL databases (e.g., Hadoop, Spark) are standard technology in big data scenarios. A recent trend for data persistence is using blockchain, which can perform with semantically annotated data.

Another trend found in a fair number of works in the literature is the use of cloud/fog computing for persistence. Cloud computing has been widely adopted with IoT solutions, providing many different services. Nevertheless, one must have to take into account the drawbacks related to security and latency issues.

4.5 Concluding remarks

This chapter discussed the importance of semantic technologies to overcome interoperability problems in healthcare systems. The impact of remote monitoring devices and sensors was also crucial to this context because of the many possibilities for data collection from the most different devices and systems. Thus, we presented EHR systems and the main approaches used to address the interoperability, highlighting the role of semantic models in this context.

After that, we discussed connected objects in the health domain and the importance of semantic models to address interoperability between devices. Interoperability has been a research challenge in the last few years, and its importance is increasing due to the also increasing number of new devices and also to the existing different EHR systems.

Semantic web tools have been widely adopted to face this problem. Despite the advances already achieved, there are just a few proposals in the literature regarding the interoperability between connected objects and EHR systems. A discussion about existing works is presented latter in the chapter. Through a use case, we intended to highlight the importance of this approach during the text.

A data-centric view was used to put semantic models as a central figure in searching for future trends. We raised some concerns from this approach that can influence new research and development opportunities to construct more efficient systems that can further benefit the health system, and especially patients.

References

[1] T. Gunter, N. Terry, The emergence of national electronic health record architectures in the United States and Australia: models, costs, and questions, Journal of Medical Internet Research 7 (2005) e3.

[2] K.H. Veltman, Syntactic and semantic interoperability: new approaches to knowledge and the semantic web, New Review of Information Networking 7 (2001) 159–183.

[3] S. Garde, P. Knaup, E. Hovenga, S. Heard, Towards semantic interoperability for electronic health records, Methods of Information in Medicine 46 (2007) 332–343.

[4] Y. Xu, D. Sauquet, P. Degoulet, M.-C. Jaulent, Component-based mediation services for the integration of medical applications, Artificial Intelligence in Medicine 27 (3) (2003) 283–304.

[5] Top 8 healthcare technology trends to watch out for in 2020 | datadriveninvestor, https://www.datadriveninvestor.com/2019/11/30/top-8-healthcare-technology-trends-to-watch-out-for-in-2020/. (Accessed 8 October 2021).

[6] K.L.-M. Ang, J.K.P. Seng, Application specific internet of things (asiots): taxonomy, applications, use case and future directions, IEEE Access 7 (2019) 56577–56590.

[7] J.N.S. Rubí, P.R. de Lira Gondim, Interoperable internet of medical things platform for e-health applications, International Journal of Distributed Sensor Networks 16 (1) (2020).

[8] S. Cho, G. May, D. Kiritsis, A semantic-driven approach for industry 4.0, in: 2019 15th International Conference on Distributed Computing in Sensor Systems (DCOSS), 2019, pp. 347–354.

[9] G. Manogaran, N. Chilamkurti, C.-H. Hsu, Emerging trends, issues, and challenges in internet of medical things and wireless networks, Personal and Ubiquitous Computing 22 (09 2018).

[10] S. Jabbar, F. Ullah, S. Khalid, M. Khan, K. Han, Semantic interoperability in heterogeneous IoT infrastructure for healthcare, Wireless Communications and Mobile Computing 2017 (2017) 9731806.

[11] M. Ganzha, M. Paprzycki, W. Pawlowski, P. Szmeja, K. Wasielewska, Semantic interoperability in the internet of things: an overview from the inter-iot perspective, Journal of Network and Computer Applications 81 (2017) 111–124.

[12] A. Dridi, S. Sassi, S. Faïz, A smart iot platform for personalized healthcare monitoring using semantic technologies, in: 29th IEEE International Conference on Tools with Artificial Intelligence, ICTAI 2017, November 6–8, 2017, IEEE Computer Society, Boston, MA, USA, 2017, pp. 1198–1203.

[13] G.M. Honti, J. Abonyi, R. Natella, A review of semantic sensor technologies in internet of things architectures, Complexity 2019 (jan 2019).

[14] S. Dash, S. Shakyawar, M. Sharma, S.R. Kaushik, Big data in healthcare: management, analysis and future prospects, Journal of Big Data 6 (2019) 1–25.

[15] Y. Cardinale, G. Freites, E. Valderrama, A. Aguilera, C. Angsuchotmetee, Semantic framework of event detection in emergency situations for smart buildings, Digital Communications and Networks (2021).

[16] V. Ntrigkogia, T. Stavropoulos, M. Vlachopoulou, I. Kompatsiaris, A state of the art survey: business cases based on semantic web technologies in healthcare, 09 2019.

[17] L. Min, Q. Tian, X. Lu, J. An, H. Duan, An openehr based approach to improve the semantic interoperability of clinical data registry, BMC Medical Informatics and Decision Making 18 (2018) 15.

[18] G. Sebestyen, A. Hangan, S. Oniga, Z. Gál, eHealth solutions in the context of internet of things, in: 2014 IEEE International Conference on Automation, Quality and Testing, Robotics, 2014, pp. 1–6.

[19] A. Palavalli, D. Karri, S. Pasupuleti, Semantic internet of things, in: 2016 IEEE Tenth International Conference on Semantic Computing (ICSC), 2016, pp. 91–95.

[20] H. Li, D. Seed, B. Flynn, C. Mladin, R. Di Girolamo, Enabling semantics in an m2m/iot service delivery platform, in: 2016 IEEE Tenth International Conference on Semantic Computing (ICSC), 2016, pp. 206–213.

[21] J. Moreira, L. Daniele, L. Ferreira Pires, K. Wasielewska, P. Szmeja, W. Pawlowski, M. Ganzha, M. Paprzycki, Towards IoT platforms' integration: semantic translations between W3C SSN and ETSI SAREF, in: Semantics-Workshop Semantic Interoperability and Standardization in the IoT (SIS-IoT), 2017.

[22] openEHR architecture overview, https://specifications.openehr.org/releases/BASE/Release-1.0.3/architecture_overview.html. (Accessed 8 October 2021).

[23] Ehr information model, https://specifications.openehr.org/releases/RM/latest/ehr.html. (Accessed 8 October 2021).

[24] Introduction to hl7 standards | hl7 international, http://www.hl7.org/implement/standards/index.cfm?ref=nav. (Accessed 13 October 2021).

[25] E. Hovenga, S. Garde, Electronic health records, semantic interoperability and politics, Electronic Journal of Health Informatics: Special Issue on Systemic Interoperability 5 (01 2010).

[26] National institutes of health (nih) | turning discovery into health, https://www.nih.gov/. (Accessed 14 October 2021).

[27] R.E. Gliklich, N.A. Dreyer, M.B. Leavy, Registries for evaluating patient outcomes: a user's guide [internet] – ncbi bookshelf, https://www.ncbi.nlm.nih.gov/books/NBK208643/, April 2014. (Accessed 14 October 2021).

[28] Iso/tr 20514, health informatics — electronic health record — definition, scope and context, https://www.iso.org/obp/ui/. (Accessed 14 October 2021).

[29] Snomed – home | snomed international, https://www.snomed.org/. (Accessed 14 October 2021).

[30] T. Sigwele, Y.F. Hu, M. Ali, J. Hou, M. Susanto, H. Fitriawan, An intelligent edge computing based semantic gateway for healthcare systems interoperability and collaboration, in: 2018 IEEE 6th International Conference on Future Internet of Things and Cloud (FiCloud), 2018, pp. 370–376.

[31] E. Adel, S. El-Sappagh, S. Barakat, M. Elmogy, A semantic interoperability framework for distributed electronic health record based on fuzzy ontology, International Journal of Medical Engineering and Informatics 12 (2020) 207.

[32] Hl7 standards – section 1c: Fhir | hl7 international, http://www.hl7.org/implement/standards/product_section.cfm?section=12. (Accessed 14 October 2021).

[33] C. Peng, P. Goswami, Meaningful integration of data from heterogeneous health services and home environment based on ontology, Sensors 19 (8) (2019) 1747.

[34] Iso – iso 13606-1:2019 – health informatics — electronic health record communication — part 1: Reference model, https://www.iso.org/standard/67868.html. (Accessed 15 November 2021).

[35] P. Muñoz, J. Trigo, I. Martinez, A. Muñoz, J. Escayola, J. García, The iso/en 13606 standard for the interoperable exchange of electronic health records, Journal of Healthcare Engineering 2 (2011) 1–24.

[36] Semantic web – w3c, https://www.w3.org/standards/semanticweb/. (Accessed 9 November 2021).

[37] D. Boscá, D. Moner, J. Maldonado, M. Robles, Combining archetypes with fast health interoperability resources in future-proof health information systems, Studies in Health Technology and Informatics 210 (2015) 180–184.

[38] V. Papež, S. Denaxas, H. Hemingway, Evaluation of semantic web technologies for storing computable definitions of electronic health records phenotyping algorithms, in: AMIA Annual Symposium Proceedings. AMIA Symposium 2017, 07 2017.

[39] H. Sun, K. Depraetere, J. De Roo, G. Mels, B. De Vloed, M. Twagirumukiza, D. Colaert, Semantic processing of ehr data for clinical research, Journal of Biomedical Informatics 58 (2015) 247–259.

[40] H. Freedman, H. Williams, M. Miller, D. Birtwell, D. Mowery, C. Stoeckert, A novel tool for standardizing clinical data in a semantically rich model, Journal of Biomedical Informatics 112 (2020) 100086.

[41] E. Adel, S. El-Sappagh, S. Barakat, J.-W. Hu, M. Elmogy, An extended semantic interoperability model for distributed electronic health record based on fuzzy ontology semantics, Electronics 10 (14) (2021).

[42] M. Compton, P. Barnaghi, L. Bermudez, R. García-Castro, O. Corcho, S. Cox, J. Graybeal, M. Hauswirth, C. Henson, A. Herzog, V. Huang, K. Janowicz, W.D. Kelsey, D. Le Phuoc, L. Lefort, M. Leggieri, H. Neuhaus, A. Nikolov, K. Page, A. Passant, A. Sheth, K. Taylor, The ssn ontology of the w3c semantic sensor network incubator group, Journal of Web Semantics 17 (2012) 25–32.

[43] Sensor model language (sensorml) | ogc, https://www.ogc.org/standards/sensorml. (Accessed 14 October 2021).

[44] M. Bermudez-Edo, T. Elsaleh, P. Barnaghi, K. Taylor, Iot-lite: a lightweight semantic model for the internet of things, in: 2016 Intl IEEE Conferences on Ubiquitous Intelligence Computing, Advanced and Trusted Computing, Scalable Computing and Communications, Cloud and Big Data Computing, Internet of People, and Smart World Congress, 2016, pp. 90–97.

[45] W. Wang, S. De, R. Toenjes, E. Reetz, K. Moessner, A comprehensive ontology for knowledge representation in the Internet of things, in: 2012 IEEE 11th International Conference on Trust, Security and Privacy in Computing and Communications, 2012, pp. 1793–1798.

[46] A. Alamri, Ontology middleware for integration of iot healthcare information systems in ehr systems, Computers 7 (4) (2018) 51.

[47] K. Janowicz, A. Haller, S.J. Cox, D. Le Phuoc, M. Lefrançois, Sosa: a lightweight ontology for sensors, observations, samples, and actuators, Journal of Web Semantics 56 (2019) 1–10.

[48] J. Calvillo-Arbizu, I. Román-Martínez, J. Reina-Tosina, Internet of things in health: requirements, issues, and gaps, Computer Methods and Programs in Biomedicine 208 (2021) 106231.

[49] oneM2M sets standards for the internet of things & m2m, https://www.onem2m.org/. (Accessed 14 October 2021).

[50] D. Lucic, A. Caric, I. Lovrek, Standardisation and regulatory context of machine-to-machine communication, in: 2015 13th International Conference on Telecommunications (ConTEL), 2015, pp. 1–7.

[51] S. Dahmen-Lhuissier, Etsi – internet of things – iot standards | machine to machine solutions (m2m), https://www.etsi.org/technologies/internet-of-things. (Accessed 14 October 2021).

[52] C. Pereira, S. Frade, P. Brandão, R. Cruz-Correia, A. Aguiar, Integrating data and network standards into an interoperable e-health solution, 2014.

[53] M. Irfan, N. Ahmad, Internet of medical things: architectural model, motivational factors and impediments, in: 2018 15th Learning and Technology Conference (L T), 2018, pp. 6–13.

[54] A.M. Rahmani, T.N. Gia, B. Negash, A. Anzanpour, I. Azimi, M. Jiang, P. Liljeberg, Exploiting smart e-health gateways at the edge of healthcare internet-of-things: a fog computing approach, Future Generation Computer Systems 78 (2018) 641–658.

[55] A. Rashed, A. Ibrahim, A. Adel, B. Mourad, A. Hatem, M. Magdy, N. Elgaml, A. Khattab, Integrated iot medical platform for remote healthcare and assisted living, in: 2017 Japan–Africa Conference on Electronics, Communications and Computers (JAC-ECC), 2017, pp. 160–163.

[56] S.K. Datta, C. Bonnet, N. Nikaein, An iot gateway centric architecture to provide novel m2m services, in: 2014 IEEE World Forum on Internet of Things (WF-IoT), 2014, pp. 514–519.

[57] D.-Hyu Kim, H. Lee, J. Kwak, Standards as a driving force that influences emerging technological trajectories in the converging world of the internet and things: an investigation of the m2m/iot patent network, Research Policy 46 (7) (2017) 1234–1254.

[58] T.N. Vidanagama, Towards realization of an iot environment: a real-life implementation of an iot environment and its analytics, in: 2017 Global Internet of Things Summit (GIoTS), 2017, pp. 1–6.

[59] Etsi – smartm2m, https://www.etsi.org/committee/smartm2m. (Accessed 14 October 2021).

[60] L. Daniele, F. den Hartog, J. Roes, Created in close interaction with the industry: the smart appliances reference (saref) ontology, in: R. Cuel, R. Young (Eds.), Formal Ontologies Meet Industry, Springer International Publishing, Cham, 2015, pp. 100–112.

[61] Saref portal, https://saref.etsi.org/. (Accessed 14 October 2021).

[62] J. Moreira, L. Ferreira Pires, M. van Sinderen, L. Daniele, Saref4health: Iot standard-based ontology-driven healthcare systems, in: Proceedings of Formal Ontology in Information Systems, 2018.

[63] S.M.R. Islam, D. Kwak, M.H. Kabir, M. Hossain, K.-S. Kwak, The Internet of things for health care: a comprehensive survey, IEEE Access 3 (2015) 678–708.

[64] C. Trojahn, R. Vieira, D. Schmidt, A. Pease, G. Guizzardi, Foundational ontologies meet ontology matching: a survey, Semantic Web 03 (2021).

[65] A. Dridi, S. Sassi, R. Chbeir, S. Faiz, A flexible semantic integration framework for fully-integrated ehr based on fhir standard, 2020, pp. 684–691.

[66] I. Villanueva-Miranda, H. Nazeran, R. Martinek, A semantic interoperability approach to heterogeneous internet of medical things (iomt) platforms, 2018, pp. 1–5.

[67] A. Rajan, V. Kumar Swathika, Ontology based public healthcare system in internet of things (iot), Procedia Computer Science 50 (2015) 99–102.

[68] D. McComb, The Data-Centric Revolution: Restoring Sanity to Enterprise Information Systems, Technics Publications, 2019.

[69] M. Fernández-López, A. Gomez-Perez, N. Juristo, Methontology: from ontological art towards ontological engineering, in: Engineering Workshop on Ontological Engineering (AAAI97), 03 1997.

[70] M.C. Suárez-Figueroa, A. Gomez-Perez, M. Fernández-López, The NeOn methodology for ontology engineering, 2012, pp. 9–34.

[71] N. Noy, D. Mcguinness, Ontology development 101: a guide to creating your first ontology, Knowledge Systems Laboratory 32 (01 2001).

[72] S. Mishra Tiwari, A. Abraham, Semantic assessment of smart healthcare ontology, International Journal of Web Information Systems 16 (2020).

[73] S. Mishra Tiwari, S. Jain, A. Abraham, S. Shandilya, Secure semantic smart healthcare (s3hc), Journal of Web Engineering 17 (2019) 617–646.

[74] J.L. Shah, H.F. Bhat, A.I. Khan, Chapter 6 – integration of cloud and iot for smart e-healthcare, in: V.E. Balas, S. Pal (Eds.), Healthcare Paradigms in the Internet of Things Ecosystem, Academic Press, 2021, pp. 101–136.

The security and privacy aspects in semantic web enabled IoT-based healthcare information systems

Ozgu Can

Ege University, Department of Computer Engineering, Bornova-Izmir, Turkey

Contents

5.1 Introduction and motivation

The recent technological changes, the widespread use of the internet, and the dynamic developments in communication technologies and sensors have created a connected world. As a result of these developments, the Internet of Things (IoT), which is accepted as the next evolution of the internet has emerged. The Oxford dictionary defines IoT as "*the connection of devices within everyday objects via the internet, enabling them to share data*" [1]. IoT provides communication among everyone and everything by extending the capabilities of the internet to enable machine-to-machine communication (M2M) [2]. Thereupon, the IoT integrates the physical world and information world to form a new way of communication. The integration of these interconnected objects is called the future internet [3]. The internet is now in the process of becoming a fully integrated future internet, and consequently, the internet will become the IoT [4].

In IoT, things that share information, communicate, and coordinate with each other are called smart things or smart devices. A smart thing is defined as *an electronic device that communicates to other devices and services and operates to some extent interactively and autonomously, where its behavior*

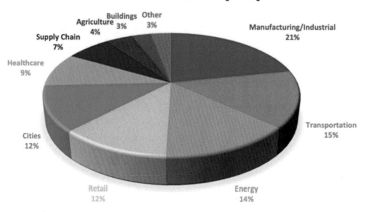

FIGURE 5.1

Top 10 IoT application areas in 2020 [11].

is characterized by loaded software and operating environment [5]. All smart concepts such as smartphones, smartwatches, smart thermostats, smart locks, smart refrigerators, smart TVs, smart medical devices, smart traffic controls, autonomous vehicles, etc. constitute the connected world system. As the number of installed IoT devices around the world was 15 billion in 2015, the predicted IoT device number is 75 billion devices for 2025 [6]. The Statista report states that the number of internet-connected devices will reach 10 billion in 2021 [7]. Cisco expects the number of connected devices to reach 500 billion by 2030 [8]. According to the IDC forecast, there will be 41.6 billion connected IoT devices or things generating 79.4 zettabytes of data in 2025 [9]. Also, the global IoT market, which was $308.97 billion in 2020 is projected to grow from $381.30 billion in 2021 to $1,854.76 billion in 2028 [10]. Fig. 5.1 shows the top ten IoT application areas in 2020 based on IoT analytics' research results [11]. As foreseen, in this connected world where everything is connected to everything, the IoT has gained increasing importance both in our personal lives and in various fields such as industry, healthcare, energy, retail, cities, supply chains, etc. Moreover, the COVID-19 pandemic has also accelerated the growth of IoT.

Today, IoT is an indispensable part of our daily lives. As in other application domains, IoT also became an emerging trend in healthcare systems. There are numerous IoT-based applications in the healthcare domain such as medical devices, smart sensors, remote monitoring, and tracking. The healthcare information system aims to reduce the cost of treatments, to improve the effectiveness and efficiency of health workers' decisions, to improve the healthcare quality and quality of personal life by ensuring qualitative, preventive, and curative healthcare to people. Within this scope, IoT offers various benefits for the goals of healthcare information systems. The benefits of IoT-based healthcare solutions include:

- to monitor patients in real-time,
- to track patients' health conditions continuously,
- to detect potential health disorders or diseases early,

- to assist remote medical assistance,
- to automate patient care,
- to improve the quality and efficiency of treatments,
- to provide patient-centric care,
- to improve patient health,
- to provide a better connection between the doctor and the patient,
- to increase patient satisfaction,
- to increase the quality of life by guiding the patient's experience,
- to take a more active role in managing personal health,
- to reduce costs,
- to reduce the time spent from the perspective of health care providers,
- to use data for medical research purposes.

The McKinsey Global Institute report states that the greatest benefit of IoT in healthcare comes from the improved efficiency in treating patients with chronic conditions [12]. The report also indicates that the treatment costs for chronic diseases constitute 60% of total healthcare spending and estimates that remote monitoring could reduce this cost by 10% to 20%. In addition to these benefits, the global pandemic has also led to a rapid increase in the use of IoT within the healthcare sector. For example, it is possible to gain medical help without physical contact and remotely track a patient's condition. Also, IoT devices are used to disinfect and sterilize rooms against COVID-19 contamination.

Consequently, IoT has improved the lifestyle of individuals and people have become more reliant on IoT-based solutions. However, these improvements have also increased the privacy and security challenges in IoT. Security professionals consider IoT as the vulnerable point for cyber attacks [13]. Thus, the security and privacy of this connected world should be ensured. The IoT security report of Palo Alto Networks [14] states that 57% of IoT devices are vulnerable to IoT attacks, 83% of devices run on unsupported operating systems, 98% of IoT traffic are unencrypted, and the challenges are growing. Fig. 5.2 shows the IoT devices that have the highest security issues [14]. The report also specifies that 26% of the threats are caused by user experience (*cryptojacking, phishing, password*), 33% of the threats are malwares (*botnet, backdoor trojan, ransomware, worm*), and 41% of the threats are exploits (*network scan, remote code execution, command injection, buffer overflow, SQL injection, zero-day, other exploits*). The top cyber threat types for each threat are shown in Fig. 5.3 [14].

Data that is collected by IoT devices are mostly sensitive and include metadata such as date, time, and location. In IoT-based healthcare solutions, devices capture and process sensitive personal health data. As seen from the IoT security report of Palo Alto Networks, IoT devices and services are vulnerable to threats and attacks. Therefore, the devices, their associated communication protocols, data and transmission of data need to be secured. On the other hand, traditional security countermeasures and privacy enforcement cannot be directly applied to IoT technologies due to their limited computing power, besides the high number of interconnected devices arise scalability issues [15]. Hence, the security and privacy needs of both patients and healthcare providers must be satisfied. Thus, the security and privacy challenges must be managed, monitored, and related measures must be taken. For this purpose, there are numerous studies presented in the literature. This chapter focuses on the semantic web-based studies for the related solutions on security and privacy aspects of IoT-based healthcare systems.

The semantic web is accepted as the extension of the current web and described by Tim Berners-Lee [16] as "The Next Web." The semantic web represents knowledge in a machine-interpretable format. Therefore, the semantic web enables to share, reuse, and integrate information, and also to infer new

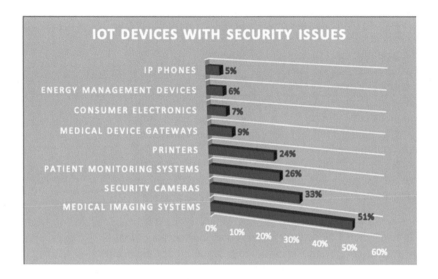

FIGURE 5.2

The IoT devices with the highest security issues [14].

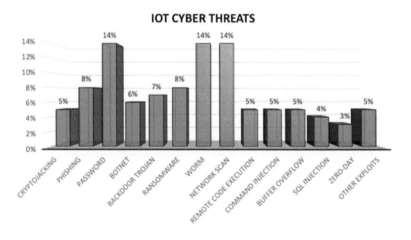

FIGURE 5.3

The top IoT cyber threats [14].

knowledge from the existing facts and rules. Thus, the semantic web technologies provide semantic interoperability and machine-to-machine communication. The semantics that is provided by the semantic web is the explicit interpretation of the domain knowledge to make machine processing more intelligent, adaptive, and efficient [17]. Semantic interoperability enables different parties to access and interpret unambiguous data [18]. The interpreted data is also used for the decision-making process. In IoT, the connected things exchange data with each other. Therefore, combining semantic web and IoT is considered a reasonable and suitable solution to promote interoperability among IoT's resources, information models, data providers, and consumers [18,19]. Besides, semantic web technologies can be used to turn the IoT data into knowledge.

The main objective of this chapter is to contribute a better knowledge and understanding of security and privacy issues in the IoT ecosystem and examine the current status in IoT-based healthcare systems. Thus, the fundamental security and privacy concepts in IoT are covered, the security and privacy concerns for the IoT healthcare systems are presented, the current semantic web-based solutions to ensure security and privacy in IoT ecosystem and IoT-based healthcare systems are examined, and challenges in the related field are discussed.

5.2 Security and privacy requirements in IoT

Security refers to the methods, tools, and personnel to protect objects against unauthorized access and use, theft, and harm. In computer science, security enables the protection of digital assets from threats and vulnerabilities. In the 1970s, security was mostly concerned with guarding the rooms. Today, with the recent developments in information and communication technologies, and as well as in IoT technology, vast volumes of data are collected, managed, processed, and shared. As a result of these developments, today's security concept deals with high-profile breaches and attacks that threaten the national security of countries and the economic prosperity of countries, businesses, and individuals.

Privacy is a core value of an individual and has been acknowledged as a fundamental right by the European Convention on Human Rights and the Universal Declaration of Human Rights [20]. In 1890, U.S. Justice Louis Brandeis called privacy *"the right to be left alone"* [21]. Despite being a moral/legal right, privacy is also perceived as the interest that individuals have in sustaining a personal space, free from interference by other people and organizations [22]. This personal space includes the privacy of the person/physical privacy (the integrity of the individual's body and the physical space of an individual), the privacy of personal behavior, the privacy of personal communications, and the privacy of personal data. In the modern world, privacy is defined as *"the ability to control the acquisition or release of information about oneself"* [23]. In the perspective of IoT, privacy is specified as the border where information from smart objects is exposed to the outside world [24]. Privacy-Enhancing Technologies (PETs) such as encryption, protocols for anonymous communications, attribute-based credentials, data masking techniques, and private search of databases are used to protect data and to prevent privacy breaches.

Sensitive data falls into three categories [25]: explicit identifiers, quasiidentifiers, and privacy attributes. Explicit identifiers represent any Personally Identifiable Information (PII), such as social security number, patient name, and contact details. PII is any information about the individual that can be used to distinguish the individual's name, social security number, date, and place of birth or any other information that is linked or linkable to an individual, such as medical, educational, and financial information [26]. Quasiidentifier is a set of attributes that can be linked with external information to reidentify an individual's identity, such as ZIP code, age, gender, and date of birth [27]. Privacy attributes represent any specific identifiable information about a person, such as any health sickness and disability. The privacy of these sensitive data must be preserved when data is published. Hence, an individual's sensitive data must be indistinguishable after the data publishing. For this purpose, data anonymization techniques that perform various operations on data before it is shared publicly are used. The purpose of data anonymization is to enhance the privacy of individuals and to ensure that data cannot be misused even if data are stolen [28]. Therefore, methods such as differential privacy [29], k-anonymity [30], l-diversity [31], and t-closeness [32] are used to provide privacy-preserving data publishing.

As IoT technology has changed people's daily lives and become indispensable, IoT devices are now vulnerable to multiple types of threats and a prime target for attackers. Also, IoT data contain personal data that should be kept confidential. Besides, security and privacy must be achieved at a satisfactory level to provide data utility effectively. Thus, the security and privacy related issues are concerned with managing information loss, the risk, and the cost of the loss. Therefore, the concepts of threat, vulnerability, attack, and risk should be understood foremost. These concepts are defined as follows:

- *Threat:* A threat is any action that exploits a vulnerability in an asset and may cause damage or harm to an asset either intentionally or unintentionally. The asset can be a person, a computer, a physical device, and information such as a database, software code, etc. Threats may result in direct loss, delays or denials, disclosure of sensitive information, modification of programs or databases, and intangible such as loss of reputation [33]. Threats can be environmental (natural disasters outside the control of humans such as earthquake, flood, storm, power outage, etc.), internal (generated by internal sources, usually an insider from the organization), external (generated from the outside of the organization) and man-made (any threat that is caused by human beings intentionally or unintentionally such as malware installation, password disclosure, unauthorized access, etc.). There is no source, list, or method that denotes all possible threats to assets. Yet, the Common Vulnerabilities and Exposures (CVE) list [34] and the National Vulnerability Database (NVD) list [35] are widely used public lists to track the known vulnerabilities and exposures. Also, MITRE ATT&CK is a knowledge base of adversary tactics and techniques based on real-world observations and used as a foundation for the development of specific threat models [36].
- *Vulnerability:* A vulnerability is a potential weakness that exploits a threat and causes unauthorized access and data breach. A security vulnerability must exist for the threat to occur. If there is a security vulnerability in a system, then there is a possibility that a threat will occur and the system is open to potential attacks. Some vulnerabilities are weaknesses and some are just side effects of other actions [37]. Vulnerabilities may result from weak passwords, software bugs, a script code injection, malicious software such as a computer virus, etc. [33].
- *Attack:* An attack is an action that uses a vulnerability to exploit a threat. A person who performs these actions is called an *attacker*. The software used to perform a malicious attack is called *malicious software (malware)*. Viruses, trojans, worms, and spywares are malwares that perform malicious actions on a system.
- *Risk:* Risk is the probability that a particular threat will be realized against a specific vulnerability to impact or harm an asset. Risk indicates a loss as a result of realizing a threat that exploits a vulnerability. The amount of harm that a threat exploiting a vulnerability can cause is called *impact (cost)* [37]. The most common risk identification is as follows:

$$Risk = Probability \times Vulnerability \times Threat$$

Security has three main principles: confidentiality, integrity, and availability, which are referred to as the CIA security triad. These principles are defined as follows:

- *Confidentiality:* Confidentiality is a principle to protect information and systems from unauthorized access. Confidentiality prevents unauthorized persons from accessing the data. Thus, information is specific only to entities that need access to information and only authorized entities can access the information and the system. Confidentiality is an important component to protect personal data and privacy. Privacy is the right of an individual to protect her personal data from unauthorized access and disclosure. Hence, personal data must be processed according to the permissions granted by the user to ensure user privacy.

- *Integrity:* Integrity prevents unauthorized modification of the data. Thus, only authorized entities can modify or change the data. Integrity is concerned with the validity and accuracy of data. Integrity ensures that only authorized entities can modify data and prevents authorized entities from making unauthorized changes to the data.
- *Availability:* Availability ensures that information and systems are accessible to authorized entities whenever they request it and used for authorized purposes. Usability is the state of being able to use the information or resource that is needed to be accessed. If the information cannot be accessed or used whenever it is needed, ensuring the confidentiality and integrity of the data will not be enough.

Each of these three principles involves a relevant protection mechanism. However, these principles are not adequate for today's security needs. Especially, the IoT ecosystem is more vulnerable to security and privacy challenges. Security issues, such as identification, authentication, authorization, access control, accountability, and nonrepudiation are also the main challenges in an IoT environment. Authentication, authorization, and accounting are also known as the *AAA* model that is used to control access to resources. The related security issues are defined as follows:

- *Identification:* Identification is the process of providing identifying information for the entity. An entity can be a person or a system. Identification uses credentials such as usernames, login ID, etc. to identify the entity.
- *Authentication:* Authentication is the process of proving that the entity is the person or entity whom she claims to be. Authentication binds the user ID to an entity. Passwords are used to verify the identity of the entity. Also, hardware/software tokens, smart cards, and biometrics are used for the authentication process. Multifactor authentication should be preferred for the authentication process. An authentication factor is a credential used for identity verification. Multifactor authentication uses different identification characteristics to identify the user and verify the user information. Thus, each additional factor increases the assurance of the authentication process. The most common example is using a password together with a code sent to the user's smartphone. The user must enter the code sent in the SMS message after entering the password to verify her identity.
- *Authorization:* Authorization is the process of assigning privileges and granting permissions/rights to an entity in order to access a resource and perform actions on that resource such as a program, system, process, or information. Privileges indicate the allowed or prohibited/restricted actions that are defined for the entity. The authorization process ensures that the user or the system has sufficient rights for the actions that will be performed on the resource and provides a correct level of access to users based on their credentials.
- *Access Control:* Access control is a mechanism that limits access to a resource and controls whether the entity has the right to access the resource or not. The access control first checks if the entity who requests to access the resource is authenticated in order to tailor the access rights to the entity. Then the access control mechanism checks the authenticated user's privileges for the actions that the user wants to perform on the resource.

- *Accountability/Auditing:* Accountability is the process of keeping a record of the actions that are performed by the user on the resources. Thus, it is ensured that users take responsibility for the actions they performed. Accountability links users to actions and an action can be traced back to the user who performed this action.
- *Nonrepudiation:* Nonrepudiation is the process to ensure that the entity cannot deny the action that the entity is performed on the resource. Nonrepudiation provides the proof that is needed to ensure the security principles of confidentiality and integrity.

5.2.1 Security and privacy challenges in IoT

IoT security is concerned with any "thing" that is connected in the ecosystem and every day the number of connected devices is increasing. Therefore, IoT is facing various security challenges due to the increasing connectedness and also the ongoing digitization process in various domains. According to the Netscout Threat Intelligence report, IoT devices are attacked within 5 minutes of being plugged into the internet [38]. The Open Web Application Security Project (OWASP) IoT project provides help to manufacturers, developers, and consumers to understand the security issues associated with the IoT and enables users to make better security decisions [39]. The OWASP IoT project also maintains a security analysis of IoT that includes vulnerabilities, threats, attack surface areas, and firmware analysis.

The main security and privacy challenges based on the generic three-layered architecture of IoT are shown in Fig. 5.4. The security of an IoT system should include the security of the whole IoT system crossing from the perception layer, network layer, and application layer.

The perception layer is the physical level of objects. Objects in this layer interact with the external world and other sensors. These physical sensors are referred as smart objects. Smart objects sense and gather data about the environment. Sensors and actuators are essential components of smart objects to collect data from the environment. The common threats against the perception layer are eavesdropping, replay attack, timing attack, jamming attack, tag removal, tag destruction, KILL command, node capture, and fake node/malicious node injection [40,41].

The network layer is also known as the transmission layer and connects to other smart things and network devices. The network layer transmits and processes sensor data. The common threats against the network layer are denial of service (DoS) attack, the man in the middle attack, sinkhole attack, Sybil attack, traffic analysis, routing attack, RFID spoofing, and RFID cloning [42,41].

The application layer is the interface between the IoT device and the network layer. The IoT device communicates the network through the application layer. The application layer offers application-specific services, handles data formatting and presentation, provides solutions such as reporting, analytics, and device control requested by the end-users. The application layer covers various applications such as smart healthcare, smart home, smart cities, smart factory, and smart vehicles. The most common threats against the application layer are phishing attack, cross-site scripting, malicious code injection, sniffing attack, malwares (virus, worm, trojan, etc.), and DoS attack [24,41].

Besides these threats, encryption attacks such as side-channel attacks and cryptoanalysis attacks are also security concerns for an IoT system. Also, privacy concerns and privacy attacks have a significant challenge in an IoT system as IoT devices gather, analyze, and transmit sensitive data across the network. The collection of massive amounts of data and powerful analyzes to transform the meaningless data into information leads to data breaches and increases the privacy concerns of users. In order to preserve privacy, data about an individual should not be available to other individuals and organizations, and the individual must have a substantial degree of control over her data and its processing

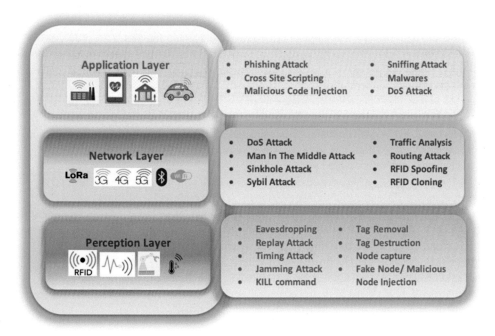

FIGURE 5.4

Threats against the 3-layered IoT architecture.

(acquisition, use, and disclosure of personal data) by other parties. In terms of privacy, identification, profiling, lifecycle transitions, linkage, privacy-violating interaction and presentation, inventory attack, localization, and tracking are the most common threats in an IoT system [24].

5.2.2 IoT security attacks and threats

The increase in the number of connected IoT devices and the complexity of IoT ecosytem introduce numerous security and privacy threats. Besides, most of the IoT devices are not designed with security issues and they do not focus enough on security and privacy. Also, a significant quantity of data moves between IoT devices and across communication networks. The main security issues in IoT are weak password protection, insufficient data protection mechanisms, insecure interfaces, poor IoT device management, weak updates, and lack of regular patches.

The IoT attack surface consists of devices, communication channels, software, and applications. Devices are the primary means that are used by attackers. Thus, the security vulnerabilities of devices could possibly allow an attacker to initiate an attack. Communication channels connect components of the IoT ecosystem. Therefore, security issues in communication channels affect the entire ecosystem. The vulnerabilities of software and applications enable attackers to compromise systems. The attack surface of devices, communication channels, software, and applications and their vulnerabilities are summarized as follows [39]:

- *Applications:* The main vulnerabilities are weak passwords, insecure password recovery mechanism, insecure data storage, username enumeration, and account lockout.
- *Device Firmware:* Firmware version display, firmware latest update dates, firmware downgrade possibility, vulnerable services, sensitive data exposures such as backdoor accounts and encryption keys are the primary vulnerabilities.
- *Device Memory:* The vulnerability comes from the third-party credentials and the sensitive data such as clear text usernames and passwords.
- *Device Network Services:* The major vulnerabilities are Denial of Service (DoS), injection, replay attack, buffer overflow, user and administrative command line interfaces, insecure password recovery mechanisms, and unencrypted services.
- *Device Physical Interface:* Firmware extraction, user and admin command line interfaces, privilege escalation, device ID, and serial number exposure are the main vulnerabilities.
- *Device Web Interface:* The vulnerabilities consist of standard web application vulnerabilities and web credential management vulnerabilities such as weak passwords, username enumeration, account lockout, and insecure password recovery mechanisms.
- *Third Party Backend APIs:* The unencrypted Personally Identifiable Information that is sent, device information leakage, and location information leakage are the main vulnerabilities.
- *Vendor Backend APIs:* Weak authentication and access controls, injection attacks, inherent trust of applications, and hidden services are the major vulnerabilities.

The IoT attack surface shows that all the major components of the IoT ecosystem can be exploited. Therefore, security and privacy issues should be considered from the design phase to the overall configuration and functionality of the entire system. For this purpose, secure settings of devices, communication channels, software, and applications should be ensured. Also, security and privacy mechanisms should be integrated in every aspect of the IoT ecosystem. Besides, the security strategies for each component of the IoT ecosystem should be built.

5.3 Security and privacy concerns in IoT-based healthcare systems

The IoT ecosystem is used in every domain and has a prominent impact on people's daily lives. Healthcare is one of the major domains that IoT solutions that are widely used. The healthcare sector is one of the fastest to adopt the IoT [43]. The healthcare industry aims to treat people and to get people healthy living habits. The healthcare industry has revolutionized and gone through various transformations from Healthcare 1.0 to 4.0 [44]. Healthcare 1.0 was doctor-centric and records were maintained manually. Healthcare 2.0 was partly technology-centric, manual records were replaced with Electronic Healthcare Records (EHRs) and monitoring devices are used to support the diagnosis. Healthcare 3.0 was patient-centric, but not capable to handle real-time big medical data [45]. Healthcare 4.0 is the era of smart healthcare in which both clinicians and patients are involved. Healthcare 4.0 uses artificial intelligence, real-time monitoring, data analytics, smart sensors, IoT, medical robots, cloud computing, fog computing, and telehealthcare technologies. In Healthcare 4.0, everybody is connected or associated through wearable devices irrespective of their locations for efficient and accurate treatment [45]. In today's smart healthcare era, the security and privacy of healthcare records are challenging due to: (i) EHR servers are being used to store patients' health records, (ii) these health records are stored at the cloud server to be accessed by only authorized users such as clinicians, hospital staff, medical and government agencies for diagnosis, treatment and data analysis, (iii) diagnosis reports are also being accessed by patients using their smart electronic gadgets, and (iv) notifications are sent to patients or emergency contacts of patients [46]. The Protenus Breach Barometer Report states that 32 million patient records in the healthcare industry have been infringed and over 3 million patient records are breached by hospital insiders between January and June 2019 [47]. Therefore, the lack of proper security measures results in multiple levels of risks to the patient's health and privacy.

The things in smart healthcare consist of heart monitoring implants, insulin pumps, smart thermometers, pacemakers, etc. These smart devices collect data, send and receive data between other devices, track information for data analysis, and keep a medical history. While smart devices increase the efficiency, accuracy, and productivity of healthcare services, they also enlarge the attack surface of healthcare. Therefore, devices are the main target of attackers. Cyber security researches show that several implantable medical devices (IMDs) are vulnerable to attacks of varying severity [48]. Hackers exploit the weaknesses of IoT devices, gain control over these devices and perform malicious activities, such as exposure of valuable data, confidential information, and botnet attacks [49]. For example, privacy and DoS attacks are conducted on Implantable Cardioverter Defibrillator (ICD) [50], software attacks are performed on implantable neurostimulators [51], security attacks are demonstrated on insulin pumps [52]. The hijacking of a medical device is called medjack. Hackers use medjack to create backdoors in hospital networks, launch targeted attacks at hospital networks, and steal medical records. Another motivation behind the medjacking is the high demand in the Darknet market for healthcare-related information [53]. Furthermore, hackers may alter the operation of these medical devices and cause these devices to operate in a life-threatening manner. Hence, the consequences of these security attacks can be lethal. As the number of devices is increasing, security vulnerabilities and additional risks are exposing. Thus, the Food and Drug Administration (FDA) requires medical device manufacturers to build security into their systems to manage postmarket cybersecurity vulnerabilities for marketed and distributed medical devices [54].

IoT devices use cloud-based solutions to store and analyze data for further decision-making. Transmission of critical medical information to the cloud without addressing security issues introduces the risk of hacking and data corruption [55]. As a result of this, data is vulnerable to unauthorized access. Thus, the assignment of different privileges to different users must be provided with an effective access control mechanism, and policy management must be enforced. Besides, identification and authentication techniques should also be used to prevent unauthorized access. An access control mechanism must also be applied for accessing the IMDs and only authorized users must access medical devices.

IoT communication environment facilitates and supports people's daily activities. On the other hand, it suffers from various security and privacy issues, such as replay attacks, the man-in-the-middle attack, impersonation, privileged-insider, remote hijacking, password guessing, DoS attacks, and malware attacks [56]. Malware botnets such as Mirai, Reaper, and Echobot lead to attacks on confidentiality, integrity, availability, and authenticity. These attacks disclose or alter the sensitive data of IoT communication, or sensitive data may not be available to authorized users [56]. In the DoS attack, the network is flooded with a lot of useless traffic caused by an attacker. The result of a DoS attack is the resource exhaustion of the targeted system due to the unavailable network. DoS attack is a security concern directed to both the network and application layers of the IoT architecture. Today's sophisticated DoS attacks offer a smoke screen to carry out attacks to breach the defensive system and data privacy of the user by deceiving the victim into believing that the actual attack is happening somewhere else [24]. The security issues against IoT architecture are categorized in three levels [57]:

1. *Low-level security issues:* Jamming adversaries, insecure initialization, low-level Sybil and spoofing attacks, insecure physical interface, sleep deprivation attack.
2. *Intermediate-level security issues:* Replay or duplication attacks, insecure neighbor discovery, buffer reservation attack, Routing Protocol for Low-Power and Lossy Networks (RPL) routing attacks, Sinkhole and Wormhole attacks, Sybil attacks on intermediate layers, authentication and secure communication, transport-level end-to-end security, session establishment and resumption, privacy violation on cloud-based IoT.
3. *High-level security issues:* Constrained Application Protocol (CoAP) security with internet, insecure interfaces, insecure software/firmware, middleware security.

The application layer of the IoT architecture gathers patients' private data, which leads to potential privacy problems. The primary privacy issue in the healthcare domain is to provide the confidentiality of patients' Electronic Medical Records (EMRs), EHRs, and Personal Health Records (PHRs). The Healthcare Information and Management Systems Society (HIMSS) defines EMR as "*an electronic record of health-related information on an individual that can be created, gathered, managed, and consulted by authorized clinicians and staff within one healthcare organization,*" EHR as "*an electronic record of health-related information on an individual that conforms to nationally recognized interoperability standards and that can be created, managed, and consulted by authorized clinicians and staff across more than one healthcare organization*" and PHR as "*an electronic record of health-related information on an individual that conforms to nationally recognized interoperability standards and that can be drawn from multiple sources while being managed, shared, and controlled by the individual*" [58]. Patients' personal data (such as demographics, social security number, and credit card) and health records are often viewed, copied, or modified without patients' consent. Data breaches through these records arise from hacking, malware, and insider threats. Consequently, these data can be modified and misused. Also, users' location data is considered as personal information. Therefore, location data is sensitive data and the release of the location data to third parties is also an important privacy issue. In IoT, data that is collected by smart medical devices often includes metadata such as location and time. Thus, such information has to be well protected to preserve personal privacy.

In the new informational ecosystem, data is power, and access to sensitive data is a profitable enterprise. The IoT-based healthcare information systems enable to collect patients' sensitive data and store these data on cloud servers or network servers to make them accessible anytime and anywhere. Also, smart devices are used to access these data. The use of smart devices, software vulnerabilities, security failures, and human error cause these databases to be accessed by unauthorized users, which leads to the exposure of sensitive data in the form of data breaches [59]. Further, phishing attacks, data theft, and malwares (such as viruses, worms, trojan horses, ransomwares, rootkits, etc.) cause breaches to occur. Besides, unauthorized access from insider attackers results in the loss, theft, or disclosure of sensitive healthcare data. The Cost of a Data Breach Report conducted by Ponemon Institute and analyzed by IBM security states that healthcare experienced a substantial increase in data breach costs year over year and in 2021 healthcare data breaches cost $9.23 million per incident and a $2 million increase over the previous year [60]. The report also indicates that 44% of the breaches exposed personal data, such as name, email, password, and healthcare data. Data breaches in medical records result in reduced patient safety, privacy violation, operational disruptions, financial loss, legal ramifications, loss of reputation, and trust.

Despite the attacks on medical devices, another security risk associated with IoT is social engineering attacks. Social engineering attack is the act of manipulating people and it is a psychological attack directly on humans using medical devices. Social engineering is focused on diverting the user from making reasonable decisions, causing the user to make irrational decisions [49]. As a result, the attacker deceives the user to give her confidential information to the attacker or to perform an action that will breach a system's information security.

Hereby, the key security and privacy concerns can be summarized as follows:

- Ensuring secure access to the patients' health records
- Access from unauthorized users to any medical device or health system
- Ransomware attacks on hospitals/healthcare organizations
- Attacks on medical devices and IMDs
- Misuse of medical devices and IMDs
- Unsecure data transmission between medical devices
- Attacks due to security flaws/bugs in devices caused by not constantly updating/patching
- The confidentiality and integrity of patient data
- Stealing patient data including health, personal, and insurance data
- Data breaches/data leakage

Finally, the security and privacy concerns in an IoT-based healthcare system are described with a motivating scenario as follows: Alice is a cancer patient. She has high blood pressure and chronic heart conditions. Now, she also suffers from Covid-19. Bob is Alice's family doctor and with his advice, Alice is wearing a wireless body sensor while she is staying at home. The sensor collects Alice's medical data such as her heart rate and blood pressure, and sends these data to Bob in order to remotely monitor her health, to detect any unwanted medical condition and to send a medical assistance to Alice's current location. Now, during her Covid-19 treatment she also shares her medical data with Eve who is a doctor in a public healthcare provider. Bob can access Alice's all medical data and add remarks to Alice's medical results. However, Eve has access permission to a restricted medical data of Alice's and can only add remarks to her medical results that are related with her Covid-19 condition. In the case of a health condition during her Covid-19 period, the smart device alerts Bob and Eve by sending Alice's health data. Then an ambulance will be dispatched to help her.

As seen in this scenario, unauthorized access and data leakage should be prevented. Therefore, it is important to define access control policies by including the additional concepts such as user roles, purposes, and time. Also, Alice can add additional people such as family members or friends to be notified in case of an emergency. However, in such a case only authorized people should monitor her medical data. In addition, malicious attempts to alter her medical data or to access wireless body sensor network should be detected.

5.4 Semantic web based solutions for security and privacy

The semantic web is known as Web 3.0. Most of the web's content is designed for humans to read, not for computers to manipulate it meaningfully [16]. The semantic Web enables to store knowledge about the content in a structured form and allows machine-interpretable representation of concepts by representing information more meaningfully [61]. Therefore, the semantic web enables the content to be understandable by both humans and machines. Semantic web allows to simplify the distribution, sharing, and exploitation of information and knowledge across multiple sources [62]. For this purpose, ontologies are used to support interoperability between systems and a common understanding of a knowledge domain. An ontology is defined as an explicit specification of a conceptualization [63]. Ontology enables the representation of the formal specification of concepts and relationships between these concepts from a domain of interest. Thus, information is given in a well-defined meaning and more effectively processed, shared, integrated, reused, and discovered.

Digital security and privacy protection have an essential importance in today's information technology environment. As information and communication technologies are progressing very fast, devices collect, process, and share all kinds of information. Therefore, it is important to meet the security and privacy requirements of digital assets. For this purpose, it is a necessity to authorize data access and define restrictions for access rights. Semantic web describes information in well-defined semantics, enables machine-to-machine interaction, provides interoperability, and automation. Thus, a semantically enriched process enables to regulate automatic access to sensitive information [64]. As stated in [62], semantic web technologies contribute to intelligent and flexible handling of privacy and security related issues by supporting information integration. Hence, several semantic web-based studies to handle security and privacy issues are presented in the literature. Within this scope, a review on security, privacy, and policy related challenges associated with semantic web technologies is presented in [62]. The existing studies propose various solutions for access control, policy management, trust management, cyber threats, consent management, and privacy concerns.

Semantic web based policy management allows to define rules for accessing a resource in order to provide users to interpret and comply with these rules. For this purpose, an Ontology Based Access Control (OBAC) to provide fine-grained policies is presented in [65,66]. A policy language is proposed in [67] to control access to resources. In [64], the main semantic web policy languages and the usage of ontologies in policy specification, conflict detection, and validation are explained in detail. An access control model that supports varying protection granularity levels and content-based access control is proposed in [68]. Mavroeidis and Bromander [69] proposes a cyber threat intelligence ontology to share threat data and threat information in an effective way. An ontology based personalized solution to preserve privacy is proposed in [28]. Moreover, ontology-based provenance management approach is presented to detect data breaches and preserve privacy [70,61]. Furthermore, several semantic web-based solutions focus on consent management [71–75], privacy standards, and regulations such as EU's General Data Protection Regulation [76] (GDPR) [77–80], and the Health Insurance Portability and Accountability Act [81] (HIPAA) [82,83] to enhance privacy preservation. Similarly, an ontology based privacy preservation approach is proposed in [84]. Besides, there are studies that focus on the privacy aspects of linked data. An investigation on the privacy challenges when personal data is published within linked data environments is presented in [85]. In [86], the potential privacy risks on linked data are introduced and discussed. The study presented in [87] proposes a personal linked data view based approach that enables users to possess and manage their data privacy in order to be aware of their personal data privacy. An automatic framework is presented in [88] to generate and publish the RDF linked data representation of cybersecurity concepts.

Consequently, semantic web enables to create accurate models for security and privacy related issues, allows meaningful interpretation of data to manage privacy, and ensures machine-readable and machine-processable representation of policies for the automation of tasks related to policy management [62]. As the semantic web brings more meaning and structure, it has the profound potential to provide effective and efficient solutions for security and privacy.

5.5 Semantic web based solutions for the security and privacy aspects in the IoT ecosystem

The distributed and heterogeneous characteristics of the IoT ecosystem lead to the interoperability challenge. The integration of semantic web and IoT enables to achieve wide-scale interoperability and move toward horizontal open systems and platforms that can support multiple applications [89]. Also, information that is semantically rich and easily accessible is integrated into the physical world, thus smart objects and digital entities are connected [90]. Hence, semantic web-based IoT solutions offer benefits in standardization, thing discovery, and search [91]. Further, building semantic web-based IoT solutions ensures a common description and understanding of the meaning of data. Thus, semantic web-based solutions enable analyzing and sharing threat data and threat information in an effective way by providing standard formats and protocols for sharing threat data, and a common understanding of the relevant concepts and terminology [69].

Semantic web-based solutions provide a well-defined meaning for the related domain information, improve the ability to exchange and use information and information processing in the IoT ecosystem. Also, semantic web represents information in a machine-understandable and machine-processable format that can be automatically treatable by algorithms. Therefore, it supports to design interoperable IoT applications. In IoT, heterogeneity, diversity, and dynamicity of devices, data and networks are the biggest challenges [92]. These challenges are achieved by semantic web technologies by defining the relationship between data, providing modular modeling and reusability, supporting interoperability, and inferring new knowledge. In the scope of security and privacy aspects of the IoT ecosystem, ontologies are used to facilitate a formalized and unified description to resolve the heterogeneity, scalability, flexibility, and interoperability challenges in the IoT security domain [92]. Ontologies also provide the data quality, robustness, precision, and redundancy requirements of a secure system [93].

In the following subsections, semantic web-based solutions to security and privacy issues in the IoT ecosystem are examined, respectively.

5.5.1 Security oriented solutions

The heterogeneous connectedness of the IoT ecosystem brings several security challenges. The growth of security threats has become more significant with the recent developments in wireless sensor networks, the growth of smart devices, and the expansion of IoT technologies. IoT devices are limited in compute, storage, and network capacity. Therefore, most of IoT devices are vulnerable to attacks and easy to hack and compromise.

IoT has several security issues that cause a variety of damage and threaten individuals' lives. Mishra et al. [92] explores the main security and privacy needs of IoT, examines semantic web technologies within IoT systems for maintaining a secure Semantic Web of Things (SWoT) and discusses security challenges such as confidentiality, availability, integrity, trustworthiness, authentication, and authorization. A reference ontology named IoTSec is developed in [94] to unify concepts and define relationships among the main basic components of risk analysis of information security. IoTSec is a quantitative security assessment for IoT environments to identify risks for the valuable assets of the organizations. An ontology-based cyber security framework based on IoTSec is proposed in [95]. For this purpose, the IoTSec ontology is extended for the prevention measures against vulnerabilities and known threats. The proposed ontology-based framework aims to run time monitoring and actuation tools to automatically detect threats to the IoT network and dynamically propose or implement suitable protection services. Another research-based on IoTSec ontology is IoTSecEv ontology [96]. The IoTSecEv ontology is a conceptual formalization of the observer's context on security preferences and links this knowledge to concepts of the IoTSec ontology. Similarly, an ontology-based framework is developed for IoT-based smart homes in [97]. The presented ontology-based security framework uses a security ontology to address data representation, knowledge, and application heterogeneity to support security and privacy preservation in the process of interactions. The ontology-based approach allows to solve the heterogeneity issues and to define security policies for both service provider and consumer.

The functional architecture of the IoT framework that incorporates secure access provision is presented in [17]. The presented framework is implemented with semantic web technologies. Thus, the interoperability of security is addressed through an ontology-based approach. The proposed architecture offers a solution for controlling secure access to services, devices, and information by creating ontology-based access control policies. Also, the interoperability of security is ensured by using semantic web technologies. The secure semantic interoperability for IoT is also studied in [98]. For this purpose, a semantic mediator component across the various IoT layers is presented to provide the required common representation and meaning of data. The main concern of the study is to control the inference operation of the reasoning components that collect, correlate, and process data that may result with the semantic attacks that exploit network or web level vulnerabilities. An IoT security ontology for smart home energy management system in smart grids to analyze and infer security issues for smart home is developed in [99]. The ontology handles threats against smart home energy management system such as tampering, DoS, spoofing, repudiation, disclosure of information, and elevation of privilege. The ontology also includes countermeasure concepts to mitigate threats such as encryption, firewall, checksum, hash, key management, credentials, and trust management.

An ontology-based security recommendation system is developed in [100]. The developed tool allows users to make well-educated decisions for taking effective security measures. The presented system utilizes semantically enriched ontology to model the IoMT components, security issues, and measures. The defined context-aware rules enable reasoning to build a recommendation system.

5.5.2 Privacy oriented solutions

The IoT provides the opportunity to collect diverse data, process, and analyze data. Privacy becomes a critical issue as the scale of the collected data is increasing day by day. In IoT, privacy is more than anonymity as profiling and data mining within any IoT scenario can form potential harm to individuals' privacy due to the automatic process of data collection, data storage, data share, and data analysis [101]. Also, the privacy mechanisms must ensure both data protection and confidentiality of personal data.

The basic privacy requirements in IoT are presented and a privacy ontology named IoT-Priv that matches these requirements is developed in [102]. The proposed ontology is an integrated solution for the generic security and privacy issues in IoT. Also, IoT-Priv handles privacy concepts regardless of the application domain. Therefore, IoT-Priv is a domainless privacy ontology that can be extended by applications due to their purposes.

A semantic web-based framework called SeCoMan is developed in [103]. The proposed framework is a solution for developing context-aware smart applications to preserve user privacy in the IoT. For this purpose, authorization policies are managed to control who can access to a given location by allowing users to share their location to the right users, at the right granularity, at the right place, and at the right time. The SeCoMan uses the semantic web to model the description of things, reason over data to infer new knowledge, and define context-aware policies.

The knowledge that is derived from the collected, stored, shared, and processed personal data involves potential privacy risks. Therefore, a privacy ontology named LIoPY is proposed in [104] to incorporate privacy legislation into privacy policies while considering several privacy requirements. The main concern of this study is preserving privacy by defining the basic concepts for the privacy requirements. The proposed ontology-based privacy-preserving in the IoT approach focuses on both controlling who can access the data collected by smart devices and addressing the privacy preservation of the whole data life cycle. The data life cycle is the whole process of collecting, storing, processing, and transmitting the collected data by smart devices. The LIoPY ontology is also used in a semantic web-based privacy-preserving API to generate a common privacy policy that reflects the user's privacy choices [105].

An ontology-based privacy-aware virtual sensor model that enforces privacy policy in IoT sensing is proposed in [106]. The proposed model optimizes the use of privacy-preserving techniques by applying the related techniques according to the virtual sensor inference intentions while preventing malicious virtual sensors to execute or access to raw sensor data. The presented study focuses on privacy preservation for IoT sensing and enforces a privacy-by-policy strategy to design privacy-sensitive IoT systems. The study uses an ontology-based approach to provide a flexible and powerful classification for personal information.

Finally, an IoT ontology is built in [107] by using General Data Protection Regulation (GDPR) to enhance privacy. The GDPR [76] is the European Union (EU) directive for the protection of natural persons with regard to the processing of personal data. The proposed ontology does not cover all the GDPR requirements related to the full data cycle, which includes data gathering, sharing, processing, retention, deletion, etc. The presented study focuses on the data gathering and sharing aspects of personal data.

5.6 Challenges and future directions for the security and privacy concerns in IoT-based healthcare systems

In smart healthcare systems, patient monitoring and administration are controlled and managed automatically without any direct human involvement [108]. This allows automatic identification and tracking of patients, real-time monitoring of patient's physiological parameters, early diagnosis of diseases, effective treatment, improving the quality of care for patients while reducing healthcare costs. The traditional healthcare systems were a closed environment within a secure network infrastructure. However, smart healthcare systems operate in an open context [108]. For example, a wearable device continuously collects data about a patient's vital signs such as heart rate, breathing, temperature, etc. This collected data is periodically transferred to a database and viewed by the patient's doctor to be used for the patient's diagnosis and treatment.

IoT wearables and smart devices offer a promising solution for objective, reliable, continuous and remote monitoring, assessment and support through ambient assisted living [109]. Thus, patients can be efficiently monitored in real-time and remotely, patient's movements and locations are tracked, clinical decision-making is supported, effective treatments, and emergency services are provided to patients especially elderly patients, Alzheimer's and dementia patients.

The traditional healthcare system's scale and quality of health and medical resources and services are insufficient in meeting personal medical needs [110]. Besides, as healthcare data has a significant value, attackers are highly motivated to carry out attacks for financial and political gains. There are increasing concerns relating to the security of healthcare data and devices. Therefore, low-level security and privacy countermeasures are unacceptable to preserve patient privacy, secure healthcare operations, and protect people's health. Moreover, there is a huge amount of data in the healthcare domain that needs to be managed, queried, and reused for patient-specific needs. For this purpose, these data should be shared between systems and services. Semantic web technology enables to build a model of a specific domain and provides to share a common understanding in order to improve the communication between people and systems. An ontology which is the core of the semantic web technology allows to define a semantically-rich knowledge base for the information management systems and integrate information coming from different sources. Therefore, semantic web technologies allow to describe the domain and device data, define rules to maintain correct inference results [111]. Thus, the health data collected from any medical device or clinic can be reused and shared not only at the point where it was produced but also by other authorized services, devices, and people. Thus, interoperability between information systems can be achieved. For this purpose, ontologies are used as information bases to build common frameworks in health information systems. Therefore, the use of semantic web-based IoT solutions is recognized as a successful approach for the healthcare domain. Semantic web-enabled IoT-based healthcare systems offer to find valuable information, to share information to ensure the right treatment, to analyze it, to transform it to knowledge in order to enable better decision-making, and to ensure the accuracy of data.

IoT-based healthcare solutions connect medical devices and sensors to gather health data in order to provide smart and connected healthcare. Patients, families, physicians, hospitals, and insurance companies benefit from this connectivity. For example, doctors monitor and track the status of their patients anytime and anywhere, the decision-making process is facilitated, a better-informed treatment is provided to patients. The sensitive data that is gathered by IoT devices cause security problems. In the healthcare domain, it is important to ensure patient privacy as well as collecting, storing, analyzing, and transferring health data accurately and securely. Semantic web also offers solutions for the security and privacy concerns introduced by the use of IoT in healthcare systems.

A semantic model for the healthcare domain with security layers interconnected with IoT devices is proposed in [112]. The proposed secure semantic smart model performs the monitoring of end-users and maintains security issues such as authentication, integrity, and confidentiality. The study summarizes threats to a healthcare system as follows: spoofing device to the system, modifying a device with corrupted data, breaking the smart device control system, injecting the incorrect information to access the system, stealing the confidential records of patients from the storage, leaking the sensitive data, inspecting the communication channel between the server and smart devices. The presented Secure-Semantic-Smart HealthCare (S3HC) framework addresses these challenges with a secure semantic web of things model that represents the IoT resources and data meaningfully and securely.

A decentralized trust model named Blockchain Decentralized Interoperable Trust framework (DIT) for Healthcare-based Internet of Things (IoHT) is presented in [113]. The DIT framework offers a solution for the following challenges in IoHT: preserving sensitive data of patients, ensuring confidentiality and integrity of patients' data against insider attacks, enhancing encryption and IoHT access control methods, support privacy preservation while circumventing pervasive tracking and profiling. The proposed blockchain-based framework supports semantic annotations for health edge layers in IoHT and secures the different stages of data-related processes by using cryptographic algorithms. As a result of this study, the proposed semantic web-based model outperforms interoperability, confidentiality, integrity, availability, mutual authentication, trustworthiness, and privacy.

A semantic web-based privacy framework is presented in [114] to allow the patient to determine the privacy risks incurred when the elements of the personal data are shared with a data consumer. The main focus of the study is privacy concerns and risks that refer to the harmful use of personal data or the harmful effect of personal data disclosure. The proposed framework allows a patient to share her personal data with a data consumer hidden behind a smart service from the IoT ecosystem. The related privacy framework determines the privacy risks entailed by this data sharing, compares these risks to the benefits to be received, and allows the patient to decide on specifying the data items that can be released along with their precision. The proposed solution offers pragmatic data sharing to patients.

A decentralized ontology-based system architecture that meets a healthcare organization's privacy needs and its enterprise security policy concerns by considering IoT trends in the healthcare domain is presented in [93]. For this purpose, a HealthCare Security and Privacy (HCSP) ontology is developed to ensure privacy. The proposed system assesses information security standards and policy compliance for data provided from connected medical devices to satisfy institutional and regulatory guidelines.

A privacy-preservation ontology named HOPPy (Holistic Ontology for Privacy-Preserving) is proposed to model the smart healthcare domain in [115]. The HOPPy defines a common privacy vocabulary to meet the privacy requirements. The presented privacy policy ontology is in accordance with patients' preferences, privacy laws, and regulations.

Consequently, security and privacy are fundamental concerns in smart healthcare. Therefore, the security and privacy issues must be addressed for each layer of the IoT architecture. However, in the literature, researches that propose semantic web-enabled solutions on the related security and privacy problems for the IoT-based healthcare systems are limited. Thus, the security and privacy issues in semantic web-enabled IoT-based healthcare information systems bring many challenges that need to be addressed in the future.

One of the challenges is ensuring the heightened access control. For this purpose, the privacy of the shared data must be preserved by preventing unauthorized access to data. The access of resources should be monitored and the unauthorized flow of information must be prevented. For example, medical wearable devices sense, collect, and share individuals' sensitive health data. Individuals must be aware of this shared data and have control over their sensitive data. Thus, unauthorized access should be prevented and fine-grained access control mechanisms need to be maintained. Semantic web technologies enable to set access restrictions at the right level of granularity based on the actual data content and to understand what pieces of information need to be treated carefully to protect privacy [116]. The access control mechanisms must also include the time and location restrictions due to the heterogeneous and dynamic structure of the IoT ecosystem. For example, while an individual allows sharing her wearable device data in weekdays, she prohibits sharing her data at weekends. Besides, conflict detection is an important issue that needs to be considered while working with access control policies.

Auditing of IoT medical devices must be enabled and detailed logs of device activities and access events must be kept. Auditing is a substantial requirement to detect anomalous, malicious, and unintentional activities. Also, using multifactor authentication increases security level by requiring more than one form of authentication such as biometric information that is specific to the patient. Thus, accessing medical devices or systems becomes more challenging for attackers.

The real spreading of IoT services requires customized security and privacy levels [15]. Therefore, another challenge is maintaining personalized security according to the privacy needs of patients. Privacy preservation in the IoT should be investigated and addressed from the perspective that privileges the interests of data owners [106]. Different IoT systems require distinct security mechanisms to avoid intrusions from the physical and cyber world [95]. Also, every individual has different privacy requirements when sharing her personal data. Personalized privacy protection must be accomplished by using the individual's personal privacy level choice. Moreover, customization is also an important issue and should be guaranteed.

Shared data is beneficial for researchers, but it may cause a privacy problem for individuals whose data has been published [84]. Thus, providing the right balance between data privacy and the need for patients and providers to interact with health data is also an important challenge while preserving privacy. Reliable and well-balanced mechanisms should be maintained for the disclosure of data without limiting the overall healthcare system's flexibility. It is critical to consider the privacy requirement of each individual independently to prevent excessive data distortion while ensuring individual privacy [117]. Further, the data that is stored in medical devices must be encrypted against data breaches. Also, data obfuscation techniques must be used to hide sensitive data of IoT medical devices.

Besides these challenges, communication between IoT devices must also be secured. The security risks associated with things-to-things communications in addition to risks relating to things-to-person communications need to be considered [101]. Also, reliable access to medical data is significant to improve the treatment process. For this purpose, end-to-end security solutions to protect communication among smart medical devices and exchange data between both ends of the communication should be developed. The end-to-end security must provide the data exchange without being read, eavesdropped, intercepted, modified, or tampered [101]. Also, auditing, log management, and patch management must be enabled. Moreover, the written code of end-points must be inspected to ensure that it is protected from reverse engineering and vulnerabilities against malwares [100].

In semantic web-based approaches, the fundamental ontological concepts must address 4W1H methodology which consists of 4W (What, When, Where, Who) and 1H (How) [118]. Most of the ontology-based approaches in IoT address What, Where, and When related competency questions, but they fail to address the Who aspect [107]. Thus, semantic web-based solutions to ensure security and privacy must consider the 4W1H methodology.

Finally, the legislation aspect of privacy must also be taken into consideration while proposing privacy-related solutions. As addressing only access control at processing time is not enough to preserve privacy, the privacy preservation mechanisms must fulfill regulations and privacy standards that define the privacy requirements. Thus, considering patients' consent and covering the regulations such as GDPR are key requirements while addressing the privacy issues. Patients have the right to know who collects, stores, and accesses their data. The term of consent is defined as restricting the disclosure of sensitive information according to the wishes of the patient [73]. Therefore, patient-oriented consent management can be used to guarantee patient privacy [119]. Also, ontology-based consent management [71,73,72,119] and ontology-based provenance solutions [61,120,70] should be investi-

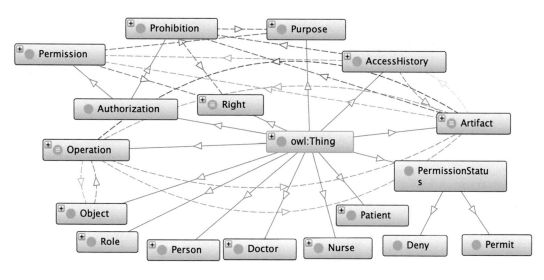

FIGURE 5.5

The structure of a privacy-preserving provenance ontology.

gated for privacy preservation in IoT-based healthcare systems. The structure of a provenance ontology that detects privacy violations in a healthcare system is presented in Fig. 5.5 [120]. As the proposed model is domain-independent, it can be used in various situations in the healthcare domain of the IoT ecosystem.

It is important to note that security is not just an add-on to existing systems, but an integral part of systems [121]. The security and privacy needs must be appropriately applied to maintain the reliability of systems and the privacy of users. IoT security is concerned with any "thing" that is connected. Therefore, the scope of IoT security should be end-to-end to support the device from the very beginning. The security and privacy challenges within IoT-based healthcare systems are primary objectives. Attacks and data breaches may threaten human life, cause patients to lose their trust, or result in loss of functioning of critical equipment within hospitals. Therefore, vulnerabilities and risks against these systems must be carefully studied, threat modeling must be developed, and all the relevant issues must be determined in order to develop sufficient security and privacy measures. In this scope, semantic-web enabled solutions for the related issues improve the efficiency and effectiveness of ensuring security and preserving privacy. Hence, the semantic web allows users to operate at abstraction levels above the technical details of format and integration, also enables autonomous and semiautonomous agents to assist in collecting, processing, and reasoning on sensor data [122]. Additionally, ontologies that play a major part in interoperability support the automatic establishment of security metrics based on explicit and reasoning information, improve the efficiency and effectiveness in security operations and help analysts to extract relevant pieces of information to characterize vulnerabilities and threats [95].

5.7 Conclusions

The IoT is a fast-growing complex network of connected sensors and devices that reaches several application domains ranging from smart homes, smart cities, smart agriculture to Industry 4.0. This rapid development of IoT offers growing opportunities for the healthcare sector. The main objective of the IoT-based healthcare systems is to provide health information to authorized parties at anytime and anywhere. As semantic web technologies are used to improve the quality of the healthcare domain and to provide interoperability between the healthcare information systems, integrating semantic web technologies into the IoT-based health information systems provides an effective and efficient solution by promoting the interoperability among resources, data providers, and consumers. However, the increasing connectedness and the ongoing digitization process in many fields convey various threats. In the healthcare domain, using reliable and timely data is critical for decision-makers at all levels of the health system. Meanwhile, maintaining the security and privacy of healthcare data is a major challenging task due to the current technological developments and the growing potential of the IoT ecosystem. Therefore, security vulnerabilities that present in the IoT-based healthcare systems should be addressed and a sufficient level of protection must be achieved. Besides, an effective IoT-based healthcare system must not compromise the fundamental security elements and privacy requirements.

Ensuring the security and privacy of the IoT-based health information systems is an important issue that requires further investigations and developments. The current researches in the literature that focus on the security and privacy-related issues for the semantic web-enabled IoT-based healthcare systems are limited. The security and privacy challenges in IoT-based healthcare systems are severe. Under the scope of the rise of IoT, the potential threats should be clearly understood and the safety of this digital future should be maintained. In this context, semantic web-based solutions to the security and privacy challenges in IoT-based healthcare systems are still open for future research.

References

[1] Oxford, Internet of Things, https://www.oxfordlearnersdictionaries.com/definition/english/internet-of-things, 2021. (Accessed 10 November 2021).
[2] Kriti Chopra, Kunal Gupta, Annu Lambora, Future internet: the Internet of Things – a literature review, in: 2019 International Conference on Machine Learning, Big Data, Cloud and Parallel Computing (COMIT-Con), 2019, pp. 135–139.
[3] Infso, Internet of Things in 2020 – Roadmap for the Future, Infso D.4 Networked Enterprise & RFID Infso G.2 Micro & Nanosystems in co-operation with The Working Group RFID of The ETP EPOSS, 2008.
[4] Lu Tan, Neng Wang, Future internet: the Internet of Things, in: 3rd International Conference on Advanced Computer Theory and Engineering (ICACTE), 2010, pp. V5-376–V5-380.
[5] Sudhi R. Sinha, Youngchoon Park, Building an Effective IoT Ecosystem for Your Business, Springer, Switzerland, 2017.
[6] Finances Online, 10 IoT trends for 2021/2022: latest predictions according to experts, https://financesonline.com/iot-trends, 2021. (Accessed 10 November 2021).
[7] Statista, Number of Internet of Things (IoT) connected devices worldwide from 2019 to 2030, https://www.statista.com/statistics/1183457/iot-connected-devices-worldwide, 2021. (Accessed 10 November 2021).
[8] Cisco, Internet of Things-At-a-Glance, https://www.cisco.com/c/en/us/products/collateral/se/internet-of-things/at-a-glance-c45-731471.pdf, 2016. (Accessed 10 November 2021).

[9] Business Wire, The growth in connected IoT devices is expected to generate 79.4ZB of data in 2025, according to a new IDC forecast, https://www.businesswire.com/news/home/20190618005012/en/The-Growth-in-Connected-IoT-Devices-is-Expected-to-Generate-79.4ZB-of-Data-in-2025-According-to-a-New-IDC-Forecast, 2019. (Accessed 10 November 2021).

[10] Fortune Business Insights, IoT market size, share & Covid-19 impact analysis, https://www.fortunebusinessinsights.com/industry-reports/internet-of-things-iot-market-100307, 2021. (Accessed 10 November 2021).

[11] IoT Analytics, Top 10 IoT application areas 2020, https://iot-analytics.com/wp/wp-content/uploads/2020/07/Top-10-IoT-applications-in-2020-min.png, 2020. (Accessed 10 November 2021).

[12] James Manyika, Michael Chui, Jacques Bughin, Richard Dobbs, Peter Bisson, Alex Marrs, Disruptive Technologies: Advances That Will Transform Life, Business, and the Global Economy, McKinsey Global Institute, San Francisco, CA, USA, 2013.

[13] Lòai Tawalbeh, Fadi Muheidat, Mais Tawalbeh, Muhannad Quwaider, IoT privacy and security: challenges and solutions, Applied Sciences 10 (12) (2020) 4102.

[14] Palo Alto Networks, State of enterprise IoT security in 2020, https://www.paloaltonetworks.com/resources/infographics/state-of-enterprise-iot-security-in-2020, 2020. (Accessed 10 November 2021).

[15] Sabrina Sicari, Alessandra Rizzardi, Luigi Alfredo Grieco, Alberto Coen-Porisini, Security, privacy and trust in Internet of Things: the road ahead, Computer Networks 76 (2015) 146–164.

[16] Tim Berners-Lee, James Hendler, Ora Lassila, The semantic web, Scientific American 284 (5) (2001) 34–43.

[17] Sarfraz Alam, Mohammad M.R. Chowdhury, Josef Noll, Interoperability of security-enabled Internet of Things, Wireless Personal Communications 61 (2011) 567–586.

[18] Ahlem Rhayem, Mohamed Ben, Ahmed Mhiri, Faiez Gargouri, Semantic web technologies for the Internet of Things: systematic literature review, Internet of Things 11 (2020) 100206.

[19] Payam Barnaghi, Wei Wang, Cory Henson, Kerry Taylor, Semantics for the Internet of Things: early progress and back to the future, International Journal on Semantic Web and Information Systems (IJSWIS) 8 (1) (2012) 1–21.

[20] Enisa, Privacy and data protection by design – from policy to engineering, https://www.enisa.europa.eu/publications/privacy-and-data-protection-by-design, 2014. (Accessed 10 November 2021).

[21] Marc van Lieshout, Linda Kool, Privacy implications of RFID: an assessment of threats and opportunities, in: IFIP International Federation for Information Processing, Vol. 262, The Future of Identity in the Information Society, Springer, Boston, 2008, pp. 129–141.

[22] Roger Clarke, Internet privacy concerns confirm the case for intervention, Communications of the ACM 42 (2) (1999) 60–67.

[23] A. Michael Froomkin, The death of privacy?, Stanford Law Review 52 (5) (2000) 1461–1543.

[24] Mark Mbock Ogonji, George Okeyo, Joseph Muliaro Wafula, A survey on privacy and security of Internet of Things, Computer Science Review 38 (2020) 100312.

[25] Sivanarayani M. Karunarathne, Neetesh Saxena, Muhammad Khurram Khan, Security and privacy in IoT smart healthcare, IEEE Internet Computing 25 (4) (2021) 37–48.

[26] Erika McCallister, Tim Grance, Karen Scarfone, Guide to Protecting the Confidentiality of Personally Identifiable Information (PII), NIST Special Publication, 2010, 800–122, 58 pp.

[27] Pierangela Samarati, Protecting respondents' identities in microdata release, IEEE Transactions on Knowledge and Data Engineering 13 (6) (2001) 1010–1027.

[28] Ozgu Can, Buket Usenmez, An ontology based personalized privacy preservation, in: 11th International Conference on Knowledge Engineering and Ontology Development (KEOD 2019), 2019, pp. 500–507.

[29] Cynthia Dwork, Differential privacy: a survey of results, in: International Conference on Theory and Applications of Models of Computation, vol. 4978, 2008, pp. 1–19.

[30] Valentina Ciriani, Sabrina De Capitani di Vimercati, Sara Foresti, Pierangela Samarati, k-Anonymity, in: T. Yu, S. Jajodia (Eds.), Secure Data Management in Decentralized Systems, in: Advances in Information Security, vol. 33, Springer, Boston, MA, 2007.

[31] Ashwin Machanavajjhala, Daniel Kifer, Johannes Gehrke, Muthuramakrishnan Venkitasubramaniam, l-Diversity: privacy beyond k-anonymity, in: IEEE 22nd International Conference on Data Engineering (ICDE'06), 2006.

[32] Ninghui Li, Tiancheng Li, Suresh Venkatasubramanian, T-closeness: privacy beyond k-anonymity and l-diversity, in: IEEE 23rd International Conference on Data Engineering 2007 (ICDE Conference 2007), 2007.

[33] John R. Vacca, Computer and Information Security Handbook, 3rd ed., Morgan Kaufmann Publishers, Cambridge, 2017.

[34] CVE, Common vulnerabilities and exposures, https://cve.mitre.org/index.html, 2021. (Accessed 10 November 2021).

[35] NVD, National vulnerability list, https://nvd.nist.gov, 2021. (Accessed 10 November 2021).

[36] MITRE, MITRE ATT&CK, https://attack.mitre.org, 2021. (Accessed 10 November 2021).

[37] David Kim, Michael G. Solomon, Fundamentals of Information Systems Security, 3rd ed., Jones & Bartlett Learning, Burlington, 2018.

[38] Netscout, Dawn of the Terrorbit Era, Netscout Threat Intelligence Report 2H 2018, 2018.

[39] OWASP, OWASP Internet of Things project, https://owasp.org/www-project-internet-of-things, 2021. (Accessed 10 November 2021).

[40] Muhammad Burhan, Rana Asif Rehman, Bilal Khan, Byung-Seo Kim, IoT elements, layered architectures and security issues: a comprehensive survey, Sensors 2018 (18) (2018) 2796.

[41] Mohamed Litoussi, Nabil Kannouf, Khalid El Makkaoui, Abdellah Ezzati, Mohamed Fartitchou, IoT security: challenges and countermeasures, Procedia Computer Science 177 (2020) 503–508.

[42] Ioannis Andrea, Chrysostomos Chrysostomou, George Hadjichristofi, Internet of Things: security vulnerabilities and challenges, in: The 3rd IEEE ISCC 2015 International Workshop on Smart City and Ubiquitous Computing Applications, 2015, pp. 180–187.

[43] Anil Chacko, Thaier Hayajneh, Security and privacy issues with IoT in healthcare, EAI Endorsed Transactions on Pervasive Health and Technology 4 (14) (2018) e2.

[44] Jigna J. Hathaliya, Sudeep Tanwar, An exhaustive survey on security and privacy issues in Healthcare 4.0, Computer Communications 153 (2020) 311–335.

[45] Rajesh Gupta, Sudeep Tanwar, Sudhanshu Tyagi, Neeraj Kumar, Mohammad S. Obaidat, Balqies Sadoun, HaBiTs: blockchain-based telesurgery framework for Healthcare 4.0, in: 2019 International Conference on Computer, Information and Telecommunication Systems (CITS), 2019, pp. 1–5.

[46] Jigna J. Hathaliya, Sudeep Tanwar, Sudhanshu Tyagi, Neeraj Kumar, Securing electronics healthcare records in Healthcare 4.0: a biometric-based approach, Computers and Electrical Engineering 76 (2019) 398–410.

[47] Protenus, Mid-year breach barometer report, https://www.protenus.com/resources/2019-breach-barometer, 2019. (Accessed 10 November 2021).

[48] Laurie Pycroft, Tipu Z. Aziz, Security of implantable medical devices with wireless connections: the dangers of cyber-attacks, Expert Review of Medical Devices 15 (6) (2018) 403–406.

[49] Syed Ghazanfar Abbas, Ivan Vaccari, Faisal Hussain, Shahzaib Zahid, Ubaid Ullah Fayyaz, Ghalib A. Shah, Taimur Bakhshi, Enrico Cambiaso, Identifying and mitigating phishing attack threats in IoT use cases using a threat modelling approach, Sensors 21 (2021) 4816, 2021.

[50] Eduard Marin, Dave Singelée, Flavio D. Garcia, Tom Chothia, Rik Willems, Bart Preneel, On the (in)security of the latest generation implantable cardiac defibrillators and how to secure them, in: Proceedings of the 32nd Annual Conference on Computer Security Applications (ACSAC'16), 2016, pp. 226–236.

[51] Laurie Pycroft, Sandra G. Boccard, Sarah L.F. Owen, John F. Stein, James J. Fitzgerald, Alexander L. Green, Tipu Z. Aziz, Brainjacking: implant security issues in invasive neuromodulation, World Neurosurgery 92 (2016) 454–462.

[52] Chunxiao Li, Anand Raghunathan, Niraj K. Jha, Hijacking an insulin pump: security attacks and defenses for a diabetes therapy system, in: IEEE 13th International Conference on e-Health Networking, Applications and Services, 2011, pp. 150–156.

[53] TrapX Research Labs, MEDJACK.4 Medical Device Hijacking, TrapX Investigative Report, 2018.

[54] FDA, Postmarket Management of Cybersecurity in Medical Devices – Guidance for Industry and Food and Drug Administration Staff, U.S. Food and Drug Administration, 2016, FDA-2015-D-5105.

[55] Jeyavel Janardhanan, T. Parameswaran, J. Mannar Mannan, U. Hariharan, Security vulnerabilities and intelligent solutions for IoMT systems, in: D.J. Hemanth, J. Anitha, G.A. Tsihrintzis (Eds.), Internet of Medical Things. Internet of Things (Technology, Communications and Computing), Springer, Cham, 2021.

[56] Mohammad Wazid, Ashok Kumar Das, Joel J.P.C. Rodrigues, Sachin Shetty, Youngho Park, IoMT malware detection approaches: analysis and research challenges, IEEE Access 7 (2019) 182459–182476.

[57] Minhaj Ahmad Khan, Khaled Salah, IoT security: review, blockchain solutions, and open challenges, Future Generation Computer Systems 82 (2018) 395–411.

[58] HIMSS, Personal health records, electronic health records key to India's national digital health mission comment letter, https://www.himss.org/resources/personal-health-records-electronic-health-records-key-indias-national-digital-health, 2021. (Accessed 10 November 2021).

[59] Adil Hussain Seh, Mohammad Zarour, Mamdouh Alenezi, Amal Krishna Sarkar, Alka Agrawal, Rajeev Kumar, Raees Ahmad Khan, Healthcare data breaches: insights and implications, Healthcare 8 (2) (2020) 133.

[60] IBM, IBM report: cost of a data breach hits record high during pandemic, https://newsroom.ibm.com/2021-07-28-IBM-Report-Cost-of-a-Data-Breach-Hits-Record-High-During-Pandemic, 2021. (Accessed 10 November 2021).

[61] Ozgu Can, Dilek Yilmazer, A novel approach to provenance management for privacy preservation, Journal of Information Science 46 (2) (2020) 147–160.

[62] Sabrina Kirrane, Serena Villata, Mathieu d'Aquin, Privacy, security and policies: a review of problems and solutions with semantic web technologies, Semantic Web 9 (2) (2018) 153–161.

[63] Thomas R. Gruber, A translation approach to portable ontologies, Knowledge Acquisition 5 (2) (1993) 199–220.

[64] Daniel Olmedilla, Security and privacy on the semantic web, in: M. Petkoviè, W. Jonker (Eds.), Security, Privacy, and Trust in Modern Data Management, in: Data-Centric Systems and Applications, Springer, Berlin, Heidelberg, 2007.

[65] Ozgu Can, Okan Bursa, Murat Osman Unalir, Personalizable ontology based access control, Gazi University Journal of Science 23 (4) (2010) 465–474.

[66] Ozgu Can, Murat Osman Unalir, Ontology based access control, Pamukkale University Journal of Engineering Sciences 16 (2) (2010) 197–206.

[67] Lalana Kagal, Tim Finin, Anupam Joshi, A Policy Based Approach to Security for the Semantic Web, The Semantic Web – ISWC 2003, Second International Semantic Web Conference, vol. 2870, Springer, 2003, pp. 402–418.

[68] Elisa Bertino, Elena Ferrari, Secure and selective dissemination of XML documents, ACM Transactions on Information and System Security 5 (3) (2002) 290–331.

[69] Vasileios Mavroeidis, Siri Bromander, Cyber threat intelligence model: an evaluation of taxonomies, sharing standards, and ontologies within cyber threat intelligence, in: 2017 European Intelligence and Security Informatics Conference (EISIC), 2017, pp. 91–98.

[70] Ozgu Can, Dilek Yilmazer, A privacy aware semantic model for provenance management, in: 8th Metadata and Semantics Research Conference (MTSR2014), vol. 478, 2014, pp. 162–169.

[71] Ozgu Can, A semantic model for personal consent management, in: 7th Metadata and Semantics Research Conference (MTSR 2013), vol. 390, 2013, pp. 146–151.

[72] Emre Olca, Ozgu Can, A meta consent model for personalized data privacy, in: 10th International on Metadata and Semantics Research (MTSR 2016), 2016.

[73] Emre Olca, Ozgu Can, Providing patient rights with consent management, in: 2nd International Symposium on Digital Forensics and Security (ISDFS'14), 2014, pp. 101–104.

[74] Anelia Kurteva, Tek Raj Chhetri, Harshvardhan J. Pandit, Anna Fensel, Consent through the lens of semantics: state of the art survey and best practices, Semantic Web (1 Jan 2021) 1–27.

[75] Yu Lin, Jie Zheng, V.I.C.O. Yongqun He, Ontology-based representation and integrative analysis of vaccination informed consent forms, Journal of Biomedical Semantics 7 (2016) 20.

[76] GDPR, General Data Protection Regulation, Regulation (EU) 2016/679 of The European Parliament and of The Council of 27 April 2016, 2016.

[77] Monica Palmirani, Michele Martoni, Arianna Rossi, Cesare Bartolini, Livio Robaldo, Legal ontology for modelling GDPR concepts and norms, in: Legal Knowledge and Information Systems, in: Frontiers in Artificial Intelligence and Applications, vol. 313, 2018, pp. 91–100.

[78] Monica Palmirani, Michele Martoni, Arianna Rossi, Cesare Bartolini, Livio Robaldo PrOnto, Privacy ontology for legal reasoning, in: A. Kó, E. Francesconi (Eds.), Electronic Government and the Information Systems Perspective, EGOVIS 2018, in: Lecture Notes in Computer Science, vol. 11032, Springer, 2018.

[79] Harshvardhan J. Pandit, Christophe Debruyne, Declan O'Sullivan, Dave Lewis, GConsent – a consent ontology based on the GDPR, in: P. Hitzler, et al. (Eds.), The Semantic Web, ESWC 2019, in: Lecture Notes in Computer Science, vol. 11503, Springer, 2019.

[80] Maryam Davari, Elisa Bertino, Access control model extensions to support data privacy protection based on GDPR, in: 2019 IEEE International Conference on Big Data (Big Data), 2019, pp. 4017–4024.

[81] HIPAA, Health information privacy, https://www.hhs.gov/hipaa/index.html, 2021. (Accessed 10 November 2021).

[82] Arkalgud Ramaprasad, Thant Syn, Khin Than Win, Ontological meta-analysis and synthesis of HIPAA, in: Proceeding of the 19th Pacific Asia Conference on Information Systems (PACIS 2014), 2014, pp. 1–10.

[83] Karuna Pande Joshi, Yelena Yesha, Tim Finin, An ontology for a HIPAA compliant cloud services, in: 4th International IBM Cloud Academy Conference (ICACON 2016), 2016.

[84] Buket Usenmez, Ozgu Can, Conceptualization of personalized privacy preserving algorithms, in: E. Garoufallou, R. Hartley, P. Gaitanou (Eds.), Metadata and Semantics Research (MTSR 2015), in: Communications in Computer and Information Science, vol. 544, Springer, Cham, 2015, pp. 195–200.

[85] David Corsar, Peter Edwards, John Nelson, Personal privacy and the web of linked data, in: Proceedings of the Workshop on Society, Privacy and the Semantic Web – Policy and Technology (PrivOn2013), CEUR, vol. 1121, 2013.

[86] Felix Ritchie, Jim Smith, Confidentiality and linked data, Paper published as part of The National Statistician's Quality Review, London, 2018.

[87] Patricia Serrano-Alvarado, Emmanuel Desmontils, Personal Linked Data: A Solution to Manage User's Privacy on the Web, Atelier sur la Protection de la Vie Privée (APVP), 2013.

[88] Arnav Joshi, Ravendar Lal, Tim Finin, Anupam Joshi, Extracting cybersecurity related linked data from text, in: 2013 IEEE Seventh International Conference on Semantic Computing, 2013, pp. 252–259.

[89] Suparna De, Yuchao Zhou, Klaus Moessner, Ontologies and context modeling for the Web of Things, in: Managing the Web of Things, Elsevier, 2017, pp. 3–36.

[90] Antonio J. Jara, Alex C. Olivieri, Yann Bocchi, Markus Jung, Wolfgang Kastner, Antonio F. Skarmeta, Semantic web of things: an analysis of the application semantics for the IoT moving towards the IoT convergence, International Journal of Web and Grid Services 10 (2/3) (2014) 244–272.

[91] Bogdan Manate, Victor Ion Munteanu, Teodor Florin Fortis, Towards a smarter Internet of Things: semantic visions, in: 2014 Eight International Conference on Complex, Intelligent and Software Intensive Systems (CISIS), 2014, pp. 582–587.

[92] Sanju Mishra, Sarika Jain, Chhiteesh Rai, Niketa Gandhi, Security challenges in semantic Web of Things, in: A. Abraham, N. Gandhi, M. Pant (Eds.), Innovations in Bio-Inspired Computing and Applications (IBICA 2018), in: Advances in Intelligent Systems and Computing, vol. 939, Springer, Cham, 2019, pp. 162–169.

[93] Hisham Kanaan, Khalid Mahmood, Varun Sathyan, An ontological model for privacy in emerging decentralized healthcare systems, in: 2017 IEEE 13th International Symposium on Autonomous Decentralized System (ISADS), 2017, pp. 107–113.

[94] Bruno A. Mozzaquatro, Ricardo Jardim-Goncalves, Carlos Agostinho, Towards a reference ontology for security in the Internet of Things, in: 2015 IEEE International Workshop on Measurements & Networking (M&N), 2015, pp. 1–6.

[95] Bruno Augusti Mozzaquatro, Carlos Agostinho, Diogo Goncalves, João Martins, Ricardo Jardim-Goncalves, An ontology-based cybersecurity framework for the Internet of Things, Sensors 18 (9) (2018) 3053.

[96] Pedro Gonzalez-Gil, Antonio F. Skarmeta, Juan Antonio Martinez, Towards an ontology for IoT context-based security evaluation, in: 2019 Global IoT Summit (GIoTS), 2019, pp. 1–6.

[97] Ming Tao, Jinglong Zuo, Zhusong Liu, Aniello Castiglione, Francesco Palmieri, Multi-layer cloud architectural model and ontology-based security service framework for IoT-based smart homes, Future Generation Computer Systems 78 (2018) 1040–1051.

[98] George Hatzivasilis, Othonas Soultatos, Darko Anicic, Arne Bröring, Konstantinos Fysarakis, George Spanoudakis, et al., Secure semantic interoperability for IoT applications with linked data, in: IEEE Global Communications Conference (GLOBECOM), 2019, pp. 1–6.

[99] Getinet Ayele Eshete, Semantic Description of IoT Security for Smart Grid, Master Thesis, University of Agder, Faculty of Engineering and Science Department of Information and Communication Technology, June 2017.

[100] Faisal Alsubaei, Abdullah Abuhussein, Sajjan Shiva, Ontology-based security recommendation for the Internet of medical things, IEEE Access 7 (2019) 48948–48960.

[101] Mahmoud Elkhodr, Seyed Shahrestani, Hon Cheung, The Internet of Things: new interoperability, management and security challenges, International Journal of Network Security & Its Applications (IJNSA) 8 (2) (2016).

[102] Mayke Ferreira Arruda, Renato Freitas Bulcão-Neto, Toward a lightweight ontology for privacy protection in IoT, in: Proceedings of the 34th ACM/SIGAPP Symposium on Applied Computing (SAC'19), 2019, pp. 880–888.

[103] Alberto Huertas Celdrán, Félix J. García Clemente, Manuel Gil Pérez, Gregorio Martínez Pérez, SeCoMan: a semantic-aware policy framework for developing privacy-preserving and context-aware smart applications, IEEE Systems Journal 10 (3) (2016) 1111–1124.

[104] Faiza Loukil, Chirine Ghedira-Guegan, Khouloud Boukadi, Aicha Nabila Benharkat, LIoPY: a legal compliant ontology to preserve privacy for the Internet of Things, in: IEEE 42nd Annual Computer Software and Applications Conference (COMPSAC), 2018, pp. 701–706.

[105] Faiza Loukil, Chirine Ghedira-Guegan, Khouloud Boukadi, Aïcha-Nabila Benharkat, Privacy-preserving IoT data aggregation based on blockchain and homomorphic encryption, Sensors 21 (7) (2021) 2452.

[106] Thiago Moreira da Costa, OPP_IoT an ontology-based privacy preservation approach for the Internet of Things, PhD Thesis, Université Grenoble Alpes, 2017, NNT: 2017GREAM003.

[107] Rachit Agarwal, Tarek Elsaleh, Elias Tragos, GDPR-inspired IoT ontology enabling semantic interoperability, federation of deployments and privacy-preserving applications, arXiv:2012.10314v1, 2020.

[108] Shantanu Pal, Michael Hitchens, Tahiry Rabehaja, Subhas Mukhopadhyay, Security requirements for the Internet of Things: a systematic approach, Sensors 20 (20) (2020) 5897.

[109] Thanos G. Stavropoulos, Asterios Papastergiou, Lampros Mpaltadoros, Spiros Nikolopoulos, Ioannis Kompatsiaris, IoT wearable sensors and devices in elderly care: a literature review, Sensors 20 (10) (2020) 2826.

[110] Nikita Malik, Sanjay Kumar Malik, Using IoT and semantic Web technologies for healthcare and medical sector, in: Ontology-Based Information Retrieval for Healthcare Systems, Wiley, 2020, pp. 91–115.

[111] Emine Sezer, Okan Bursa, Ozgu Can, Murat Osman Unalir, Semantic web technologies for IoT-based health care information systems, in: The Tenth International Conference on Advances in Semantic Processing (SEMAPRO 2016), 2016, pp. 45–47, 2016.

[112] Sanju Mishra Tiwari, Sarika Jain, Ajith Abraham, Smita Shandilya, Secure Semantic Smart HealthCare (S3HC), Journal of Web Engineering 17 (8) (2019) 617–646.

[113] Eman M. Abou-Nassar, Abdullah M. Iliyasu, Passent M. El-Kafrawy, Oh-Young Song, Ali Kashif Bashir, et al., DITrust chain: towards blockchain-based trust models for sustainable healthcare IoT systems, IEEE Access 8 (2020) 111223–111238.

[114] Mansour Naser Alraja, Hanadi Barhamgi, Amjad Rattrout, Mahmoud Barhamgi, An integrated framework for privacy protection in IoT-applied to smart healthcare, Computers and Electrical Engineering 91 (2021) 107060.

[115] Driss El Majdoubi, Hanan El Bakkali, Souad Sadki, Asmae Leghmid, Zaina Maqour, HOPPy: holistic ontology for privacy-preserving in smart healthcare environment, in: 2021 Fifth World Conference on Smart Trends in Systems Security and Sustainability (WorldS4), 2021, pp. 248–253.

[116] Thomas Thurner, How semantic web serves the security and e-health domain, https://2021-eu.semantics.cc/how-semantic-web-serves-security-and-e-health-domain, 2021. (Accessed 10 November 2021).

[117] Ozgu Can, Personalised anonymity for microdata release, IET Information Security 12 (4) (2018) 341–347.

[118] Garvita Bajaj, Rachit Agarwal, Pushpendra Singh, Nikolaos Georgantas, Valérie Issarny, 4W1H in IoT semantics, IEEE Access 6 (2018) 65488–65506.

[119] Okan Bursa, Emine Sezer, Ozgu Can, Murat Osman Unalir, Using FOAF for interoperable and privacy protected healthcare information systems, in: 8th Metadata and Semantics Research Conference (MTSR2014), vol. 478, 2014, pp. 154–161.

[120] Ozgu Can, Dilek Yilmazer, Improving privacy in health care with an ontology-based provenance management system, Expert Systems 37 (1) (2020) e12427.

[121] Gokhan Polat, Fadi Sodah, Security issues in IoT challenges and countermeasures, ISACA Journal 2 (2019) 1–7.

[122] Michael Compton, Payam Barnaghi, Luis Bermudez, et al., The SSN ontology of the W3C semantic sensor network incubator group, Web Semantics: Science, Services and Agents on the World Wide Web 17 (2012) 25–32.

Knowledge-based system as a context-aware approach for the Internet of medical connected objects

6

Antonella Carbonaro

Department of Computer Science and Engineering – DISI, University of Bologna, Cesena, Italy

Contents

6.1 Introduction

The era of the Internet of Things (IoT), with a huge number of IoT devices already in everyone's lives, is bringing about major changes in several important domains such as Smart Cities, e-health, security, intelligent transportation, infrastructure management, etc. One of the most significant aspects of the IoT is the interconnection of things, which provides interconnected systems of different services and applications. Huge amounts of data are generated every day, the knowledge of which is extremely valuable. The most advanced IoT systems try to assess a situation based on the knowledge expressed by the data in order to enable services to make intelligent decisions [1].

However, one of the most important challenges of existing IoT technologies is that of interoperability as data is based on predetermined formats without following vocabularies to describe interoperable data [2]. The basic structure of IoT is machine-to-machine communication, as IoT is based on a wide variety of different heterogeneous systems and technologies and there is no standardized language for data representation and processing. For example, sensor measurements are required to be distributed and analyzed by other devices or sensors and not to be usable by humans for more sophisticated processing. This has contributed to a large number of IoT systems that are incompatible. Thus, it is very difficult to extract knowledge from the huge number of data provided by the IoT applications every second. These measurements should follow standards so that the data can be transferred from one machine to another.

The IoT revolution has made a considerable contribution also to the expansion and empowering of e-healthcare services [3,4]. In fact, smart remote and mobile healthcare applications are making an enormous shift in the health and social care workforce efficiency as well as patients' wellbeing, creating several tools for collecting and managing data effectively and providing several solutions to e-healthcare challenges: data, information, and knowledge can be used in real-time to support effective integration of prevention, treatment, and recovery services across healthcare services. For example, blood pressure, body temperature, heartbeat, respiratory rate, oxygen saturation, blood glucose level, wrist pulse signal, galvanic skin response, magnetoencephalogram, electrooculography, electromyogram, electrocardiogram, and electroencephalogram are health-related parameters that can be collected in huge quantities by medical connected objects (MCOs) to realize smart healthcare applications. These MCOs are heterogeneous in terms of deployment contexts, computing capabilities, and communication protocols, as they are designed by different manufacturers. Accordingly, the exchanged information is heterogeneous in formats and units and does not have the same coding format. In this context, ensuring semantic interoperability becomes difficult [5]. The knowledge era allows to improve e-healthcare systems through the management of the knowledge base and the automatic support of decisions. In this perspective, e-healthcare applications are now exchanging not only an enormous volume of data but also an important quantity of information and a large knowledge base.

Once again, interoperability is the key: beyond the issues related to different international healthcare systems and organizations, a substantial lack of cohesive data models in Electronic Health Records (HERs) and poor interoperability highlight a weakness in e-health infrastructures [6], made even more severe if we consider that these data can be of essential importance in daily activities and research, e.g., on the effectiveness of drugs and therapeutic strategies. Thus, semantic interoperability is becoming a crucial feature and it is hard to imagine a healthcare or clinical system architecture without it [7,8]. The MCO ontologies appear a valuable resource to improve the data contained in the information model and to support service management operations [9]. Knowledge graph (KG) models are becoming commonly used models in healthcare systems providing a flexible approach to integrate data and knowledge and assisting in inferring meaning [10]. Often KG systems use reasoning mechanisms in order to ensure active and assisted monitors of patients [11].

The main goal of this work is to propose a KG-based context-aware approach for the Internet of MCOs that consists in:

- defining a semantic representation of MCOs with their data (characteristics, capabilities, deployment contexts, measurements, etc.) and contexts to resolve the semantic heterogeneity problem;
- integrating knowledge about the observed patients such as symptoms, treatments, events, risks;
- integrating high-value medical knowledge;
- ensuring the good functioning of the employed MCOs to guarantee sustainable and effective patient monitoring;
- facilitating the interaction between doctors and users through the development of a semantic-based patient monitoring system.

The chapter is organized as follows. First, Section 6.2 describes different relevant data that should be considered to model knowledge in a knowledge-based system for e-health and highlights main requirements. Then, to provide a solid foundation, Section 6.3 introduces the most interesting and high-value knowledge representations in the e-health domain. Next, Section 6.4 describes KGs, a symbolic abstraction used for encoding a knowledge base, a collection of statements having the form of interlinked subject–predicate–object triples. Section 6.5 proposes the KG framework using examples to highlight the different data collected in the integrated domain model. Finally, Section 6.6 closes the discussion and points out future directions.

6.2 **Knowledge-based system in health**

An e-healthcare system has to consider and manage information from different sources and contexts: data collected in clinical environments, data collected in daily living environments and high-value domain knowledge. Data collected from the actual use of digital health systems represent a significant potential for patients, epidemiology, and health services. They are highly heterogeneous and generally implemented with a focus on workflow and documentation, not on data quality or consistency, and often do not offer a patient-centered overall picture. Electronic patient records are rich and complex and contain hundreds of attributes: data on a patient's medical history, demographic data, diagnoses, medications, allergies, radiological images, laboratory test results, etc. They are characterized by ambiguous semantics and quality standards resulting from different collection processes between sites. These data are collected in a clinical environment, e.g., laboratories and hospitals. Data collected in daily living environments are generally unstructured, often obtained from unlabeled information derived from health conversations (e.g., discharge letters, pathology reports, etc.) and dynamic data from wearables and MCOs (e.g., data streams like glycemic index, food intake, and sports activities performed) that continuously transmit information about a person's fitness and health using mobile health technologies [12,13]. Thus, integration between hospital and proximity care is made explicit to include telemedicine. HL7's Fast Healthcare Interoperability Resources (FHIR) is emerging as a popular standard for health data exchange and the development of new applications. It supports technologies such as RDF and SPARQL that can provide effective solutions to enable interoperability and a common language between healthcare systems and can lead to the disambiguation of information through the adoption of various available terminologies and ontologies and forms of reasoning about healthcare data. The representation of FHIR resources is mostly through JSON objects, while the representation of ontologies is in RDF serializations. Thus, RDF graphs are suitable to combine information about patients, observations, and corresponding drugs with the extension of knowledge about drugs and relationships between objects. Furthermore, specific medical knowledge is increasingly standardized and freely available, also in a machine-readable format. This high-value knowledge (in categories such as epidemiology, symptomatology, diagnosis, treatment, medication, nutrition) should be carefully considered and linked to map information onto an integrated domain model by providing support for logical reasoning. Section 6.3 describes the most up-to-date and high-value knowledge representations in the e-health domain. Knowledge modeling is an effective way to organize and utilize the three described types of dispersed knowledge, and it is also an important step in constructing knowledge-based applications. The ultimate goal of knowledge modeling is to organize the scattered knowledge from different data sources to form a unified knowledge model which computers can process for knowledge management or other applications. In this regard, the main goal of this work is to propose a knowledge-based system as a context-aware approach for the Internet of MCOs that consists in:

- modeling a comprehensive data- and process-oriented conceptual approach for knowledge representation;
- proposing a dynamic data interchange and processing system able to:
  - receive and execute contextual queries and focuses,
  - produce and consume KGs,
  - support automatic feeding and contextual querying.

The framework should enable the transition from a document model to a data- and process-oriented one, as a bridge between knowledge that avoids new implementations at each integration and/or modification and allows the integration of new ontologies, data lakes, vocabularies, data sets, etc. It should be not a simple set of document repositories, but a system for the interchange and processing of data and services for the citizen and for the health workers involved in the care process. Considering that every medical information has a context, the result of the queries should be a view for the specific use case that changes according to the medical needs. These views are essential to filter the huge amount of information in the different datastores. For example, there is more information to consider if the medical query is about the long-term effects of taking a drug; conversely, if an application is needed to alert the patient that he/she is taking drugs to lower his/her heart rate when his heart rate drops below a certain threshold, only fewer observations are needed (too many hours have passed since taking the drug, the heart rate has increased above a certain threshold).

The chapter highlights the power of using KG technologies, to provide the formal tools and languages for knowledge representation and reasoning about health data, in order to implement the described approach promoting semantic interoperability and comparing real-world clinical practice with accepted best practices and guidelines. In particular, the chapter highlights the use of KGs considering data collected in daily living environments and specific medical knowledge, exposing them in a structured format and forming a context-aware resource graph.

6.3 Context modeling using knowledge graphs

Smart ontology-based systems for e-health will likely rely on diverse sources of ontology knowledge both implicitly and explicitly [14,15]. The adoption of semantic technologies and logical formalisms for explaining e-health systems is in line with the 'breaking the black box' trend, who foresees the return of rule-based approaches in combination with description logic (DL) techniques to obtain newly transparent models [16]. These explainable knowledge-enabled systems include a domain knowledge representation in the application domain, have mechanisms to incorporate patient context, are interpretable, and host explanation structures that generate user-understandable, context-aware, and provenance-enabled explanations of the functioning of the e-health system and the knowledge used [17]. The most interesting and high-value knowledge representations in the e-health domain are described below.

The Disease Ontology is a rich knowledge base of human diseases managed by the Institute for Genome Sciences at the University of Maryland School of Medicine initially developed in 2003. It aims to provide the biomedical community with consistent, reusable, and sustainable descriptions of human disease terms and phenotypic characteristics and disease concepts from the related medical vocabulary. It semantically integrates other vocabularies through cross-mapping, e.g., Disease Ontology terms in MeSH, ICD, NCI's thesaurus, SNOMED, and OMIM.

Orphanet is considered the largest existing database for rare diseases. Currently, it has data for more than 7000 rare diseases. Each disease is given a unique and stable identifier, the ORPHAcode. The Orphanet rare disease nomenclature comprises heterogeneous entities organized in descending order: disease groups, diseases, subtypes. A disease included in the database may represent a disorder, a malformative syndrome, a clinical syndrome, a morphological or biological abnormality, or a particular clinical situation (during the course of a disease). Disorders are organized into groups and subdivided into clinical, etiological, or histopathological subtypes. It also offers direct information on specialized centers, diagnostic tests, patient associations, research projects, and registries and biobanks.

In 2013 Orphanet and the European Bioinformatics Institute (EMBL-EBI) created ORDO. It is an OWL ontology derived directly from Orphanet and updated twice a year with periodic extractions. It plays a central role in many rare disease projects (similar to DBpedia in the LOD graph). It models rare diseases epidemiological data (age of onset, prevalence, mode of inheritance) and relationships between the disease and its genetic cause (if known). It provides references to the International Classification of Diseases 10 (ICD-10), SNOMEDCT, Medical Subject Headings (MeSH), Medical Dictionary for Regulatory Activities (MedDRA), Online Mendelian Inheritance in Man (OMIM) and Universal Protein Resource (UniProt), Ensembl, Reactome, and Genatlas. ORDO 2.9 (the latest version at the time of writing) consists of 14,559 classes and 205,428 annotation assertion axioms.

While ORDO is mainly used to name diseases (e.g., "idiopathic achalasia"), HPO is used to describe the clinical phenotype observed in a patient (e.g., "muscle weakness"). The combination of HPO together with Orphanet has always been considered a promising resource for the automated classification of rare diseases. The developers of ORDO and HPO have recently been working on the integration of both ontologies, annotating Orphanet phenome types with appropriate HPO terms. The result of this interoperability effort is called the HPO ORDO Ontological Module (HOOM) and its first version was released in 2018. The computable descriptions of human disease using HPO phenotypic profiles have become a key element in several algorithms being used to support genomic discovery and diagnostics.

Founded in 2007 by nine countries, SNOMED International is a nonprofit organization of 40 member countries, including 22 in Europe, which owns and maintains SNOMED CT, the world's most comprehensive clinical terminology. With SNOMED CT it is possible to record medical data more accurately, to exchange patient data both within the healthcare team and with patients, both locally and across borders. SNOMED CT can also be used in healthcare data and analytics platforms, clinical research, applied research, and other research activities to improve healthcare. It represents the world's most comprehensive clinical terminology for electronic health data exchange with over 350,000 concepts and clinical findings such as signs and symptoms and tens of thousands of surgical, therapeutic, and diagnostic procedures. SNOMED CT also describes concepts representing body structures, organisms, substances, pharmaceuticals, physical objects, physical forces, samples, etc. Twenty-two European countries already participate in SNOMED and others are added regularly (for example, the German Federal Institute for Drugs and Medical Devices [BfArM] has announced its membership of SNOMED International for national use of SNOMED CT in 2021). SNOMED International produces and maintains mappings to other coding systems, classifications, and terminologies. Some mappings are with GMDN, ICD-10, ICD-O, MedDRA, Orphanet (from 2021). The SNOMED CT to Orphanet map will provide the content of rare diseases in the international version, also providing a link table generated by Inserm as part of the SNOMED International–Inserm collaboration agreement. BioPortal is the world's most comprehensive repository of biomedical ontologies. In May 2021 it contained nearly 900 ontologies, over 13 million classes, more than 36,000 properties, and over 55 million mappings.

DrugBank Online is a comprehensive, free-access online database containing information on drugs and drug targets. It is both a bioinformatics and cheminformatics resource, combining detailed drug data (i.e., chemical, pharmacological, and pharmaceutical) with comprehensive drug target information (i.e., sequence, structure, and pathway). It is widely used by the pharmaceutical industry, medicinal chemists, pharmacists, physicians, students, and the general public. In 2021 DrugBank expanded coverage to seven additional regions (Austria, Colombia, Indonesia, Italy, Malaysia, Thailand, and Turkey).

The latest version of DrugBank online contains approximately 15,000 drug entries, including 2695 approved small-molecule drugs, 1470 approved biologics (proteins, peptides, vaccines, and allergens), 131 nutraceuticals, and over 6654 experimental (discovery phase) drugs. In addition, 5259 nonredundant (i.e., drug target/enzyme/transporter/carrier) protein sequences are linked to these drug entries. Each entry contains more than 200 data fields with half of the information dedicated to chemical data and the other half dedicated to the drug target or protein. It offers connectivity with SNOMED-CT, MedDRA, ICD-10, etc.

If it is true that one ontology is not sufficient to describe the various information that characterizes an e-health system, it is also true that there are several competitive ontologies to describe concepts in the same domain. Thus, there is an overlap problem that does not facilitate the goal of Linked Data. In fact, while noteworthy efforts have been devoted towards syntactic and semantic interoperability as described, the way in which individual-level data are collected and coded can be extremely different even between institutions within the same country. Therefore, data often require a long and costly preprocessing phase, which consists of mapping variables between coding standards and releases before being shared with others to contribute to multicenter studies.

6.4 Knowledge graphs

KG represents a collect of interlinked descriptions of entities, where descriptions have a formal structure and each entity represents part of the description of the entities related to it (forming a network). KGs combine characteristics of several data management paradigms and can be understood as database (for structured queries support), graph (for network data structure), and knowledge base (for formal semantics representation). However, a knowledge graph is not like any other database; it is supposed to provide new insights, which can be used to infer new things about the world. In recent years, KGs have become the base of many context-aware modeling systems, representing a structured representation of facts, consisting of entities, relationships, and semantic descriptions [18]. Entities can be real-world and abstract concepts, relationships represent the relation between entities, and semantic descriptions refer to entity and relationship properties with a well-defined meaning. Data about entities, relationships, and descriptions are represented through the Resource Description Framework (RDF) [19] language. Because KGs are based on standards for the identification, retrieval, and representation of information and knowledge, and scattered entities are interconnected by links, it is possible to crawl the entire data space, fuse data from different sources, and provide expressive query capabilities over aggregated data, similarly to how a local database is queried today [20]. For this purpose, the Simple Protocol and RDF Query Language is the standard language for querying, combining, and consuming structured data in a similar way SQL does this by accessing tables in relational databases. Since Linked Data is exclusively based on open web standards, data consumers and domain experts can use generic tools to access, analyze, and visualize data. KG technology means being able to connect different types of data in meaningful ways. So, a KG is necessarily built on semantics.

KGs have the ability to project information into a multidimensional conceptual space where similarity measures along different dimensions can be used to group together related concepts. This allows for an integrated solution that not only identifies the meanings of entities, clustering them into a unified knowledge layer, but also correlates concepts to allow for inference generation and insights. The KG architecture is based on a layered approach, and each layer provides a set of specific functionalities.

Semantic layers, on the top of the stack, include ontology languages, rule languages, query languages, logic, reasoning mechanisms, and trust. Ontologies, as a source of formally defined terms, play an important role within knowledge-intensive contexts such as the one described in this chapter. Ontologies can be reused, shared, and integrated across applications, and aim at capturing domain knowledge in a generic fashion and provide a commonly agreed understanding of the domain [21]. Ontologies constitute the backbone of KGs expressing concepts and relationships of a given domain and specify complex constraints on the types of resources and their properties. It can be said that a KG is a way an ontology could be represented. The real divide between ontology and knowledge graph has nothing to do with size or semantics, but rather the very nature of the data. KGs are fact-oriented, while ontologies are schema-oriented. Unlike KGs, in domain ontologies the focus is not on data (or facts), but on a highly expressive description (disjointness, cardinality, restrictions) of the concepts, the relations between them, and useful annotations (synonyms, definitions, comments, design choices, etc.). An ontology is metadata/schema, whereas the KG is the data itself. An ontology usually deals with concepts, not instances of concepts. So, not every RDF graph is a KG.

Rule languages allow writing inference rules in a standard way that can be used for automatic reasoning. First-order logic and DL [22] are frequently used to support the reasoning system which can make inferences and extract new insights based on the resource content relying on one or more ontologies. The reasoning is the process of extracting new knowledge from an ontology and its instance base and represents one of the most powerful features, especially for dynamic and heterogeneous environments. A semantic reasoner is a software system whose primary goal is to infer knowledge that is implicitly stated by reasoning upon the knowledge explicitly stated, according to the rules that have been defined [23,24]. Reasoners are also used to validate the ontology, that is, they check the consistency, satisfiability, and classification of its concepts to make sure that the ontology does not contain any inconsistencies among its term definitions.

6.5 Integrated domain model

In the previous sections we introduced how the use of KG can provide standardized frameworks for the concept representation in e-health systems exposing in a structured format the different representations converted to standard RDF formats and forming a context-aware resource graph. The framework represents data collected in clinical environments, data collected in daily living environments, and high-value domain knowledge. So, in our approach, the resource graph can be further expanded to include concepts, relationships, and data from other ontologies and linked open projects. For example, we can add a triple:

$$\text{schema:angina}_p ectorisowl : sameAssnomedct : 194828000$$

to indicate that the concept "angina pectoris" in the Schema.org vocabulary has the same meaning as that of "snomedct:194828000" which is the ischemic heart disease concept under the SNOMED-CT clinical ontology, also related to the class "Angina co-occurrent and due to coronary arteriosclerosis" and the subclass "Preinfarction syndrome and related to Family history: Angina in first degree female relative less than 65 years".

Similarly, angina concept can be related to DBpedia (the central interlinking hub of the Web of Data containing millions of RDF links to other Web data sources), Mesh (Medical Subject Headings, a comprehensive controlled vocabulary for the purpose of indexing life sciences journal articles and books), or Wordnet (a fairly large on-line lexical reference system offering broad coverage of general lexical English relations) by adding the following triples:

$$\text{"schema:angina}_p ectorisowl : sameAsdbpedia : 552599"$$

$$\text{"schema:angina}_p ectorisowl : sameAsmeshld : D000787"$$

$$\text{"schema:angina}_p ectorisowl : sameAswordnet : 14197107"$$

$$\text{"schema:angina}_p ectorisowl : sameAswordnet : 14131521"$$

which represent "disease of the throat or fauces marked by spasmodic attacks of intense suffocative pain" and "the heart condition marked by paroxysms of chest pain due to reduced oxygen to the heart" concepts.

So doing, we expanded the resource graph by linking it with other health ontologies, LOD, linked open health data, and hierarchy concept graphs containing hyponyms and hypernyms. Accordingly, they can be accessed and queried in a uniform way using standard languages. Data visualization in a personalized manner is now possible through a Web dashboard. The homogenized data are now available for statistical analysis, for example to monitor the percentage of individuals aged 15 or above who had cancer, or are overweight and obese, and also to survey health conditions and recourse to health services. Once the data is represented as RDF and exposed through a SPARQL endpoint we can combine them with other data belonging to a different LOD portal, because the different storage modalities are irrelevant from a SPARQL query perspective. For instance, we could carry out a selective survey of risk factors affecting the health status of all families living in a specific area by cross-checking data extracted from devices with those from the registry of families in the territory. The interpretation and the processing of the resulted knowledge can be furthermore additionally enhanced owing to the SWRL to propose rules for different goals, for example to verify the proper functioning of the connected objects and the validity of the detected data and to provide the adequate service for patients [25]. Knowledge reasoning over KGs, which is intimately bound up with RDFS and OWL, is closely related to ontology. The reasoning method based on ontology mainly uses the more abstract frequent patterns, constraints, or paths to infer. We used KG intelligent reasoning capabilities to perform domain knowledge reasoning through modeled domain knowledge and rules. KG reasoning methods infer unknown relations from existing triples, providing efficient correlation discovery ability for resources in KGs and completing KGs. The framework is able to also include data collected by the sensors in daily living environments. When the system starts receiving events from the sensors, the KG is automatically updated and a set of inference rules is applied to infer the context of a specific patient, e.g., for the purpose of selecting the appropriate care service and selecting the appropriate interaction device. For example, the following rule allows us to detect the position of a subject by activating a proximity sensor; if the sensor's status is "on state", the rule establishes that the subject assigned to the sensor is in the proximity of the object on which the sensor is positioned. This is a general rule that may be useful for the recognition of activities other than the one under examination.

Rule Sensor_detection: Sensor $(?y)^\wedge$ hasType $(?y,$ sensor_type $)^\wedge$ assignedTo $(?y, ?x)^\wedge$ hasCurrentState $(?y,$ on_state $)^\wedge$ deployedIn $(?y, ?r)$ -> detectedIn $(?x, ?r)$

The position of a patient detected by the previous rule allows the selection of the appropriate device and could be enriched with other criteria related to the patient profile and preferences. For example, if the patient is detected in the bathroom and the sensor in the bathroom detects a presence exceeding a certain threshold time, then an alert can be triggered and notified to both the patient and the closest caregiver. Thanks to the rules described above, enabling the reasoner makes it possible to infer various information on the position and status of the specific patient. In particular, its position has been automatically inferred both at the level of proximity to the object and at the level of room and home thanks to the Sensor_detection rule and partOf ontology properties; we also have information about the activities he is performing, and if one of them has generated an alert, also about which correct activity he should perform. Finally, the assistant devices that will be activated to manage the service are inferred. Moreover, consider that the sensors are characterized by numerous properties such as date range, frequency, date and time, etc. We can set SWRL rules to automatically verify the values of vital signs detected by the sensors, with respect to the interval that they must assume to validate the operation of the sensor. The following rule performs the validity of sensor vital signs:

> **Rule Validity_vital_signs:** IoT(?o)$^\wedge$ Sensing-device(?s)$^\wedge$ contains (?o, ?s)$^\wedge$ Measurement(?m)$^\wedge$ detects(?o,?m)$^\wedge$ hasvalue(?m,?v)$^\wedge$ hasmaxValue(?o,?maxv)$^\wedge$ hasminValue(?o,?minv)$^\wedge$ swrlb:greaterThanOrEqual (?v,?minv)$^\wedge$ swrlb:lessThanOrEqual (?v,?maxv) -> validity(?m, true)

We established a broad workflow that takes a stream of low-level-encoded IoT information instances, transforms them to domain-level OWL concepts, and reasons with them to generate knowledge. The use of rules is important to capture and modulate some characteristic aspects of the domain allowing to describe relations that cannot be described using DL. Many efforts have been made to address the challenges of knowledge representation and its related applications. However, there remain several open problems and promising future directions; one of the most crucial is that related to scalability, especially on large KGs. There is a trade-off between computational efficiency and model expressiveness, with a limited number of works applied to more than one million entities [26].

6.6 **Discussion and conclusions**

In this chapter, we have presented the application of KG technologies to manage rapidly changing and highly interconnected relevant information in healthcare. In the context of connected medical objects and freely available knowledge bases, we used knowledge reasoning on KGs to provide correlation discovery capability for resources in heterogeneous KGs. We used the implicit information obtained through the reasoning techniques to improve the context representation and proposed the creation of knowledge-based queries and rules to provide automatic health decision support. Inference mechanisms in ontology-based knowledge systems allow for the representation of additional attributes that cannot be naturally inferred using traditional ontology models. Reasoning about enriched information can lead to interesting considerations and conclusions. The research field of ontology-based reasoning KG is still in an early stage of development and faces many challenges. In the future, it will be increasingly important to explore the possibilities related to explicit knowledge reasoning to address the inherent complexity that characterizes e-healthcare systems. For example, also the current COVID pandemic made clear that the e-health community demands for unified frameworks for sharing and exchanging digital (epidemiological) data and facilitating the flow of information between health workers, stakeholders, policymakers, and the public. In addition, a promising field of research that is evolving very rapidly involves incorporating additional information like textual descriptions to semantically enrich KG representations. While there is a great effort on embedding KGs, that is, binary relational structures, not much is known about embedding relations of higher arities, like events, without breaking them down into binary incidence structures (set of pairwise edges) to express complex concept units. To this end, recent developments in deep learning and natural language processing have enabled intelligent ways to uncover structured, concise, and unambiguous knowledge.

References

[1] Y. Chen, IoT, cloud, big data and AI in interdisciplinary domains, Simulation Modelling Practice and Theory (2020).

[2] M. Noura, M. Atiquzzaman, M. Gaedke, Interoperability in internet of things: taxonomies and open challenges, Mobile Networks and Applications 24 (3) (2019) 796–809.

[3] B. Farahani, F. Firouzi, K. Chakrabarty, Healthcare IoT, in: Intelligent Internet of Things, Springer, Cham, 2020, pp. 515–545.

[4] S. Krishnamoorthy, A. Dua, S. Gupta, Role of emerging technologies in future IoT-driven Healthcare 4.0 technologies: a survey, current challenges and future directions, Journal of Ambient Intelligence and Humanized Computing 1 (47) (2021).

[5] S. Jabbar, F. Ullah, S. Khalid, M. Khan, K. Han, Semantic interoperability in heterogeneous IoT infrastructure for healthcare, Wireless Communications and Mobile Computing (2017).

[6] A. Dagliati, A. Malovini, V. Tibollo, R. Bellazzi, Health informatics and EHR to support clinical research in the COVID-19 pandemic: an overview, Briefings in Bioinformatics 22 (2) (2021) 812–822.

[7] G.S. Patange, Z. Sonara, H. Bhatt, Semantic interoperability for development of future health care: a systematic review of different technologies, in: Proceedings of International Conference on Sustainable Expert Systems, Springer, Singapore, 2021, pp. 571–580.

[8] R. Burse, M. Bertolotto, D. O'Sullivan, G. McArdle, Semantic interoperability: the future of healthcare, in: Web Semantics, Academic Press, 2021, pp. 31–53.

[9] A. Chatterjee, A. Prinz, M. Gerdes, S. Martinez, Automatic ontology-based approach to support logical representation of observable and measurable data for healthy lifestyle management: proof-of-concept study, Journal of Medical Internet Research 23 (4) (2021) e24656.

[10] P.P. Jayaraman, A.R.M. Forkan, A. Morshed, P.D. Haghighi, Y.B. Kang, Healthcare 4.0: a review of frontiers in digital health, Wiley Interdisciplinary Reviews: Data Mining and Knowledge Discovery 10 (2) (2020) e1350.

[11] X. Chen, S. Jia, Y. Xiang, A review: knowledge reasoning over knowledge graph, Expert Systems with Applications 141 (2020) 112948.

[12] F. Piccinini, G. Martinelli, A. Carbonaro, Accuracy of mobile applications versus wearable devices in long-term step measurements, Sensors 20 (21) (2020) 6293.

[13] F. Piccinini, G. Martinelli, A. Carbonaro, Reliability of body temperature measurements obtained with contactless infrared point thermometers commonly used during the COVID-19 pandemic, Sensors 21 (11) (2021) 3794.

[14] J.K. Tarus, Z. Niu, G. Mustafa, Knowledge-based recommendation: a review of ontology-based recommender systems for e-learning, Artificial Intelligence Review 50 (2018) 21–48.

[15] S. Riccucci, A. Carbonaro, G. Casadei, An architecture for knowledge management in intelligent tutoring system, in: IADIS International Conference on Cognition and Exploratory Learning in Digital Age, CELDA, 2005, pp. 473–476.

[16] R. Reda, F. Piccinini, A. Carbonaro, Semantic modelling of smart healthcare data, in: Proceedings of SAI Intelligent Systems Conference, Springer, Cham, 2018, pp. 399–411.

[17] A. Carbonaro, F. Piccinini, R. Reda, Semantic description of healthcare devices to enable data integration, in: Information Technology – New Generations, Springer, Cham, 2018, pp. 627–630.

[18] S. Ji, S. Pan, E. Cambria, P. Marttinen, P.S. Yu, A survey on knowledge graphs: representation, acquisition and applications, arXiv preprint, arXiv:2002.00388, 2020.

[19] O. Lassila, F. van Harmelen, I. Horrocks, J. Hendler, D.L. McGuinness, The semantic web and its languages, IEEE Intelligent Systems & Their Applications 15 (6) (2000) 67–73.

[20] T. Heath, C. Bizer, Linked Data: Evolving the Web into a Global Data Space, Synthesis Lectures on the Semantic Web: Theory and Technology, 2011, pp. 1–136, 1(1).

[21] A. Andronico, A. Carbonaro, L. Colazzo, A. Molinari, Personalisation services for learning management systems in mobile settings, International Journal of Continuing Engineering Education and Life Long Learning 14 (4–5) (2004) 353–369.

[22] M. Krotzsch, F. Simancik, I. Horrocks, A description logic primer, arXiv preprint, arXiv:1201.4089, 2012.

[23] R. Reda, A. Carbonaro, Design and development of a linked open data-based web portal for sharing IoT health and fitness datasets, in: Proceedings of the 4th EAI International Conference on Smart Objects and Technologies for Social Good, 2018, pp. 43–48.

[24] R. Reda, F. Piccinini, G. Martinelli, A. Carbonaro, Heterogeneous selftracked health and fitness data integration and sharing according to a linked open data approach, Computing 1 (23) (2021).

[25] S. Riccucci, A. Carbonaro, G. Casadei, Knowledge acquisition in intelligent tutoring system: a data mining approach, in: Mexican International Conference on Artificial Intelligence, Springer, 2007, pp. 1195–1205.

[26] S. Ji, S. Pan, E. Cambria, P. Marttinen, S.Y. Philip, A survey on knowledge graphs: representation, acquisition, and applications, IEEE Transactions on Neural Networks and Learning Systems (2021).

Toward a knowledge graph for medical diagnosis: issues and usage scenarios

Antonio De Nicola[a], Rita Zgheib[b], and Francesco Taglino[c]

[a]*ENEA, Centro Ricerche Casaccia, Rome, Italy*
[b]*Canadian University Dubai, Dubai, United Arab Emirates*
[c]*IASI-CNR, Rome, Italy*

Contents

7.1 Introduction

Availability of a large amount of data in the medical sector and an increasing computational infrastructure to exploit them allow the definition of new advanced services for medical diagnosis to support the work of medical doctors and to increase the quality of life of patients. Telemedicine, ambient assisted living, and a better management of electronic health records are some of the new generation of services that need to be further developed in the next years. Most of the issues that are faced by medical doctors and ICT practitioners defining these services are related to their level of trust and reliability. How to guarantee them is still in an open issue to be addressed.

To this regard, practitioners have defined a plethora of standards and specifications to be used for better management of information and an increased level of trust of the adopting services. For instance, SNOMED CT [1] and ICD-10 [2] are standardized specifications that are widely used in clinical contexts. However, further issues arise due to the lack of interoperability among these different standards, and specifications adopted in the medical domain. Ontologies are recognized as an effective instrument to deal with the interoperability problem and the human disease ontology (DOID) [3] and the Infectious

Disease Ontology (IDO) [4] are only some of the many proposals in the area. However, existing ontologies cover different aspects of the medical domain and have different objectives spanning from clinical diagnosis to a better organization of electronic medical records. To deal with such issues, we propose a knowledge graph for medical diagnosis leveraging and aligning existing largely used standards and ontologies. In detail, we present some of the typical issues to be faced and aligned them by focusing on ICD-10, SNOMED CT, DOID, and SYMP ontology [5].

Then we discuss some scenarios of usage for the envisioned knowledge graph. We address how a knowledge graph can benefit interoperability of electronic health records, can support clinicians in their work and patients in increasing the quality of their lives by means of a new generation of telemedicine services, and can be an added value for medical insurance services.

The rest of the paper is organized as it follows. Section 7.2 presents the related work in the area. The strategy to develop the knowledge graph for medical diagnosis is presented in Section 7.3. The issues for engineering the knowledge graph are presented in Section 7.4. Then, the envisioned usage scenarios are described in Section 7.5. Finally, Section 7.6 draws some conclusions and future research directions.

7.2 Related work

Patient medical records are more comprehensive and available than ever before. Refinements and system upgrades are continually enhancing clinic workflows and doctor-patient interactions. Encoding medical information is a complex task because vocabularies always involve a trade-off between completeness and usability, just as with an English dictionary. Many research projects succeeded to propose a medical data vocabulary and taxonomy that are used in the hospitals [6]:

- CPT (Current Procedural Terminology)[1] is a standard vocabulary used for procedures, and radiology tests, laboratory tests that physicians or their staff perform in office settings. It does not include diagnoses or conditions, including the intensity of outpatient visits. The American Medical Association (AMA) founded the first version of CPT and released a new version every 4 years.
- Health Level 7 (HL7)[2] is one of the several American National Standards Institute (ANSI)—accredited Standards Developing Organizations (SDOs) operating in the healthcare area. It is dedicated to provide a comprehensive framework and related standards for the exchange, integration, sharing and retrieval of electronic health information that support clinical practice and the management, delivery and evaluation of health services.
- UMLS[3] [7] (The Unified Medical Language System) is a repository of biomedical vocabularies developed by the US National Library of Medicine. The UMLS integrates over 2 million names for some 900,000 concepts from more than 60 families of biomedical terminologies and 12 million relations among these concepts. Vocabularies combined in the UMLS Metathesaurus include

---

[1] https://www.ama-assn.org/amaone/cpt-current-procedural-terminology.
[2] https://www.hl7.org/.
[3] http://umlsks.nlm.nih.gov.

the NCBI taxonomy,[4] Gene Ontology,[5] the Medical Subject Headings (MeSH),[6] OMIM,[7] and the Digital Anatomist Symbolic Knowledge Base.[8]

- SNOMED CT [1] is a standardized, multilingual vocabulary of clinical terminology built to encode the concepts and corresponding meanings that are used in health information and to support the efficient clinical recording of data to enhance patient care. It provides codes, terms, synonyms, and definitions used in clinical documentation and reporting. All SNOMED CT concepts are organized into taxonomic (IS-A) hierarchies; for example, *"viral pneumonia"* IS-A *"infectious pneumonia,"* *"infectious pneumonia"* IS-A *"pneumonia,"* and *"pneumonia"* IS-A *"lung disease."* It provides the core general terminology for electronic health records (EHR) most commonly in clinics office, not hospitals. But it is hard to build tools to make SNOMED CT easy to navigate. Also, not all systems use it. SNOMED CT can be mapped to other coding systems, such as ICD-9 and ICD-10, which help facilitate semantic interoperability.

- ICD-10 [2] is the 10th revision of the International Classification of Diseases, which is a medical classification created by the US National Center for Health Statistics (NCHS) and the Centers for Medicare and Medicaid Services (CMS). All updates and official release are handled by the CDC (Centers for Disease Control and Prevention).[9] It contains codes for diseases, signs, symptoms, abnormal findings, complaints, social circumstances, and external causes of injury or diseases. It is a standard diagnostic tool for health management and is widely used by most hospitals today. ICD-9 is the previous version of ICD-10.

Data representation and information homogeneity between different medical institutions is a primary challenge for automatic diagnosis and disease detection systems. Ontologies have been widely used in real-world applications domains from healthcare and life science to banking and government to improve classification accuracy and information modeling. Most of the ontologies are expressed in the well-known Web Ontology Language (OWL) [8]. Semantic reasoners such as Pellet [9] are all widely adopted to produce ontology-based automatic classification systems. Medical ontologies are valuable and effective methods of representing medical knowledge. In this direction, they are much stronger than biomedical vocabularies since they enable reasoning and inference. The literature review refers to several OWL ontologies for representing diseases.

- The human Disease Ontology (DOID) [3] is an open-source ontology developed to promote the integration of biomedical data that is correlated with human disease. The latest release of DOID [10] includes 17,563 disease terms. For some terms, it also includes references to UMLS MESH, NCI, SNOMEDCT, ICD9, and ICD10.

- Symp [5] is an ontology that describes human symptoms. Each disease has several symptoms associated with it. The authors built an algorithm that aligns SYMP and DOID ontologies to relate diseases and their symptoms. This alignment helps the automation of disease detection as demon-

---

4 https://www.ncbi.nlm.nih.gov/taxonomy.
5 http://geneontology.org/.
6 https://www.nlm.nih.gov/mesh/meshhome.html.
7 https://www.ncbi.nlm.nih.gov/omim/.
8 https://pubmed.ncbi.nlm.nih.gov/9452983/.
9 https://www.cdc.gov/.

strated in [11], but not all diseases are connected to symptom concepts. SYMP has been integrated in DOID ontology, and the latest version[10] links a few diseases to their symptoms.

- The Infectious Disease Ontology (IDO) [4] was developed to assist the analysis of the COVID-19 disease. It includes three extensions: IDO Virus (VIDO), the Coronavirus Infectious Disease Ontology (CIDO),[11] and an extension of CIDO focusing on COVID-19 (IDO-COVID-19). IDO covers the entities relevant to infectious diseases generally and not specific contagious diseases associated with particular pathogens. Its coverage ranges across biological scales (gene, cell, organ, organism, population), disciplinary perspectives (biological, clinical, epidemiological), and successive stages along the chain of infection (host, reservoir, vector, pathogen) [12]. At the heart of the IDO Core, there is the term disease, which is imported from the Ontology for General Medical Science (OGMS) [13].
- The Smart Health Care Ontology for healthcare (SHCO)[12] is an ontology created to describe a patient's episode in an intelligent environment where IoT medical devices are implemented. The ontology describes the patient's information, vitals, diagnosis, the connected health devices, disease, and the involved actors such as the nurse and the doctors.

In the literature, specific ontologies have been developed to describe the concepts related to a specific disease. For instance, many ontologies have been proposed to model information related to COVID-19 supporting the organization and representation of COVID-19 case data on a daily basis. Examples of these ontologies are CODO [14], an ontology for collection and analysis of covid data, Kg-COVID-19,[13] and Linked COVID-19 Data: Ontology.[14] The cardiovascular ontology [15] completes DOID ontology with cardiovascular taxonomy and rules to determine the end point of a disease course, and to locate the material basis of a cardiovascular disease. Other specific disease ontologies are the neurological disease ontology [16], fibriotic disease lung [17], and cancer visualization ontology [18].

The presented terminologies and ontologies are used today by the healthcare systems. However, the usage of the terminologies can vary from hospital to hospital and from one country to another. Therefore, the usage of different codes makes data sharing and processing hard tasks. To address this issue is a challenging activity that represents the motivation for many projects that are working on integrating and mapping codes from different terminologies. For instance, the DOID ontology includes disease and medical vocabularies from MeSH, ICD, NCI's thesaurus, SNOMED CT, and OMIM. In [19], the authors have integrated some ICD-10 concepts, ICD-O3 cancer diagnosis terminologies and SNOMED CT. The objective of this project is to improve the reuse of oncology data that is limited because of the heterogeneity of terminologies. The project presented in [20] focuses on harmonizing the WHO (World Health Organization) classifications and SNOMED CT within the ICD-10 version. A mapping effort between the terms shows that the WHO terms are semantically and hierarchically different from the terms of SNOMED CT. The study suggests some collaboration between the entities to enhance EHR. All of the integration frameworks efforts until today are still at their early stage; they are either focusing on specific diseases (heart diseases, oncology, etc.) or they are focusing on integrating some terminologies.

---

[10] https://bioportal.bioontology.org/ontologies/DOID/?p=summary.

[11] https://bioportal.bioontology.org/ontologies/CIDO.

[12] https://tiwari2019.github.io/myrepo/OnToology/Healthcare.owl/documentation/index-en.html.

[13] https://github.com/Knowledge-Graph-Hub/kg-covid-19.

[14] https://zenodo.org/record/3765375#.XraWJmgzblU.

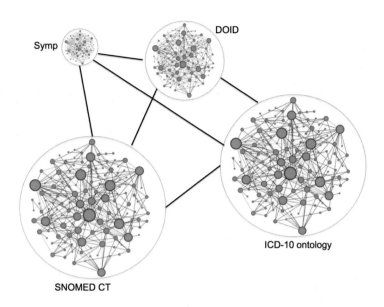

FIGURE 7.1

Backbone of the knowledge graph for medical diagnosis.

7.3 A knowledge graph for medical diagnosis

The plethora of existing medical standards, vocabularies, taxonomies, and ontologies presented in Section 7.2 are a significant step toward an effective digitalization aimed at supporting medical doctors. However, hospitals use standards mainly with the goal of indexing and retrieving electronic medical records. Conversely, an automatic medical diagnosis would require ontologies, such as DOID, to perform reasoning on available knowledge. Unfortunately, they are not widely used in hospitals. Furthermore, the exchange of electronic medical records between different hospitals using different standards could face interoperability issues, for instance, due to different names assigned to medical terms, different conceptual structures, or even different units of measure.

According to [21], a knowledge graph is *"a graph of data intended to accumulate and convey knowledge of the real world, whose nodes represent entities of interest and whose edges represent relations between these entities."* A knowledge graph for supporting medical diagnosis should define the concepts and the relationships pertaining to this domain of interest. Our idea is that this knowledge graph should leverage already existing knowledge developed by the medical community. To this aim, we propose to align and merge existing ontologies and standards into a unique knowledge structure. As a first step toward the development of the knowledge graph for medical diagnosis here, we present the main issues in aligning some of the most representative medical ontologies. In particular, we focus on DOID and SYMP ontologies, which address diagnosis of human diseases, and SNOMED CT and ICD-10, which provide the most comprehensive classifications of medical terms.

The backbone of the envisaged knowledge graph is sketchily depicted in Fig. 7.1. In our proposal, the DOID ontology plays a key role since it puts together symptoms and diagnosis. It should be noted that it already includes elements from the above-mentioned specifications but the resulting integration is still preliminary.

7.4 Issues for ontology alignment

Ontology matching is the operation of finding correspondences between semantically related entities of ontologies, which determines an alignment between ontologies [22]. Here, we present some of the issues identified to match the DOID and SYMP ontologies, and DOID and ICD-10. Indeed, we also assume social validation [23,24] by domain experts [25] of the resulting alignments as an open issue to be tackled to guarantee trust and reliability of the envisaged medical services leveraging the knowledge graph; however, here, we focus on the technical aspects of ontology matching. The social validation problem will be treated in a future work.

7.4.1 Alignment between DOID and SYMP ontologies

A precondition for semantics-based medical diagnosis is the availability of an ontology where all the concepts representing the diseases are connected to the concepts representing the corresponding symptoms. The DOID ontology includes 8084 diseases[15] but only 409 are connected to symptoms, which are imported from the SYMP ontology, by means of the object property named *"has_symptom."* For instance, the disease named *"intestinal tuberculosis"* is connected to *"nausea," "bleeding," "vomiting,"* and *"gastrointestinal bleeding."* This means that more than 94% of diseases included in the ontology are not connected to any symptom. Hence, the amount of work for domain experts to connect diseases with symptoms could be exhausting. However, this can be reduced by analyzing the names and definitions of the ontology concepts representing diseases. In fact, 5067 out of the 7675 diseases (66.02%) without symptoms have a definition in the ontology. In detail, the following cases occur:

- A term describing a symptom is contained in the disease label of a disease included in DOID. For instance, *"coma"* is a symptom of the disease named *"hepatic coma."*
- A term describing a symptom is contained in a disease definition. Hence, it is possible to search for textual patterns in disease definitions. Fig. 7.2 shows how a label of a symptom from the SYMP ontology is searched in the definitions of the diseases contained in the DOID ontology. In detail, the symptom named *"cough"* is contained in the definition of the disease named *"respiratory syncytial virus infectious disease."* Accordingly, the pair disease-symptom can be connected together through the object property *"has_symptom."*

7.4.2 Issues for ICD-10 and DOID-SYMP alignment

Mapping DOID and SYMP ontologies is essential to building the knowledge graph for medical diagnosis. However, mapping ICD-10 to both ontologies is another challenge that must be tackled. In fact, ICD-10 is used by most hospitals to classify and code all diagnoses, symptoms, and procedures

---

[15] This number considers only diseases that are not further specified in the ontology. The used DOID ontology was downloaded on August 21, 2020.

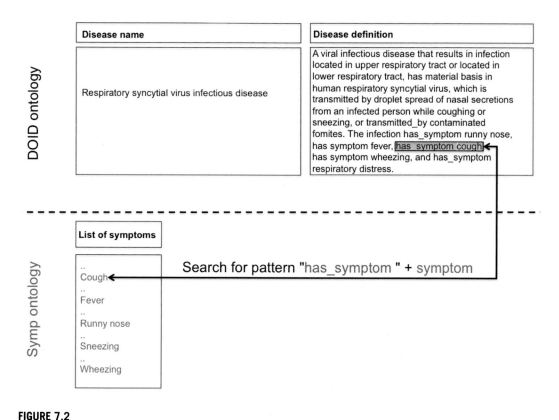

FIGURE 7.2

Search of SYMP symptoms in the definitions of DOID diseases.

in IT systems. In contrast, DOID aligned with SYMP provides the required understanding of the diseases and their relations to the symptoms. The SYMP ontology includes 874 symptoms, but no one has been mapped to an ICD-10 code. This means that all the symptoms in the SYMP ontology must be mapped to the ICD-10 codes. Only 182 symptoms are connected and cross-referenced to ICD-9, with the object property "hasDBXRef." The same terms are also connected to other databases such as SyOID and UMLS. For instance, the symptom "abdominal pain" has the following cross-references "ICD9CM2005:789," "SyOID:2880," "UMLS:C0000737." ICD-9 is an older version of ICD-10 that is not used anymore by hospital information systems. But, the CDC has provided official mapping between ICD-9 and ICD-10; hence, ICD-9 could be used to map some symptoms to ICD-10.

From the other side, the DOID ontology includes 8084 diseases that have been extracted from different databases such as NCI, MESH, and SNOMED CT. 3522 diseases have been cross-referenced to medical databases, including ICD-10 codes with the object property "hasDBXRef." Hence, around 60% of the terms must be mapped to ICD-10 codes. From the other side, the remaining non-ICD-10 referenced diseases include 442 mapped to ICD-9, and 3912 diseases mapped to SNOMED CT terms and others. Besides ICD-9, SNOMED CT is a very well-known and used ontology. An active commu-

ICD9	ICD-10
• 7841 throat pain • 78833 mixed incontinence (female) (male)	• R070 Pain in throat • N3946 Mixed incontinence

FIGURE 7.3

An example of mapping symptoms from ICD-9 to ICD-10.

ICD9	ICD-10	
7813 lack of coordination	R270, R278, R279,	Ataxia, unspecified, Other lack of coordination, Unspecified lack of coordination

FIGURE 7.4

An example of mapping symptoms from ICD-9 to multiple codes in ICD-10.

nity runs it, and a mapping between SNOMED CT and ICD-10 is officially released. In conclusion, mapping DOID and SYMP to ICD-10 requires mapping three types of terms:

- diseases/symptoms referenced with ICD-9 codes;
- diseases/symptoms referenced with SNOMED CT codes;
- diseases/symptoms that are not referenced by any database.

Based on the previously mentioned terms, three ways are possible in order to map DOID and SYMP terms to ICD-10:

1. Using the CDC conversion files: ICD-9 consists of a tabular list containing a numerical list of the disease code numbers in tabular form, an ordered index to the disease records, and a classification system for surgical, diagnostic, and therapeutic procedures. CDC provides materials and guidelines for converting ICD-9 to ICD-10. For instance, the file 018_I9gem consists of a tabular list containing the ICD-9 in the first column and the corresponding ICD-10 code in the second column. An example of the mapping codes is shown in Fig. 7.3. This mapping presents a challenge related to the fact that ICD-10 includes more terms than ICD-9. This is due to the detailed description of the diseases in the new ICD-10. This fact has led to finding in some cases where one ICD-9 code can be mapped into multiple ICD-10 codes as shown in Fig. 7.4. This case can be solved by adding all the ICD-10 codes as children nodes of the originally general term. In any case, based on the practitioners' diagnosis, either they choose the specific code or the general one. No data is missed in that case.

2. Using the SNOMED CT ontology: In DOID, 3,912 diseases are mapped to SNOMED CT terms. SNOMED CT is considered one of the most comprehensive terminologies in healthcare. In order to support automation of clinical data and improve reimbursement, SNOMED CT proposed a SNOMED CT to ICD-10-CM map (referred to as "the Map") as well as a collection of materials

and guidelines for the same purpose.[16] Moreover, the I-MAGIC (Interactive Map-Assisted Generation of ICD Codes) algorithm can be used as well to map the SNOMED CT to ICD-10-CM in real-time. Using SNOMED CT to map DOID terms to ICD-10 is a good solution, even though it is accompanied by hard programming and technical work.

3. Performing string matching: A further and complementary strategy for mapping SYMP and ICD-10 entries is based on the exploitation of textual information present in both the resources. In the SYMP ontology, all of the concepts have a label, which is the name for referring to the symptom. Furthermore, 249 symptoms are associated with a natural language description, i.e., a sentence explaining the meaning of the symptom, which is defined via the *http://www.w3.org/2002/07/owl# IAO_0000115* property, whereas 156 concepts are associated with a synonym via the *#hasExactSynonym*[17] property. Finally, 76 concepts have both a description and a synonym.

Concerning ICD-10, beside the label, most of the concepts have a *short description* and *approximate synonyms,* which are a natural language definition, and alternative terms for referring to a concept, respectively.

The key idea is to apply natural language techniques to compare textual information from SYMP and ICD-10, in order to identify mappings between entries from the two resources.

7.5 Usage scenarios for the medical diagnosis knowledge graph

In this section, we illustrate three real-world cases that can benefit from semantic solutions and the adoption of an integrated knowledge graph in the medical sector. The first scenario is about the Electronic Health Records (EHR) where semantic interoperability is still a key open issue. The second scenario is about telehealth and telemedicine where semantics-based reasoning systems can help to support a diagnosis formulation remotely. Finally, the third scenario is about medical insurance, where semantic reasoning and the semantic integration and interoperability between business process management systems from different institutions, e.g., hospitals and insurance companies can improve and speed up processing of medical policies.

7.5.1 Electronic health records interoperability

The Electronic Health Record (EHR) is the digital version of the medical history of a patient. Currently, it is having a great impact on the medical sector. In particular, the global electronic health records market size is expected to reach USD 35.1 billion by 2028, registering a compound annual growth rate (CAGR) of 3.7% over the forecast period, according to a report by Grand View Research, Inc.[18]

One of the key features of an EHR is that health information can be created and managed by authorized providers in a digital format capable of being shared with other providers across more than one healthcare organization. However, this cross-organizational sharing of information is possible only if EHR interoperability is fully implemented.

---

[16] https://www.nlm.nih.gov/research/umls/mapping_projects/snomedct_to_icd10cm.html.

[17] IRI: http://www.geneontology.org/formats/oboInOwl#hasExactSynonym.

[18] https://www.grandviewresearch.com/press-release/global-electronic-health-records-market.

Interoperability is complex, and goes beyond the ability to move information from a system to another. According to the Healthcare Information and Management Systems Society,[19] interoperability can be classified into three levels:

- Foundational, which enables one EHR system to receive data from another system but it does not require that the receiver is able to interpret it.
- Structural, which enables data to be exchanged between information systems and interpreted at the data field level.
- Semantic, which requires that two systems can exchange, interpret, and use information.

Hence, two EHR systems are truly interoperable if they are able to exchange and then use the data. For this to occur, the transferred content must contain standardized coded data so that the receiving system can interpret it. In particular, interoperability at semantic level requires that involved organizations and institutions agree on standards for representing different health-related information and that such standards are somehow integrated. However, lack of standardized data is still an issue, limiting the ability to share data electronically for patient care [26].

In this scenario, a knowledge graph for medical diagnosis can play a significant role for addressing the desired semantic interoperability in the clinical sector. Knowledge graphs deal with the representation of the structure and the semantic relations and, due to their formal nature, they can facilitate the definition of nonambiguous computational vocabularies, and enable solutions for data interoperability. Furthermore, the knowledge graph can be applied for classification tasks as for instance by Fries et al. [27], where ontologies are used in combination with rules generated by medical experts in order to label patient records.

For example, a patient has their health records at a local hospital that uses ICD-10 codes for diseases and diagnoses and CPT for medical procedures and services (EHR$_1$). The patient is on vacation and falls ill. Hence, she goes to visit a clinic that uses SNOMED-CT as clinical terminology (EHR$_2$). First, the patient may not be able to provide all details of her medical history, which can make a difference for the doctor charged with his care. Second, the diagnosis saved under EHR$_2$ will not be shared with her local hospital. A knowledge graph that aligns different clinical terminologies and standards such as SNOMED-CT and ICD-10 provides interoperable electronic health records. It also allows the electronic sharing of patient information between other EHR systems and healthcare providers across organizational boundaries.

7.5.2 Automatic reasoning in telemedicine

Telemedicine is the delivery of health care information across distances. As such, it encompasses the whole range of medical activities including diagnosis, treatment and prevention of diseases, as well as continuing education of healthcare providers and consumers, and research and evaluation [28]. Telemedicine will shift care from hospitals and clinics to homes and mobile devices. This transition will revolutionize the provision of health services in the same manner as home banking and online shopping is doing in other sectors.

---

[19] https://www.himss.org/.

The number of people using telemedicine has increased steadily over the years. For instance, according to the American Telemedicine Association[20] more than half of all US hospitals has a telemedicine program, whereas the UK's National Health Service Long Term Plan[21] claimed that "digitally enabled care will go mainstream." Furthermore, current reports estimate that telemedicine's influence will continue to swell over the next few years, too. In fact, the global telehealth market is expected to reach $266.8 billion by 2026, showing a CAGR of 23.4% between 2018 and 2026. Telemedicine also has emerged as a critical technology to bring medical care to patients while attempting to reduce the transmission of COVID-19 among patients, families, and clinicians [29].

Common telemedicine services include: virtual visits, which shorten the wait for an appointment and ensure you get healthcare wherever you are located. In a typical telemedicine interaction, patients send health-related information to medical doctors, as for instance, blood pressure and blood sugar, but also images of a wound, eye, or skin condition, as well as medical records filed by another doctor, such as X-rays. Consequently, the doctors can send information to manage people's care at home, as for instance, notifications or reminders to take medication, new suggestions for improving diet and mobility, etc.

Health ontologies, such as the Telehealth Smart Home ontological model [30], have already proven to be effective to support telemedicine [31]. A knowledge graph for medical diagnosis, where diseases and symptoms are connected, can enable engineering of reasoning systems capable to support doctors in analyzing data remotely captured through medical devices and in prising the adequate decision as, for instance, when a patient has to be routed to the right specialist. Hence, in this scenario, the envisaged usage for the knowledge graph for medical diagnosis is to support automatic diagnosis based on the symptoms of a patient. Semantic similarity is an artificial intelligence technique that makes it possible to compute how much two semantically annotated resources (e.g. an electronic health record or the description of a disease) are similar. Usually, semantic similarity reasoning leverages either only the taxonomy [32], or even other kinds of relations defined in an ontology [33]. It can be used to estimate the similarity value between a set of symptoms reported in the electronic health record of a patient and the symptoms characterizing a disease. The most likely diseases are those with the highest similarity values with respect to the patient's symptoms. Indeed, the knowledge graph presented in this paper gathers both knowledge about diseases and that about symptoms, included, respectively, in the DOID and Symp ontologies. In addition, chatbots supported by semantic inference systems could directly interact with patients and reasoning on a mix of background knowledge (e.g., knowledge graphs) and patient specific data in order to respond to specific questions from the patients themselves.

7.5.3 Medical insurance management

A medical insurance helps to cover the expenses related to examinations, surgeries, and any other health treatment. Also, medical insurances represent a big economic market. In fact, the global healthcare insurance market size was estimated at 2.4 trillion in 2019 and is expected to expand at a CAGR of 6.7% from 2020 to 2027.[22] The market is majorly driven by the high cost of healthcare, rising prevalence of chronic diseases as well as increasing disposable income. Even the COVID-19 has created

---

[20] https://www.americantelemed.org/.

[21] https://www.longtermplan.nhs.uk/.

[22] https://www.grandviewresearch.com/industry-analysis/healthcare-insurance-market.

a positive impact on the healthcare insurance industry as more and more people have started investing in healthcare plans. There has been a rise of 50.0% in the queries related to health policies due to the global COVID-19 pandemic. In order to speed up processing of medical insurance policies, semantic interoperability can be crucial. Insurance organizations and systems need to communicate with medical organizations and systems. The correct and unambiguous identification of terms describing pathologies and medical treatments is fundamental. This is true both at data and process level. In fact, in a future of increasing automation systems, different organizations need not only to exchange data in a seamless way but also to enable processes to interact reducing the human intervention as much as possible. The adoption of a knowledge graph that includes diseases and treatments can help in automatically understanding the validity of a request of reimbursement.

For instance, imagine that a request of reimbursement has been received by an insurance company. The informative system of the company analyzes the medical prescription accompanying the request and finds out that the patient needs for a "nose reconstruction with skin graft." Accessing the underlying knowledge graph, and in particular the taxonomic organization of treatments, the system is able to understand that *reconstruction nose with skin graft* is a specific case of *rhinoplasty*, since the former is a subclass of the latter, and in turn, it is a *plastic surgery*. According to the policies of the insurance company, a plastic surgery cannot be reimbursed if not motivated properly. Then, since no further documentation has been provided, the request is automatically rejected by the system.

Furthermore, a knowledge graph where diseases and treatments are integrated and linked can enable systems to automatically detect whether new health problems are related to previously existing diseases. In fact, in this case, according to the specific insurance policy, reimbursement requests can be rejected or partially accepted. For instance, suppose that a person sends a request of reimbursement about an *arthroscopy of the left knee with meniscus repair* to her insurance company. If the information system of the company is based on a knowledge graph linking diseases and treatments, it can understand that the *arthroscopy of left knee with meniscus repair* is a kind of procedure for addressing knee-related problems, as for instance a *knee injury*. Then, if the history record maintained by the company about the person says that in the distant past, she had a *left knee injury*, the company can decide to reject the request or at least to ask for additional medical examinations.

7.6 Conclusion

This article introduced several terminologies in the healthcare system. We discussed the importance of medical ontologies to understand the concepts of the medical domain better, for instance, DOID ontology aligned with the SYMP ontology provides a clear knowledge about diagnosis and their symptoms. We proposed to use ICD-10 as a key for automation, reasoning on clinical data, and disease diagnosis. A careful survey of the current research on disease diagnosis ontologies reveals that the alignment of DOID and SYMP ontology effectively describes diseases and related symptoms. However, this alignment is not enough for disease diagnosis; ICD-10 codes must be referenced in the ontology to align with the clinical terms and codes. This alignment employs multiple challenges related to mapping DOID terms with SYMP terms from one side. From the other side, connecting DOID and SYMP terms to ICD-10 is another challenge that must be tackled. In this article, we presented potential solutions to face these challenges: text matching and use of official transitions file from ICD-9 and SNOMED CT to ICD-10. Finally, we presented the challenges and potential of using semantic technologies in healthcare. As future work, we intend to consider other ontologies, standards, and specifications for the alignment with the medical diagnosis knowledge graph.

References

[1] SNOMED CT website, https://www.nlm.nih.gov/healthit/snomedct/index.html. (Accessed 29 June 2021).

[2] International classification of diseases (icd), 10th revision, clinical modification website, https://www.cdc.gov/nchs/icd/icd10cm.htm. (Accessed 29 June 2021).

[3] L.M. Schriml, C. Arze, S. Nadendla, Y.W.W. Chang, M. Mazaitis, V. Felix, et al., Disease ontology: a backbone for disease semantic integration, Nucleic Acids Research 40 (D1) (2012) D940–D946.

[4] S. Babcock, J. Beverley, L. Cowell, B. Smith, The infectious disease ontology in the age of COVID-19, 05 2020.

[5] O. Mohammed, R. Benlamri, S. Fong, Building a diseases symptoms ontology for medical diagnosis: an integrative approach, in: The First International Conference on Future Generation Communication Technologies, IEEE, 2012, pp. 104–108.

[6] E. Coiera, Guide to Health Informatics, CRC Press, 2015.

[7] O. Bodenreider, The unified medical language system (umls): integrating biomedical terminology, Nucleic Acids Research 32 (suppl_1) (2004) D267–D270.

[8] OWL web ontology language, https://www.w3.org/TR/owl-features/. (Accessed 29 June 2021).

[9] E. Sirin, B. Parsia, B.C. Grau, A. Kalyanpur, Y. Katz, Pellet: a practical owl-dl reasoner, Journal of Web Semantics 5 (2) (2007) 51–53.

[10] W.A. Kibbe, C. Arze, V. Felix, E. Mitraka, E. Bolton, G. Fu, et al., Disease ontology 2015 update: an expanded and updated database of human diseases for linking biomedical knowledge through disease data, Nucleic Acids Research 43 (D1) (2015) D1071–D1078.

[11] R. Zgheib, S. Kristiansen, E. Conchon, T. Plageman, V. Goebel, R. Bastide, A scalable semantic framework for IoT healthcare applications, Journal of Ambient Intelligence and Humanized Computing (2020) 1–19.

[12] L.G. Cowell, B. Smith, Infectious disease ontology, in: Infectious Disease Informatics, Springer, 2010, pp. 373–395.

[13] R.H. Scheuermann, W. Ceusters, B. Smith, Toward an ontological treatment of disease and diagnosis, Summit on Translational Bioinformatics 2009 (2009) 116.

[14] B. Dutta, M. DeBellis, Codo: an ontology for collection and analysis of covid-19 data, arXiv preprint, arXiv:2009.01210, 2020.

[15] A. Barton, A. Rosier, A. Burgun, J.F. Ethier, The cardiovascular disease ontology, in: FOIS, 2014, pp. 409–414.

[16] M. Jensen, A.P. Cox, N. Chaudhry, M. Ng, D. Sule, W. Duncan, et al., The neurological disease ontology, Journal of Biomedical Semantics 4 (1) (2013) 1–10.

[17] C.J. Ryerson, T.J. Corte, J.S. Lee, L. Richeldi, S.L. Walsh, J.L. Myers, et al., A standardized diagnostic ontology for fibrotic interstitial lung disease. An international working group perspective, American Journal of Respiratory and Critical Care Medicine 196 (10) (2017) 1249–1254.

[18] A. Esteban-Gil, J.T. Fernández-Breis, M. Boeker, Analysis and visualization of disease courses in a semantically-enabled cancer registry, Journal of Biomedical Semantics 8 (1) (2017) 1–16.

[19] J.N. Nikiema, V. Jouhet, F. Mougin, Integrating cancer diagnosis terminologies based on logical definitions of SNOMED CT concepts, Journal of Biomedical Informatics 74 (2017) 46–58.

[20] G. Damante, D.F. Gongolo, S. Brusaferro, V. Della Mea, The integration of WHO classifications and reference terminologies to improve information exchange and quality of electronic health records: the SNOMED–CT ICF harmonization within the icd-11 revision process.

[21] A. Hogan, E. Blomqvist, M. Cochez, C. D'amato, G.D. Melo, C. Gutierrez, et al., Knowledge graphs, ACM Computing Surveys 54 (4) (2021), https://doi.org/10.1145/3447772.

[22] P. Shvaiko, J. Euzenat, Ontology matching: state of the art and future challenges, IEEE Transactions on Knowledge and Data Engineering 25 (1) (2011) 158–176.

[23] A. Burton-Jones, V.C. Storey, V. Sugumaran, P. Ahluwalia, A semiotic metrics suite for assessing the quality of ontologies, Data & Knowledge Engineering 55 (1) (2005) 84–102, https://doi.org/10.1016/j.datak.2004.11.010.

[24] A. De Nicola, M. Missikoff, R. Navigli, A software engineering approach to ontology building, Information Systems 34 (2) (2009) 258–275, https://doi.org/10.1016/j.is.2008.07.002.

[25] A. De Nicola, M. Missikoff, A lightweight methodology for rapid ontology engineering, Communications of the ACM 59 (3) (2016) 79–86, https://doi.org/10.1145/2818359.

[26] Defining the Provider Data Dilemma: Challenges, Opportunities, and Call for Industry Collaboration, Tech. Rep., Council for Affordable Quality Healthcare, 2016.

[27] J.A. Fries, E. Steinberg, S. Khattar, S.L. Fleming, J. Posada, A. Callahan, et al., Ontology-driven weak supervision for clinical entity classification in electronic health records, Nature Communications 12 (2017) (2021), https://doi.org/10.1038/s41467-021-22328-4.

[28] R. Wootton, J. Craig, V. Patterson, Introduction to Telemedicine, CRC Press, 2017.

[29] B. Calton, N. Abedini, M. Fratkin, Telemedicine in the time of coronavirus, Journal of Pain and Symptom Management 60 (1) (2020) e12–e14, https://doi.org/10.1016/j.jpainsymman.2020.03.019.

[30] F. Latfi, B. Lefebvre, C. Descheneaux, Ontology-based management of the telehealth smart home, dedicated to elderly in loss of cognitive autonomy, in: C. Golbreich, A. Kalyanpur, B. Parsia (Eds.), Proceedings of the OWLED 2007 Workshop on OWL: Experiences and Directions, Innsbruck, Austria, June 6–7, 2007, in: CEUR Workshop Proceedings, vol. 258, CEUR-WS.org, 2007, http://ceur-ws.org/Vol-258/paper42.pdf.

[31] A. De Nicola, M.L. Villani, Smart city ontologies and their applications: a systematic literature review, Sustainability 13 (10) (2021), https://doi.org/10.3390/su13105578.

[32] A. De Nicola, A. Formica, M. Missikoff, E. Pourabbas, F. Taglino, A comparative assessment of ontology weighting methods in semantic similarity search, in: 11th International Conference on Agents and Artificial Intelligence (ICAART 2019), Vol. 2, 2019, pp. 506–513.

[33] R. Quintero, M. Torres-Ruiz, R. Menchaca-Mendez, M.A. Moreno-Armendariz, G. Guzman, M. Moreno-Ibarra, Dis-c: conceptual distance in ontologies, a graph-based approach, Knowledge and Information Systems 59 (1) (2019) 33–65.

CHAPTER

A naturopathy knowledge graph and recommendation system to boost the immune system

KISS: Knowledge-based Immune System Suggestion

Amelie Gyrard[a] and Karima Boudaoud[b]

[a]*M3 (Machine-to-Machine Measurement), Paris, France*
[b]*University of Nice Sophia Antipolis, Sophia Antipolis, France*

Contents

"Let food be thy medicine and medicine be thy food."

Hippocrates

8.1 Introduction

Lancet Planetary Health 2019 [1] published that it will be the year of nutrition: links between food systems, human health, and the environment. Because of the COVID-19 world-wide pandemic, there is a need for any complementary solutions to boost the immune system. Matt Ritchell's immune system book [2] encourage better sleep, physical exercise, meditation, and nutrition. Nowadays, healthy lifestyle, fitness, and diet habits have become central applications in our daily life: (1) "Healthy diet is the most effectual approach to prevent disease. Food and victuals is a key to have good health" according to [3], and (2) Chopra and Tanzi encourage a balanced diet [4].

Increasingly, **Recommendation Systems (RS)** [5,6] are invading our lives (e.g., YouTube for videos, Amazon for products, and Netflix for movies). Existing RS system surveys do not address suggestions to boost the immune system yet [7] (machine learning-based RS published in 2018) [8–10]. Well-being RS (published in 2019) [11] illustrates the emerging research need. The are several kinds of recommendation systems: (1) Content-based (CB) comes from information retrieval and information filtering, (2), Collaborative filtering (CF) predicts item utility for a particular user based on the items previously rated by other users, (3) **knowledge-based** (that we are interested in this book chapter), and (4) hybrid approaches.

Investigating well-being applications for a healthy lifestyle is time-consuming for users and requires an eagerness to learn. Boulos et al. [12] highlight there are missing links between exiting food ontologies. There is a need to ease the integration of those food ontologies and enrich them with the necessary knowledge to boost the immune system since none of them address it yet.

There is a need to design a naturopathy knowledge graph to be employed within a knowledge-based recommender system to boost the immune system. The naturopathy knowledge graph integrates ontology-based food projects to reuse past expertise and disseminate FAIR principles [13] by encouraging researchers to share their reproducible experiments by publishing online their ontologies, data sets, rules, etc. The set of the ontology codes shared online can be automatically processed; if the ontology code is not available yet, the scientific publications describing the food ontologies are semiautomatically processed with Natural Language Processing (NLP) techniques to feed the food reasoning engine to build the naturopathy recommender system.

We combine multidisciplinary approaches: naturopathy describing foods and its benefits, and smart health using IoT (see our other book chapters [14,15], more focused on this topic) and AI technologies (knowledge-based systems).

We designed the following **Research Questions (RQ)**:

- **RQ1**: Is there any food that can help boost the immune system?
- **RQ2**: Are there reusable ontologies describing food?
- **RQ3**: How to deduce meaningful information (e.g., data analytics) for food data? How to design the well-being recommendation system to encourage a diet that boosts the immune system?

Contributions (C):

- **C1**: Food ontology catalog (and its associated tools) is designed to encourage past projects to share their expertise implemented as ontologies; it addresses RQ2 and is explained in Section 8.2.2, Section 8.3.1, and Table 8.1.
- **C2**: A unified food knowledge based focused on naturopathy to improve well-being and boost the immune system; it addresses RQ1 and is explained in Section 8.3.2.
- **C3**: A reasoning engine is designed to provide more sophisticated suggestions and address requirements such as allergies, diets, etc.; it addresses RQ3 and is explained in Section 8.3.3.

Structure of the chapter: Related work reviewed in Section 8.2. The naturopathy knowledge graph and recommendation system to boost the immune system is described in Section 8.3. The conclusion and future work are provided in Section 8.4.

8.2 Related work: food knowledge graphs and recommendation systems

This section classifies existing work into tree main categories: (1) Food knowledge graphs (ontologies and data sets) in Section 8.2.1, (2) food-based recommender system in Section 8.2.2, (3) food information extraction in Section 8.2.3, and (4) shortcomings of the literature study in Section 8.2.4.

8.2.1 Food knowledge graphs: ontologies and data sets

DBpedia [16] (semantic version of Wikipedia) references food and their nutrients, but there is no emphasis on which ones can boost the immune system. **Edamam**[1] is a food knowledge base that highlights nutrients while cooking recipe. Boulos et al. [12] review food ontologies. We enrich the food ontology catalog with more projects referenced by the authors even if the ontologies are not available yet (see Table 8.1). **OntoFood**[2] ontology (available in the BioPortal biomedical ontology catalog) is integrated with SWRL nutrition rules for diabetic patients. FoodOn (Griffiths et al. [17]) design the farm-to-fork food ontology to describe food safety, food security, agricultural, and animal husbandry practices linked to food production, culinary, nutritional and chemical ingredients, and processes. **Food Product Ontology** (Kolchin et al. [18]) is an ontology that describes food products: product, price, store, and company data. **Open Food Facts**[3] is an open source food database: users can search for food's nutritional information and compare products from around the world [12]. The food industry use Open Food Facts to track, monitor, and plan food production. **AGROVOC**[4] [19] is a widely used in agriculture, fisheries, forestry, and food. **FoodWiki** (Celik et al. [20]) classifies foods, nutritional information, and the recommended daily intake. Cantais et al. [21] design a food ontology to provide guidance for diabetes patient. Chang-Shing et al. [22] design a web-based system helping tourists plan menus based on nutrition analysis and food composition.

Conclusion: None of those works provide explicit descriptions on food that boost the immune system.

8.2.2 Food recommender systems

Personal Health KG for diet (that includes an ontology) (Seneviratne et al. [23]) consider user's temporal personal health data (and mentioned our Personalized Health KG past work [24]) to personalize dietary recommendations for diabetic patients. Food recommendations are provided to answer such questions: (1) What should I eat for breakfast? (2) What foods can I eat if I have a dairy allergy? (3) What can I substitute for food Y? (patient's taste preferences).

  ProTrip (Subramaniyaswamy et al. [25]) is an IoT health nutrition-centric and ontology-based tourism recommender system based on hybrid filtering mechanisms (collaborative, content-based, and knowledge-based). ProTrip adapts user's food preferences through a questionnaire-based survey: allergic constraints, preferred type of food (vegetarian, nonvegetarian, egg containing), sweet and spice level, and availability according to climatic conditions and local areas. The recommender system considers chronic diseases such as asthma, cold, kidney stones, diabetes, haemophilia, hyperlipidaemia,

---

[1] https://www.edamam.com/.
[2] https://bioportal.bioontology.org/ontologies/OF/?p=summary.
[3] https://vest.agrisemantics.org/content/open-food-facts-food-ontology.
[4] http://agroportal.lirmm.fr/ontologies/AGROVOC.

hypertension, hypothyroidism, migraine, rheumatoid arthritis, ulcer, and wheezing. ProTrip uses the Apache Mahout recommendation engine library (SlopeOne recommender schemes). The recommender system is evaluated with participants from the International Symposium on Big Data and Cloud Computing Challenges 2015: 76% of participants disclose the accuracy of the recommendations as a 5-star rating for relevance and 80% of the users feel ProTrip as a health-centric recommender system.

Smart Recommender System of Hybrid Learning (SRHL) (Nouh et al. [11]) suggests healthy food for personalized well-being and prevents diseases. The hybrid RS (content-based and collaborative filtering), uses unsupervised machine learning algorithms. The recommendations take into account time, activity, location, monetary costs, ingredients, health, nutritional value, availability, and the effects of combining the ingredients.

The Iranian snack knowledge-based recommender system mobile application for type II diabetic patients (Norouzi et al. [26] [27]) also provides constraint-based reasoning and a roulette wheel algorithm. Recommendations take into consideration patients' interests, Iranian culture, dietary habits, and seasons. No ontology is employed within this knowledge-based RS. The RS has been evaluated with nutritionists who also provided diets.

Personalized Dietary Recommendation For Travelers (Karim et al. [28]) integrates an ontology that combines food, nutrition, and travel concepts. Since food and nutrition concepts are already modeled within the ontology, it would be highly relevant to reuse the knowledge. Furthermore, eight SWRL rules executed with the Pellet reasoner have been developed for providing recommendations (e.g., this is a dish for breakfast/lunch/dinner, an obese patient, heart patient, and healthy person).

Diet Food Recommendation System (DFRS) for Diabetic Patients (Kumar et al. [3]), hybrid RS (content and collaborative-based), suggests dishes for the type of diet menu according to the total nutrition to be taken daily. It comprises: data preprocessing, weight tuning, diet planning, food ontology construction, SOM training, and K-means clustering.

Health food recommendation system (Ge et al. [29,30]) for mobile Android platforms suggest recipes by taking into consideration users' health (e.g., chronic diseases such as obesity and diabetes) and users' preferences. This RS does not use ontologies.

Al-Nazer et al. [31] highlight that queries such as "Does eating bananas prevent diabetes?" are not easily answered by web search engines. Semantic web technologies are employed for understanding the users' queries and structuring the information on the web. Personal preferences, heath constraints (pregnant, smoker, diseases, allergies, medical history), culture, and religions are considered for food recommendation. The user's profile ontology collect user name, gender, weight, height, blood type, skin color, and Body Mass Index (BMI), and food search history. Four hundred fifty-three queries are collected from different sources such as domain experts, users through surveys and health consumer websites and categorized into: food-centric, nutrition-centric, recipe-centric, disease-centric, body part-centric, and body function-centric.

NutElCare Ontology-based RS (Espin et al. [32]) suggests healthy diet plans for the elderly. The ontology defines the user profile: Body Mass Index (BMI), time of year and geographical environment, physical activity, and the level of mastication and swallowing of the user. NulElCare collects and represents nutritional information to provide nutrition tips for the elderly. NulElCare retrieves reliable and complete nutritional information from expert sources, either humans (e.g., nutritionists, gerontologists, bromatologists) or computerized (e.g., information systems, nutritional databases, World Health Organization (WHO)) and Spanish Society of Patental and Enteral Nutrition SENPE recommendations. NutElCare uses the AGROVOC FAO thesaurus. Espin et al. address the problem of heterogeneity of the representation of information that prevents the information from being reused by other processes or applications.

    FOODS ontology and web-based food-menu RS (Snae et al. [33]) is addressed for patients with diabetes in Thailand. The RS is based on PIPS (Personalised Information Platform for health and life services).[5] FOODS recommends recipes by considering ingredient nutritional factors and seasonal availability of specific ingredients, according to user profiles (e.g., age, and diet during the week).

Conclusion: None of those work design a naturopathy recommender systems to boost the immune system.

8.2.3 Food information extraction with natural language processing: named-entity recognition

FoodOntoMap (Popovski, Eftimov et al. [1]) extract food concepts from recipes, using semantic tags from their food ontology. 22,000 recipes from Allrecipes are gathered from five recipe categories: appetizers and snacks, breakfast and lunch, dessert, dinner, and drinks. It is claimed there is no annotated corpus with food concepts, and there are only a few rule-based food named-entity recognition systems for food concepts extraction. We will demonstrate that Table 8.1 helps in finding, classifying, and reusing existing ontologies. **FoodIE** (Popovski, Eftimov et al. [34])[6] is a rule-based food-named Named-Entity Recognition (NER) method for food Information Extraction (IE) from unstructured recipe data. The rule engine executes computational linguistics and semantic information rules describing the food entities. FoodIE comprises four steps: (1) Food-related text preprocessing (e.g., remove quotation marks, white space, ASCII transliteration, standard mathematical decimal notation), (2) Text POS-tagging and post-processing of the tag data set, (3) Semantic tagging of food tokens in the text, and (4) Food-named entity recognition. The data set comprises 200 recipes that are processed and evaluated (100 recipes analyzed to build the rule engine, 100 new recipes for test phase). Recipes are from Allrecipes (https://www.allrecipes.com/) and MyRecipes (https://www.myrecipes.com/); it highlights there is no standardized format for the recipe description. FoodBase[7] is released by Popovski and Eftimov et al.

Conclusion: Those works apply NLP techniques on recipes, rather than food ontologies or scientific publications describing ontologies.

8.2.4 Shortcomings of the literature study

We summarize the following limitations of the literature study:

- Doing the systematic literature review is a time-consuming task. There is a need to share the literature review innovatively (e.g., a knowledge repository for food supported by tools) to ease the work of future researchers. The set of ontologies are classified in Table 8.1.

---

[5] http://www.csc.liv.ac.uk/~floriana/PIPS/PIPSindex.html.

[6] https://github.com/GorjanP/foodie.

[7] http://cs.ijs.si/repository/FoodBase/foodbase.zip.

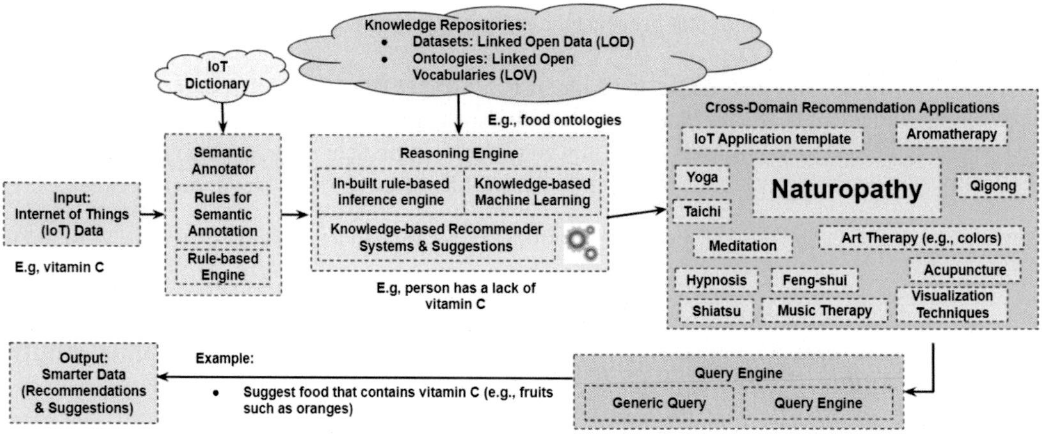

FIGURE 8.1

Well-being recommender system architecture [36]. This chapter focuses on naturopathy to boost the immune system.

- A lot of ontologies cannot be exploited since they are not accessible online. There is a need to disseminate best practices (e.g., following FAIR principles [35]) to ease the task of computers to analyze ontologies and extract meaningful domain knowledge automatically.
- There is a lack of prototypes/experiments (e.g., web service, web application) that can be easily reproduced or tested to understand the applications and illustrate the limitations clearly.
- NLP techniques are applied on recipes, rather than food ontologies or scientific publications describing ontologies.
- Explicit descriptions on food that boost the immune system are missing within food ontologies.
- **There is no naturopathy recommender system to boost the immune system.**

8.3 Naturopathy knowledge graph and recommendation system to boost immune system: knowledge-based immune system suggestion

This section explains the knowledge-based naturopathy recommender system to boost the immune system. It corresponds to the "Naturopathy" component (subcomponent within the green box called the 'cross-domain recommendation applications") within the well-being recommender system architecture [36], as depicted in Fig. 8.1.

8.3.1 Collecting food ontologies: LOV4IoT-food ontology catalog

To deduce meaningful information from IoT data produced by health devices to monitor patients [14], we need common sense knowledge. We searched on Google and Google Scholar for a set of specific key phrases which (1) starts with ontology-based, (2) finishes with ontology, or (3) starts with semantic-based, knowledge-based, knowledge-graph, or related synonyms. For instance, for the food domain, the key phrases are as follows: food, naturopathy, healthy diet, nutrients, cooking, etc.

Ontologies enable us to share and reuse knowledge by designing concepts and relationships within a specific domain [37]. We intended knowledge catalogs to cover various topics (emotion, food, fitness, obesity, sleep, stress, and depression) [36] relevant to IoT (in this chapter, we focused on food).

Most relevant scientific publications are classified in Table 8.1 when that information is provided:

- **Ontology** eases the reuse of the domain expertise already designed in previous projects. The column *Ontology Availability (OA)* explicitly describes if the ontology code is accessible online (URIs mentioned in the publication or after corresponding with the authors).
- **Sensors** used and measurement type (e.g., RFID embedded on food).
- The **reasoning** employed to analyze data. Sometimes, rules can be reused to interpret data in other applications using the same data efficiently. *Reasoning* column within Table 8.1 references the reasoning mechanism employed within the projects.
- **Cited publications** enriched our domain knowledge repository with scientific papers to prove the veracity of facts mentioned in our recommender system.

Our knowledge repository is the result of a continuous enrichment of the LOV4IoT knowledge repository [38] since 2012 (that references almost 800 ontology-based IoT projects). It is an innovative solution to share the Systematic Literature Review (SLR) (SLR guidelines [39]) as a tool rather than a survey paper. The repository is continuously updated, inspired from Agile software development methodologies, to enrich it with additional domains, use cases and knowledge. We take into consideration the latest publications, surveys, and we carefully analyze their reference sections that can introduce complementary topics and key scientific publications.

Food knowledge catalog: The food knowledge base, which classifies a set of ontologies describing food, recipes, etc. (see Table 8.2 for demo URLs). It references more than 53 ontology-based projects (in October 2021), as referenced in Table 8.1. The ontologies available online are classified on the top, then the ontologies are classified by the year of publications. The food ontology data set provides a subset with only 12 ontologies, which shared their ontology code online for semiautomatic knowledge extraction.

Visualizing Food Ontologies: To ease a quick discovery of the ontology content, the WebVOWL tool [67] helps to visualize ontologies available within LOV4IoT-Food and that are preselected if they can be loaded without errors. The set of issues while loading the ontology is described in [68] and shared to the WebVOWL team. Listing 8.1 shows an example of ontology metadata required with WebVOWL [67] for a better visualization. Best practices that encourage ontology quality and interoperability are quickly explained in Section 8.3.4 and summarized in Table 8.3.

8.3.2 Naturopathy knowledge graph: extracting and integrating food ontologies and data sets

The naturopathy knowledge graph is built from the set food ontologies mentioned in Table 8.1. Since the ontology-based food projects were not focused on providing suggestions to boost the immune system, we enrich our knowledge base with the content from a set of books about "alimentation as a third medicine" [69] or "aliments to cure" [70].

Table 8.1 Ontology-based food projects to design the integrated food knowledge base and reasoning engine. The ontologies available online are classified on the top, then the ontologies are classified by year of publications.

Authors	Year	Project	OA	Reasoning
Dooley, Griffiths et al. [17,40]	2018 2016	FoodOn – Universal Food Ontology	✓	✗
Peroni et al. [41]	2016	20 food ontologies (fish, honey, etc.)	✓	✗
Celdran et al. [42]	2016	Supermarket and location ontologies	✓	✓ Semantic rules, SWRL
Gyrard et al. [43]	2015	Naturopathy ontology and data set	✓	✓ Jena rule-based engine
Kolchin et al. [18]	2013	Food product ontology	✓	✗
Sabou et al. [44]	2009	SmartProducts: Food and recipe	✓	✓ owl:Restriction rules
Calore, Pernici et al. [45]	2007	Foodshop case study	✓	✓ owl:Restriction rules
Gaia	–	Restaurant	✓	✓ owl:Restriction rules
Tropical fruits	–	Tropical fruits	✓	✓ owl:Restriction rules
–	–	Pizza	✓	✓ owl:Restriction rules
–	–	Wine	✓	✓ owl:Restriction rules
Mooney	–	Restaurant	✓	✗ No owl:Restriction rules
–	–	Beverage	✓	✗ No owl:Restriction rules
–	2000	Beer	✓	✗
Seneviratne et al. [23]	2021	Personal Health KG for diet	✗	✓
Subramaniyaswamy et al. [25]	2019	Personalized food recommendation	✗	✓ RS
Espin et al. [46]	2016	Nutrition Diet RS for elderly people	✗	✓ RS, SWRL rules, Pellet
Boulos et al. [12]	2015	Survey paper: Food ontologies for IoT	✗	✗
Chi et al. [47]	2015	Disease dietary consultation system	✗	✓ Rule-based
Karim et al. [28]	2015	Personalized dietary RS for travelers	✗	✓ RS, SWRL, Pellet
Celik et al. [20]	2015	FoodWiki, mobile safe food consumption	✗	✓ > 30 SWRL rules, Pellet
Pizzuti et al. [48,49]	2014	Food track and traceability ontology	✗	✓ Pellet
Su et al. [50,51]	2014	Personalized fitness, diet plan, food	✗	✓ SPIN rules, SPARQLMotion
Tumnark et al. [52]	2013	Personalized dietary RS for weightlifting	✗	✓ RS, SWRL, Pellet, SQWRL
Chen et al. [53]	2013	Ontology-based diet RS	✗	✓ RS, Jena rule, fuzzy, knapsack
Curiel et al. [54]	2013	Users and restaurant ontologies	✗	✓ Jena OWL reasoner
Miao et al. [55]	2013	User preferences for personalized RS	✗	✓ Bayesian model, SWRL
Suksom et al. [56]	2013	Personalized food RS	✗	✓ RS, fuzzy inference
Yasavur et al. [57]	2013	Health, food, activity, beverage	✗	✗
Vadivu et al. [58]	2010	Natural food, chemicals and diseases	✗	✗
Fudholi et al. [59,60]	2009	Menu RS	✗	RS, SWRL, SQWRL, fuzzy
Gu et al. [61]	2009	Fridge, food, smart home, RFID	✗	– Cannot access any PDF
Snae et al. [33]	2008	FOODS, food ontology	✗	✗
Sachinopoulou et al. [62]	2007	Personal health and wellness ontology	✗	– Cannot access any PDF
Li et al. [63]	2007	Food ontology for diabetes diet care	✗	– Cannot access any PDF
Chen et al. [64]	2006	Cocktail (drink) RS and mood	✗	✓ DL RACER OWL reasoner, RS
Ribeiro et al. [65]	2006	Ontology for cooking	✗	✗
Cantais et al., [21]	2005	Food/diet/product for diabetes control	✗	✓ Ontology reasoners (Racer, Pellet)
OpenFoodFacts [66]	–	Food products from around the world	✗	✗

Legend: Recommender System (RS), Ontology Availability (OA).

```
1   <owl:Ontology rdf:about="&naturopathy;">
2       <rdf:type rdf:resource="http://purl.org/vocommons/voaf#Vocabulary"/>
3       <dc:title xml:lang="en">The naturopathy ontology</dc:title>
4       <dc:description xml:lang="en">The naturopathy ontology describes relationships between foods, their nutrients, their
            color, the mood, the diseases, the diets and the allergies.</dc:description>
5       <dc:creator rdf:datatype="http://www.w3.org/2001/XMLSchema#string">Amelie Gyrard</dc:creator>
6       <dcterms:modified rdf:datatype="http://www.w3.org/2001/XMLSchema#date">2021−10−10</dcterms:modified>
7       <dcterms:issued rdf:datatype="http://www.w3.org/2001/XMLSchema#date">2013−09−12</dcterms:issued>
8       <owl:versionInfo rdf:datatype="http://www.w3.org/2001/XMLSchema#decimal">1.2</owl:versionInfo>
9       <vs:term_status>Work in progress</vs:term_status>
10      <vann:preferredNamespacePrefix>naturopathy</vann:preferredNamespacePrefix>
11          <vann:preferredNamespaceUri>http://sensormeasurement.appspot.com/naturopathy#</vann:preferredNamespaceUri>
12          <foaf:homepage>http://lov4iot.appspot.com/?p=lov4iot−food</foaf:homepage>
13  </owl:Ontology>
```

Listing 8.1: Naturopathy ontology metadata required with WebVOWL for a better visualization.

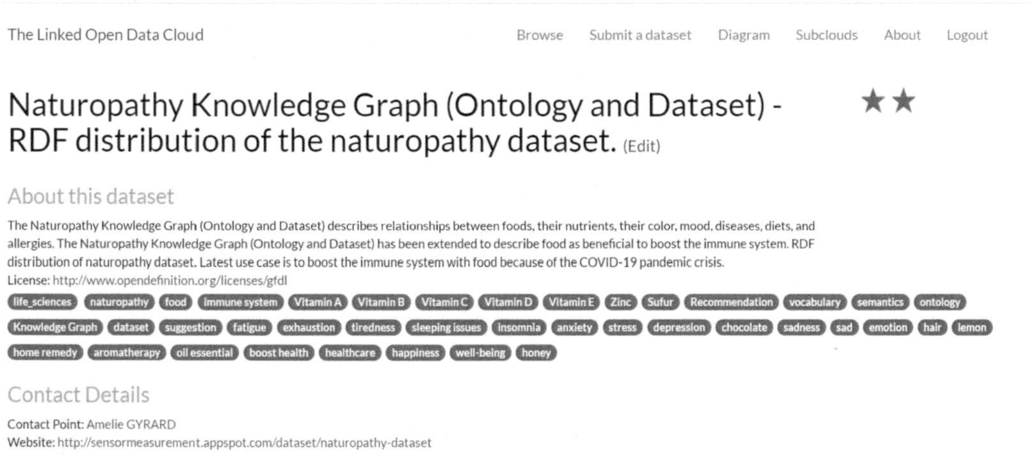

FIGURE 8.2

The naturopathy data set referenced on the Linked Open data (LOD) cloud.

The **naturopathy ontology** is referenced on the LOV4IoT-Food ontology catalog and the **naturopathy data set** is referenced on the Linked Open data (LOD) cloud as depicted in Fig. 8.2 (see Table 8.2 for URLs).

Mapping with other ontologies and knowledge graphs: Food-related knowledge is mapped as much as possible with the DBpedia ontology (`rdf:type Food`) and DBpedia resources, as demonstrated in Listing 8.2.

Knowledge extraction from food ontologies: We semiautomatically analyzed the ontologies with Natural Language Processing techniques (NLP): (1) ontology code (RDF/XML), and (2) scientific publications describing the ontologies. When the ontologies can be processed, we selected a set of specific keywords (e.g., vitamin) to compare knowledge provided by ontologies.

```
1  <naturopathy:Food rdf:about="RoyalJelly">
2  <rdfs:label xml:lang="fr">Royal Jelly</rdfs:label>
3  <rdfs:label xml:lang="en">Gelee royale</rdfs:label>
4  <rdf:type rdf:resource="&dbpedia_ont;Food"/>
5  <owl:sameAs rdf:resource="http://dbpedia.org/resource/Royal_jelly"/>
6  <m3:isRecommendedFor rdf:resource="BoostImmuneSystem"/>
7  <rdfs:comment xml:lang="en">Royal Jelly is recommended when tired and to boost the immune system.</rdfs:comment>
8  <txn:hasImage rdf:resource="http://sensormeasurement.appspot.com/images/FoodNaturopathyRS/RoyalJelly.png"/>
9      </naturopathy:Food>
```

Listing 8.2: Food-related knowledge is mapped to DBpedia.

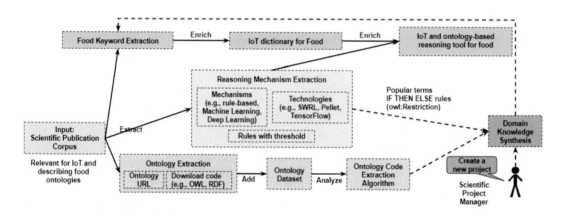

FIGURE 8.3

Knowledge extraction architecture from food ontologies.

From a set of publications on food ontology-based projects, we extract knowledge such as which sensors are employed, which reasoning mechanisms are used to interpret sensor measurements, is there an ontology available and reusable, etc. (as depicted in Fig. 8.3).

Ontologies are compared with each other to extract a common pattern to later generate a unified and federated food knowledge graph. As an example, we automatically retrieve key phrases (see Fig. 8.4) if they mention **food related keywords** such as: "protein," "food," "product," "additive," "ingredient," "nutrient," "calorie," "fat," "fish," "oil," "meat," "vegan," "vegetarian," "recipe," "weight," "age," "sex," "height," "diet," "patient," "person," "fiber," "vitamin," "carbohydrate." Since there is no corpus for such tasks, we have semimanually built the gold data set as depicted in Fig. 8.4. Each column corresponds to an ontology, and each row represents a food-related term.

We prioritize terms relevant to describe why food type and its composition (e.g., Vitamin C) is good to boost the immune system.

Applic	Concept/Pro	Pernici20	Naturopa	Kolchin20	SmartPro	SmartPro	SmartPro	Haussma	Pizza	Wine	SmartPro	FruitsTr
Food	erson patien	Not found	patient	http://sche	Not found	Not found	Not found	Not found	Not found	Not found	Not found	Not foun
Food	ingredient	Not found	Not found	ingredient	ingredientp	ingredientp	ingredient	ingredient	Not found	Not found	defaulting	Not foun
Food	food	materiala	food	food	veganfoo	veganfoo	veganfood	food	food	othertomatobas	food	Not foun
Food	additive	additive	Not found	foodadditive	additive	additive	Not found	Not found	Not found	Not found	Not found	Not foun
Food	product	meatprod	Not found	productors	product	product	meatprod	chemicalf	Not found	Not found	smartpro	Not foun
Food	meat	meatprod	meat	Not found	meatprod	meatprod	meatprod	meatfromc	meatypizza	redmeat	Not found	Not foun
Food	oil	oil	essentialoil	Not found	peanutoil	peanutoil	Not found	Not found	Not found	Not found	Not found	Not foun
Food	essential oil	Not found	essentialo	Not found	Not found	Not found	Not found	Not found	Not found	Not found	Not found	Not foun
Food	fat	milkbutter	Not found	Not found	fat fats	fat fats	nearlyfatfr	Not found	Not found	Not found	fat	Not foun
Food	fruit	fruitandve	fruit	Not found	fruit	fruit	cranberryf	Not found	fruittoppin	fruit	Not found	fruituse
Food	trient nutriti	foodstuffin	nutrient	Not found	nutrient nu	nutrientpo	Not found	Not found	Not found	Not found	nutrient nu	Not foun
Food	vitamin	Not found	vitamin	Not found	vitaminc	Not found	Not found	Not found	Not found	Not found	Not found	vitamin
Food	fish	freshfish	shellfish	Not found	shellfish	shellfish	Not found	Not found	fishtopping	fish	Not found	Not foun
Food	cheese	freshchees	cheese	Not found	creamche	creamche	cheeseca	Not found	cheesetop	cheesenutsdes	Not found	Not foun
Food	getable legu	vegetable	vegetable	Not found	leafvegeta	leafveget	Not found	Not found	vegetabletc	Not found	Not found	Not foun

FIGURE 8.4

Food gold data set: knowledge extraction from food ontologies that highlight the need to collect and integrate knowledge from complementary ontologies.

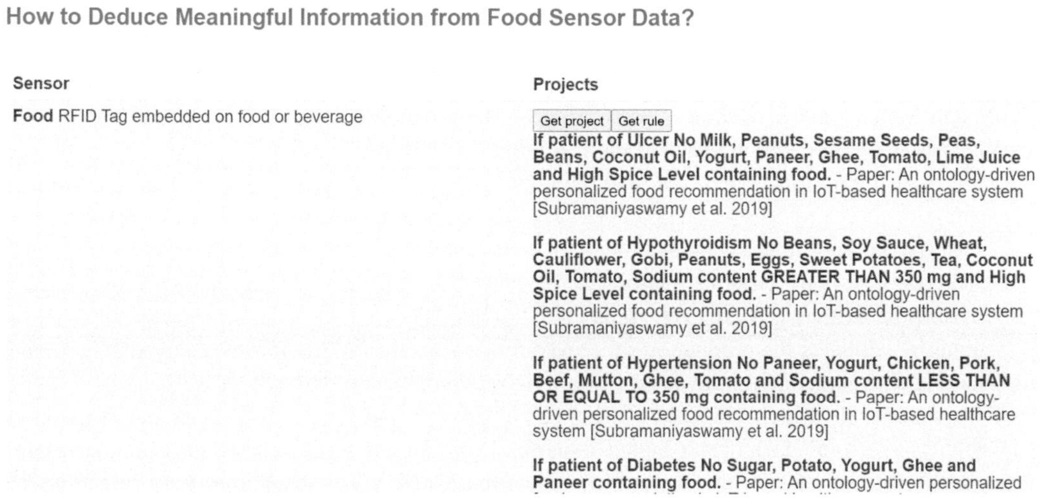

How to Deduce Meaningful Information from Food Sensor Data?

Sensor

Food RFID Tag embedded on food or beverage

Projects

Get project　Get rule

If patient of Ulcer No Milk, Peanuts, Sesame Seeds, Peas, Beans, Coconut Oil, Yogurt, Paneer, Ghee, Tomato, Lime Juice and High Spice Level containing food. - Paper: An ontology-driven personalized food recommendation in IoT-based healthcare system [Subramaniyaswamy et al. 2019]

If patient of Hypothyroidism No Beans, Soy Sauce, Wheat, Cauliflower, Gobi, Peanuts, Eggs, Sweet Potatoes, Tea, Coconut Oil, Tomato, Sodium content GREATER THAN 350 mg and High Spice Level containing food. - Paper: An ontology-driven personalized food recommendation in IoT-based healthcare system [Subramaniyaswamy et al. 2019]

If patient of Hypertension No Paneer, Yogurt, Chicken, Pork, Beef, Mutton, Ghee, Tomato and Sodium content LESS THAN OR EQUAL TO 350 mg containing food. - Paper: An ontology-driven personalized food recommendation in IoT-based healthcare system [Subramaniyaswamy et al. 2019]

If patient of Diabetes No Sugar, Potato, Yogurt, Ghee and Paneer containing food. - Paper: An ontology-driven personalized

FIGURE 8.5

Rule discovery for food with SLOR-food.

8.3.3 Knowledge-based immune system suggestion: ontology-based food recommendation to boost the immune system

Food reasoning discovery with SLOR-Food: Rules are collected and can be defined in our recommendation system to personalize food suggestions according to patient restrictions such as diets and allergies, as depicted in Fig. 8.5.

Our knowledge-based recommender system [71] integrates a set of ontologies (as summarized in Table 8.1), data sets and other sources such as scientific publications, and books related to the naturopathy domain.

The recommender system provides suggestions by using the property `<m3:isRecommendedFor rdf:resource="AAA"/>` as introduced in Listing 8.2. We automatically retrieve foods that boost the immune system with the property `<m3:isRecommendedFor rdf:resource="BoostImmuneSystem"/>`.

We designed the **Naturopathy recommender system to boost the immune system**: Natural products, such as herbs, prebiotic, probiotics, and selective medical diets, help for a healthy lifestyle [72]. The "Clinical naturopathy: an evidence-based guide to practice" book [72] addresses numerous diseases and syndromes (e.g., food allergy/intolerance, asthma, hypertension, stroke, anxiety, depression, insomnia). Additional fields, such as aromatherapy [73,74] are highly relevant.

The naturopathy application is extended with explicit links to describe food boosting the immune system (some explanations and figures are hereafter, otherwise in Appendix 8.A):

- **Food suggestion to boost the immune system** (Fig. 8.6). Numerous ingredients and nutrients that boost the immune system can be retrieved for various reasons explained hereafter (e.g., ingredients that contain Vitamin C).
- Food with **vitamins** to boost the immune system: Vitamin A (Fig. 8.9), Vitamin C (Fig. 8.7), Vitamin D (Fig. 8.10), and Vitamin E (Fig. 8.11).

  Vitamins such as vitamin C are well known to boost the immune system, as highlighted in Fig. 8.7. "Vitamin C is one of the strongest vitamins to increase immunity. Foods that have Vitamin C like hot peppers, have more Vitamin C than oranges. Many fruits (e.g., kiwi, papaya, strawberry) and vegetables (e.g., broccoli) are packed with Vitamin C."[8] Similarly, other vitamins such as Vitamin A (Fig. 8.9), Vitamin D (Fig. 8.10), and Vitamin E (Fig. 8.11) are excellent to boost the immune system.

  "Foods that are high in carotenoids compounds are: carrots, sweet potatoes, pumpkin, cantaloupe, and squash. The body turns carotenoids into vitamin A, that have an antioxidant effect to help strengthen the immune system against infection."[9] (Fig. 8.9).

  "Vitamin D regulates the production of a protein that selectively kills infectious agents, including bacteria and viruses. Vitamin D alters the activity and number of white blood cells, known as T 2 killer lymphocytes, which can reduce the spread of bacteria and viruses. Winter-associated vitamin D deficiency—from a lack of sun-induced vitamin D production—can weaken the immune system, increasing the risk of developing viral infections that cause upper respiratory tract infections."[10] (Fig. 8.10).

  "Vitamin E is a powerful antioxidant that helps your body fight off infection. Almonds, peanuts, hazelnuts, sunflower seeds, spinach, and broccoli are all high in vitamin E."[11] (Fig. 8.11).

---

[8] https://www.wptv.com/news/region-n-palm-beach-county/north-palm-beach/foods-to-eat-that-will-boost-your-immune-system-during-the-coronavirus.

[9] https://health.clevelandclinic.org/eat-these-foods-to-boost-your-immune-system/.

[10] https://edition.cnn.com/2020/03/25/health/immunity-diet-food-coronavirus-drayer-wellness/index.html.

[11] https://health.clevelandclinic.org/eat-these-foods-to-boost-your-immune-system/.

- Food with **zinc** to boost the immune system (Fig. 8.12). "Zinc reduces the length of a cold. Zinc also has antiviral properties to prevent coronavirus entry into cells and decrease the severity of the virus."[12] (Fig. 8.12).
- Food with **probiotics** to boost the immune system (Fig. 8.13). "Probiotics like yogurt keep your gut healthy, which is directly tied to your immune system."[13] (Fig. 8.13).
- Food with **beta carotene** to boost the immune system (Fig. 8.14). "Beta carotene gets converted to vitamin A, which is essential for a strong immune system. It helps antibodies respond to toxins and foreign substances. Sweet potatoes, carrots, mangoes, apricots, spinach, kale, broccoli, squash, and cantaloupe contain beta carotene."[14] (Fig. 8.14).
- Food with **sulfur** to boost the immune system (Fig. 8.15). For instance, "Garlic contains Allicin which breaks down into sulfur containing compounds boost white blood cells, an important part of the immune system."[15]
- Food or aromatherapy reducing **stress** (Fig. 8.16), **anxiety** (Fig. 8.18), and **depression** (Fig. 8.17). "Aromatherapy through inhalation optimizes the mood and reduces effects of anxiety, depression, and stress" [74].
- Food **fatigue, exhaustion, and tiredness** (Fig. 8.19).
- Food reducing **sleeping issues and insomnia** (Fig. 8.20). For instance, Roman chamomille is recommended if there are sleeping issues/insomnia.
- **Aromatherapy** with essential oils boosts the immune system and clean the air (Fig. 8.21), based on a research review published in 2021 [73] and books such as [74]. "Aromatherapy through inhalation reduces physical disorders associated with immune system dysfunction" [74].
- **Meditation** and hypnosis videos to boost health. Techniques such as Mindfulness-Based Stress Reduction (MBSR) from researchers such as Jon Kabat-Zinn [75,76] are scientifically proven with a cohort of more than 18,000 patients, and specific research on mindfulness meditation benefits on brain and the immune system with Jon Kabat-Zinn and Richard Davidson [77,78].

8.3.4 Evaluation

Technical evaluation: As a technical evaluation, the naturopathy data set is referenced on the Linked Open data (LOD) cloud as explained above. We also synthesize 16 rules to disseminate ontology best practices in Table 8.3 to encourage researchers to share more their food knowledge [68] to ease reusability and better interoperability. For each rule, we provide examples of bad practices and best practices to help beginners in their learning journey in a set of slides entitled "Step-by-step tutorial to improve the ontology quality, dissemination, reuse."[16] This work is an enhancement of our previous work [79].

---

[12] http://potentialenergytraining.com/2020/03/19/boost-your-immunity-through-nutrition/?utm_source=rss&utm_medium=rss&utm_campaign=boost-your-immunity-through-nutrition.

[13] https://www.wptv.com/news/region-n-palm-beach-county/north-palm-beach/foods-to-eat-that-will-boost-your-immune-system-during-the-coronavirus.

[14] https://edition.cnn.com/2020/03/25/health/immunity-diet-food-coronavirus-drayer-wellness/index.html.

[15] http://potentialenergytraining.com/2020/03/19/boost-your-immunity-through-nutrition/?utm_source=rss&utm_medium=rss&utm_campaign=boost-your-immunity-through-nutrition.

[16] Slides step-by-step tutorial to improve the ontology quality, dissemination, reuse, etc. Semantic Web Best Practices: https://goo.gl/Rg4cGr.

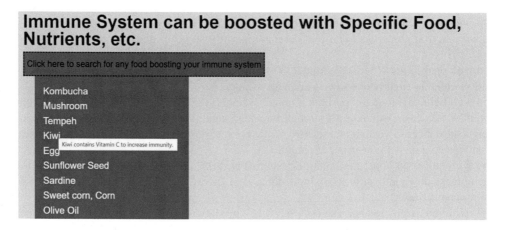

FIGURE 8.6

Food suggestions to boost the immune system.

Immune System can be boosted with Food Containing Vitamin C

"Vitamin C is one of the strongest vitamins to increase immunity. Foods that have Vitamin C like hot peppers, have more Vitamin C than oranges. So many fruits (e.g., kiwi, papaya, strawberry) and vegetables (e.g., broccoli) are packed with Vitamin C."

Click here to search for any food that contains Vitamin C to boost your immune system

Spinach
Lemon, Citrus
Pineap| Lemon contains Vitamin C.
      | Vitamin C is one of the strongest vitamins to increase immunity.
Brussels Sprout
Grapefruit
Bell Pepper, Pepper
Strawberry
Blueberry
Broccoli
Sauerkraut, Cooked cabbage

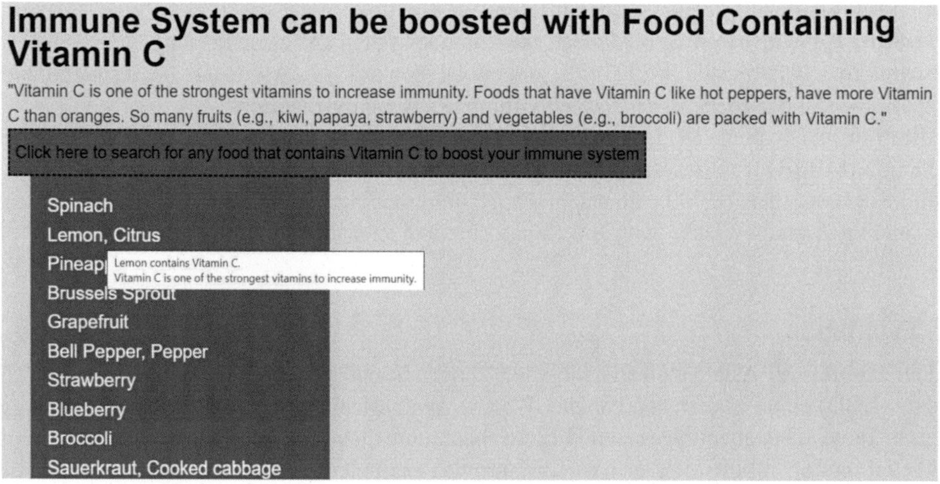

FIGURE 8.7

Foods containing **Vitamin C** are beneficial to boost the immune system.

LOV4IoT ontology catalog evaluation has been evaluated since it comprises almost 800 ontology-based IoT projects (in this chapter, we focus on food ontologies). The evaluation form and results are available.[17] The result encourages us to pursue the ontology classification work to cover more and more domains.

---

[17] http://lov4iot.appspot.com/?p=evaluation.

Table 8.2 Set of online naturopathy demonstrators: ontology, data set, ontology catalog, rule discovery, and full scenarios.

Tool name	Tool URL
LOV4IoT-Food Ontology-based IoT Project Catalog	http://lov4iot.appspot.com/?p=lov4iot-food
LOV4IoT-Food Web Service and Dumps (for Developers)	http://lov4iot.appspot.com/?p=queryFoodOntologiesWS
SLOR-Food Rule Discovery	http://linkedopenreasoning.appspot.com/?p=slor-food
Boost Immune System Full Scenarios	http://sensormeasurement.appspot.com/?p=boostImmuneSystem
Naturopathy Full Scenarios	http://sensormeasurement.appspot.com/?p=naturopathy
Naturopathy Data set	http://sensormeasurement.appspot.com/dataset/naturopathy-dataset
Naturopathy Data set on LOD	https://lod-cloud.net/dataset/Naturopathy
Naturopathy Ontology	http://sensormeasurement.appspot.com/naturopathy#

Ontology	Domain	URL	WebVOWL	ISSUE to solve	Last Tested
NaturopathyGyrard2014Health	Food	http://sensormeasurement.appspot.com/naturopathy	OK		February 2021
Kolchin2013Food	Food	http://purl.org/foodontology#	OK		February 2021
Esnaola2019FoodAgricultureF	Food	https://iesnaola.github.io/PFEEPSA/PFEEPSA/ontology/o	OK		February 2021
Celdran2014FoodPrivacySecurity	ivacy/Security/Fo	http://reclamo.inf.um.es/secoman/owl/SupermarketOntology.c	OK		February 2021
Celdran2014FoodPrivacySecurity	curityPrivacy/Sec	http://reclamo.inf.um.es/secoman/owl/LocationOntology.owl	OK		February 2021
Pernici2008Food	Food	http://web.tiscali.it/lchkl/ontology/FoodOntology.owl#	OK		February 2021
Griffiths2016Food	Food	https://raw.githubusercontent.com/FoodOntology/foodon/m	OK		February 2021
Griffiths2016Food	Food	https://raw.githubusercontent.com/FoodOntology/foodon/m	OK		February 2021
Griffiths2016Food	Food	https://raw.githubusercontent.com/FoodOntology/foodon/m	OK		February 2021
SmartProducts2012Food	Food	http://projects.kmi.open.ac.uk/smartproducts/ontologies/v2	Loaded ontology with warnings		February 2021
SmartProducts2012Food	Food	http://sensormeasurement.appspot.com/ont/food/smartProducts/recipes.owl			February 2021
SmartProducts2012Food	Food	http://projects.kmi.open.ac.uk/smartproducts/ontologies/v2	Loaded ontology with warnings		February 2021
SmartProducts2012Food	Food	http://projects.kmi.open.ac.uk/smartproducts/ontologies/v2	Loaded ontology with warnings		February 2021
GaiaRestaurant	Food	http://wise.vub.ac.be/ontologies/restaurant.owl		dropbox file	February 2021
Fruits Tropicaux	Food	http://sensormeasurement.appspot.com/ont/food/fruitsTropicaux			February 2021
Wine	Food	http://www.w3.org/TR/2003/PR-owl-guide-20031215/wine			February 2021
RestaurantMooney	Food	https://files.ifi.uzh.ch/ddis/oldweb/ddis/fileadmin/ont/nli/restaurant.owl			February 2021
Pizza	Food	http://smi-protege.stanford.edu/repos/protege/protege4/incubator/org.obolibrary.hide/trunk/un			February 2021
BBCFood2014	Food	https://www.bbc.co.uk/ontologies/fo/1.1.ttl	Loaded ontology with warnings		February 2021
Beverage	Food	http://rdfs.co/bevon/0.8/rdf	Failed to load ontology		February 2021
Haussmann2019FoodKG	Food	https://raw.githubusercontent.com/foodkg/foodkg.github.io/master/ontologies/dgo.owl			
Haussmann2019FoodKG	Food	https://raw.githubusercontent.com/foodkg/foodkg.github.io/master/ontologies/WhatToMake_Individuals.rdf			
Haussmann2019FoodKG	Food	https://raw.githubusercontent.com/foodkg/foodkg.github.io/master/ontologies/WhatToMake_FoodOn.rdf			
Haussmann2019FoodKG	Food	https://raw.githubusercontent.com/foodkg/foodkg.github.io/m	Loaded ontology with warnings		February 2021
HeflinBeer2000	Food	http://www.cs.umd.edu/projects/plus/SHOE/onts/	Failed to loa	URL documentation f	February 2021
Peroni2016Food	Food	http://www.essepuntato.it/static/tmp/0	Failed to loa	404 page not found	February 2021
BioPortalFoodOntology2015	Food	http://bioportal.bioontology.org/ontologies/FOOD_ONTOL(	Failed to loa	The page you are look	February 2021

FIGURE 8.8

Evaluation of food ontologies loadable with WebVOWL.

Challenges to automate tasks when dealing with ontologies are depicted in Fig. 8.8.[18] If the ontologies are shared online, sometimes the ontologies can be loaded by tools (e.g., ontology visualization tools or semantic web framework for ontology analysis).

---

[18] Available online for better visibility: shorturl.at/ADIQY.

Table 8.3 Ontology best practices: check list summary [68].

Rule number	Description	Difficulty
Rule 1	Finding a good ontology name	*
Rule 2	Finding a good ontology name space	**
Rule 3	Sharing your ontology online	**
Rule 4	Adding ontology metadata	**
Rule 5	Adding rdfs:label, rdfs:comment, dc:description for each concept and property	*
Rule 6	All classes start with an uppercase and properties with a lowercase.	*
Rule 7	Submitting your ontology to ontology catalogs	**
Rule 8	Reusing and linking ontologies	***
Rule 9	Dereferenceable URI copy paste the name space URL of your ontology in a web browser to get the code	**
Rule 10	Checking syntax validator	*
Rule 11	Adding ontology documentation	*
Rule 12	Adding ontology visualization	*
Rule 13	Improving ontology design	***
Rule 14	Improving dereferencing URI and content negotiation	***
Rule 15	Ontology can be loaded with ontology editors (e.g., Protege)	**
Rule 16	Registering your ontology on prefix catalogs	*

8.4 Conclusion and future work

We designed the naturopathy knowledge graph (ontology and data set) employed within a recommender system to boost the immune system, called the Knowledge-based Immune System Suggestion (KISS). The naturopathy knowledge graph acquires knowledge from more than 50 ontology-based food projects that we classify within the LOV4IoT-Food ontology catalog. The naturopathy data set is referenced on the Linked Open data (LOD) cloud. The LOV4IoT-Food ontology catalog supports researchers with: (1) the Systematic Literature Survey, which is a time-consuming task and requires an eagerness to learn and investigate existing projects, (2) FAIR principles to encourage researchers to share their reproducible experiments by publishing online ontologies, data sets, rules, etc. The set of ontology codes available online can be automatically processed; if the ontology code is not available, the scientific publications describing the food ontologies are semiautomatically processed with Natural Language Processing (NLP) techniques to feed the food reasoning engine to build the naturopathy recommender system.

Short-term challenges: LOV4IoT is relevant for the IoT community. The results are encouraging to update the data set with additional domains and ontologies. LOV4IoT leads to the AIOTI (The Alliance for the Internet of Things Innovation) IoT ontology landscape survey form[19] and analysis

---

[19] https://ec.europa.eu/eusurvey/runner/OntologyLandscapeTemplate.

result,[20] executed by the Standard WG—Semantic Interoperability Expert Group. It aims to help industrial practitioners and nonexperts to answer those questions: Which ontologies are relevant in a certain domain? Where to find them? How to choose the most suitable? Who is maintaining and taking care of their evolution?

Mid-term challenges: Automatic knowledge extraction from ontologies and scientific publication describing the ontology purpose is challenging, as highlighted in our AI4EU Knowledge Extraction for the Web of Things (KE4WoT) Challenge. The challenge encourages to reuse the expertise designed by domain experts and make the domain knowledge usable, interoperable, and integrated by machines. We released the set of ontologies, as dumps, web services, tutorials, and make them available for the challenge.

Long-term challenges: To improve the veracity and the evaluation of the recommender, involving domain experts such as dietitians, nutritionists, naturopaths, Traditional Chine Medicine doctors, Ayurvedic doctors, etc. would enhance the naturopathy knowledge graph, by proving more of the facts.

The recommender system can be extended by considering additional domains such as aromatherapy with essential oils, Bach Flower Remedies, mindfulness, and activities such as Yoga, etc. Emphasis on the emotional aspect will be done (e.g., fear, pessimism, sadness) since it impacts the immune system. We investigated psycho-physiology [80] research field to prove such facts. Collecting emotion ontologies has been investigated within the ACCRA project (robots for healthy ageing that provide well-being applications) [81].

8.5 Disclaimer

It is obvious that the World Health Organization (WHO) advice (e.g., for COVID-19) must be followed as well.

Appendix 8.A Demonstrators

Figs. 8.9–8.21 have been described in the scenarios from Section 8.3.3.

---

[20] https://bit.ly/3fRpQUU.

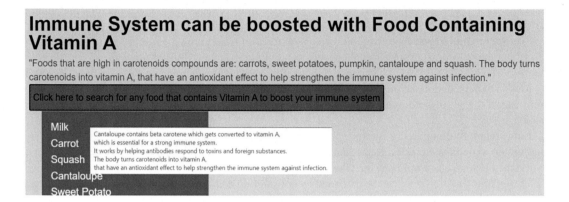

FIGURE 8.9

Food containing **Vitamin A** are beneficial to boost the immune system.

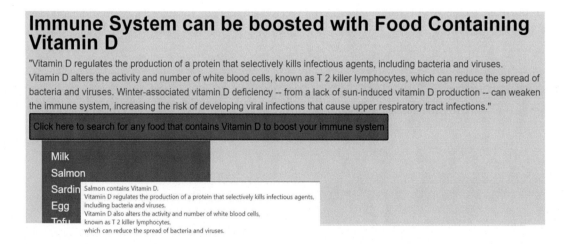

FIGURE 8.10

Food containing **Vitamin D** are beneficial to boost the immune system.

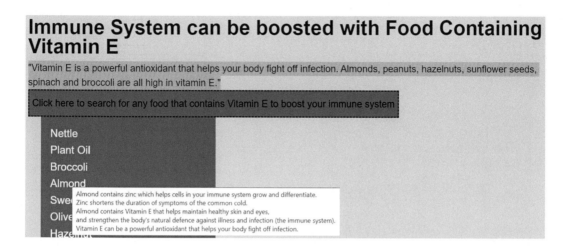

FIGURE 8.11

Food containing **Vitamin E** are beneficial to boost the immune system.

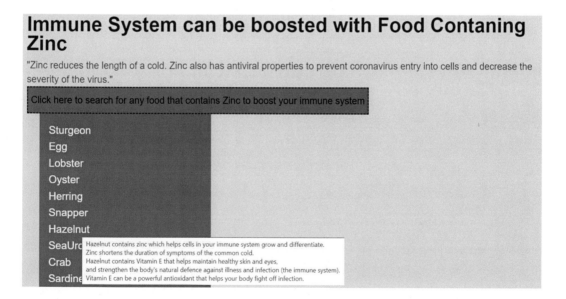

FIGURE 8.12

Food containing **zinc** are beneficial to boost the immune system.

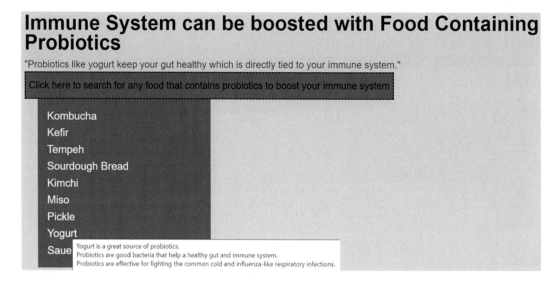

FIGURE 8.13

Food containing **probiotics** are beneficial to boost the immune system.

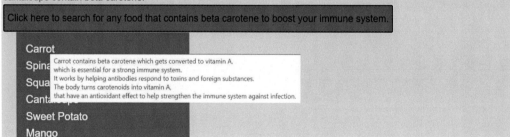

FIGURE 8.14

Food containing **beta carotene** are beneficial to boost the immune system.

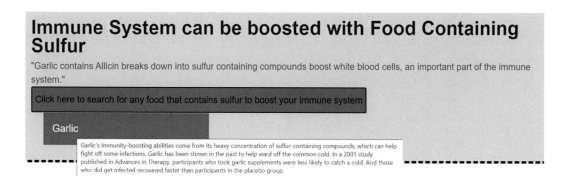

FIGURE 8.15

Food containing **sulfur** are beneficial to boost the immune system.

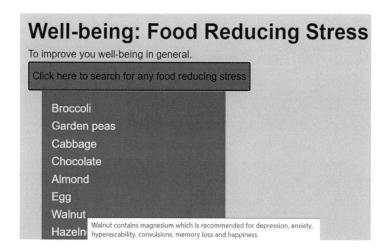

FIGURE 8.16

Food reducing stress.

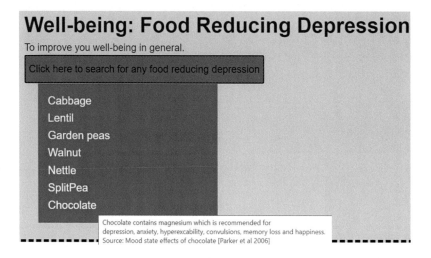

FIGURE 8.17

Food reducing depression.

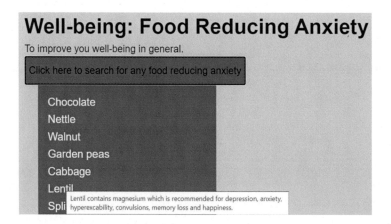

FIGURE 8.18

Food suggestion to reduce anxiety.

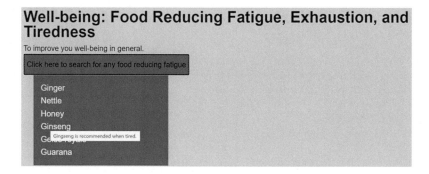

FIGURE 8.19

Food suggestion to reduce fatigue.

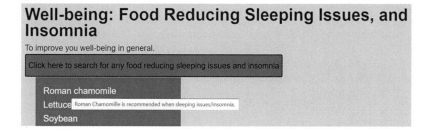

FIGURE 8.20

Food suggestion to encourage better sleep.

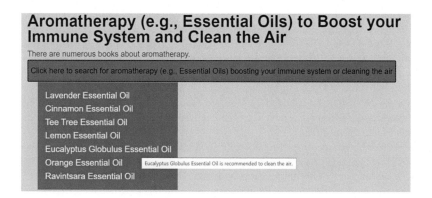

FIGURE 8.21

Aromatherapy (e.g., essential oils) to boost the immune system and clean the air.

References

[1] G. Popovski, B. Korousic-Seljak, T. Eftimov, FoodOntoMap: linking food concepts across different food ontologies, in: KEOD, 2019.

[2] M. Ritchell, An Elegant Defense: The Extraordinary New Science of the Immune System: A Tale in Four Lives, 2019.

[3] B. Raj Kumar, K. Latha, DFRS: diet food recommendation system for diabetic patients based on ontology, International Journal of Applied Engineering Research (2015).

[4] D. Chopra, Supergenes, Self, 2017.

[5] D. Jannach, M. Zanker, A. Felfernig, G. Friedrich, Recommender Systems: An Introduction, Cambridge University Press, 2010.

[6] C.C. Aggarwal, et al., Recommender Systems, Springer, 2016.

[7] I. Portugal, et al., The use of machine learning algorithms in recommender systems: a systematic review, Expert Systems with Applications (2018) (IF: 3.768 in 2017).

[8] J. Lu, D. Wu, M. Mao, W. Wang, G. Zhang, Recommender system application developments: a survey, Decision Support Systems (2015) (IF: 3.565 in 2017).

[9] J. Bobadilla, F. Ortega, A. Hernando, A. Gutiérrez, Recommender systems survey, Knowledge-Based Systems (2013) (IF: 4.396 in 2017).

[10] G. Adomavicius, A. Tuzhilin, Toward the next generation of recommender systems: a survey of the state-of-the-art and possible extensions, IEEE Transactions on Knowledge and Data Engineering (2005) (IF: 2.775 in 2017).

[11] R. Nouh, et al., A smart recommender based on hybrid learning methods for personal well-being services, MDPI Sensors (2019) (IF: 3.031 in 2018).

[12] M. Boulos, A. Yassine, S. Shirmohammadi, C. Namahoot, M. Brückner, Towards an internet of food: food ontologies for the internet of things, MDPI Future Internet Journal (2015).

[13] Coming to terms with fair ontologies.

[14] A. Gyrard, A. Kung, SAREF4EHAW-compliant knowledge discovery and reasoning for IoT-based preventive healthcare and well-being, in: Semantic Models in IoT and e-Health Applications, Elsevier, 2022.

[15] A. Gyrard, U. Jaimini, M. Gaur, S. Shekarpour, K. Thirunarayan, A. Sheth, Reasoning over personalized healthcare knowledge graph: a case study of patients with allergies and symptoms, in: Semantic Models in IoT and e-Health Applications, Elsevier, 2022.

[16] C. Bizer, J. Lehmann, G. Kobilarov, S. Auer, C. Becker, R. Cyganiak, S. Hellmann, DBpedia – a crystallization point for the web of data, Journal of Web Semantics (2009).

[17] E.J. Griffiths, D.M. Dooley, P.L. Buttigieg, R. Hoehndorf, F.S. Brinkman, W.W. Hsiao, FoodON: a global farm-to-fork food ontology. The development of a universal food vocabulary, in: ICBO/BioCreative, 2016.

[18] M. Kolchin, D. Zamula, Food product ontology: initial implementation of a vocabulary for describing food products, in: Conference of Open Innovations Association FRUCT, 2013.

[19] C. Caracciolo, A. Stellato, S. Rajbahndari, A. Morshed, G. Johannsen, Y. Jaques, J. Keizer, Thesaurus maintenance, alignment and publication as linked data: the AGROVOC use case, International Journal of Metadata, Semantics and Ontologies (2012).

[20] D. Çelik, FoodWiki: ontology-driven mobile safe food consumption system, Scientific World (2015) (IF: 1.219 in 2013).

[21] J. Cantais, D. Dominguez, V. Gigante, L. Laera, V. Tamma, An example of food ontology for diabetes control, in: Workshop on Ontology Patterns for the Semantic Web, International Semantic Web Conference (ISWC, A-Rank Conference), 2005.

[22] C.-S. Lee, M.-H. Wang, H.-C. Li, W.-H. Chen, Intelligent ontological agent for diabetic food recommendation, in: International Conference on Fuzzy Systems (World Congress on Computational Intelligence), IEEE, 2008.

[23] Oshani Seneviratne, Jonathan Harris, Ching-Hua Chen, Deborah L. McGuinness, Personal health knowledge graph for clinically relevant diet recommendations, arXiv preprint, arXiv:2110.10131.

[24] A. Gyrard, M. Gaur, K. Thirunarayan, A. Sheth, S. Shekarpour, Personalized health knowledge graph, in: 1st Workshop on Contextualized Knowledge Graph (CKG) co-located with International Semantic Web Conference (ISWC), 8–12 October 2018, Monterey, USA, 2018, pp. 8–12.

[25] V. Subramaniyaswamy, G. Manogaran, R. Logesh, V. Vijayakumar, N. Chilamkurti, D. Malathi, N. Senthilselvan, An ontology-driven personalized food recommendation in IoT-based healthcare system, Journal of Supercomputing (2019).

[26] S. Norouzi, M. Nematy, H. Zabolinezhad, S. Sistani, K. Etminani, Food recommender systems for diabetic patients: a narrative review, Reviews in Clinical Medicine (2017).

[27] S. Norouzi, A.K. Ghalibaf, S. Sistani, V. Banazadeh, F. Keykhaei, P. Zareishargh, F. Amiri, M. Nematy, K. Etminani, A mobile application for managing diabetic patients' nutrition: a food recommender system, Archives of Iranian Medicine (2018).

[28] S. Karim, U.U. Shaikh, Q. Rajput, Ontology-based personalized dietary recommendation for travelers, in: Southern Association for Information Systems Conference (SAIS), 2015.

[29] M. Ge, F. Ricci, D. Massimo, Health-aware food recommender system, in: Conference on Recommender Systems (RecSys, B-Ranked Conference), ACM, 2015.

[30] M. Ge, M. Elahi, I. Fernaández-Tobías, F. Ricci, D. Massimo, Using tags and latent factors in a food recommender system, in: International Conference on Digital Health, 2015.

[31] A. Al-Nazer, T. Helmy, M. Al-Mulhem, User's profile ontology-based semantic framework for personalized food and nutrition recommendation, Procedia Computer Science (2014).

[32] V. Espín, M.V. Hurtado, M. Noguera, K. Benghazi, Semantic-based recommendation of nutrition diets for the elderly from agroalimentary thesauri, in: International Conference on Flexible Query Answering Systems (B-Ranked Conference), Springer, 2013.

[33] C. Snae, M. Bruckner, FOODS: a food-oriented ontology-driven system, in: International Conference on Digital Ecosystems and Technologies (DEST C-Ranked Conference), IEEE, 2008.

[34] G. Popovski, S. Kochev, B. Korousic-Seljak, T. Eftimov, FoodIE: a rule-based named-entity recognition method for food information extraction, in: ICPRAM, 2019.

[35] M.D. Wilkinson, M. Dumontier, I.J. Aalbersberg, G. Appleton, M. Axton, A. Baak, N. Blomberg, J.-W. Boiten, L.B. da Silva Santos, P.E. Bourne, et al., The FAIR guiding principles for scientific DataManagement and stewardship, Scientific Data (2016) (IF: 5.305 in 2017).

[36] A. Gyrard, A. Sheth, IAMHAPPY: Towards An IoT Knowledge-Based Cross-Domain Well-Being Recommendation System for Everyday Happiness, 2020.

[37] T.R. Gruber, Toward principles for the design of ontologies used for knowledge sharing?, International Journal of Human-Computer Studies (1995).

[38] A. Gyrard, C. Bonnet, K. Boudaoud, M. Serrano, LOV4IoT: a second life for ontology-based domain knowledge to build semantic web of things applications, in: IEEE International Conference on Future Internet of Things and Cloud, 2016.

[39] G. Rizzo, F. Tomassetti, A. Vetro, L. Ardito, M. Torchiano, M. Morisio, R. Troncy, Semantic enrichment for recommendation of primary studies in a systematic literature review, Digital Scholarship in the Humanities (2017).

[40] D.M. Dooley, E.J. Griffiths, G.S. Gosal, P.L. Buttigieg, R. Hoehndorf, M.C. Lange, L.M. Schriml, F.S. Brinkman, W.W. Hsiao, Foodon: a harmonized food ontology to increase global food traceability, quality control and data integration, npj Science of Food (2018).

[41] S. Peroni, G. Lodi, L. Asprino, A. Gangemi, V. Presutti, FOOD: food in open data, in: International Semantic Web Conference (A-rank conference), Springer, 2016.

[42] A.H. Celdran, F.J.G. Clemente, M.G. Pérez, G.M. Pérez, SeCoMan: a semantic-aware policy framework for developing privacy-preserving and context-aware smart applications, IEEE Systems Journal (2016) (IF: 4.337 in 2017).

[43] A. Gyrard, Designing Cross-Domain Semantic Web of Things Applications, Ph.D. thesis, Telecom ParisTech, Eurecom, 2015.

[44] M. Sabou, J. Kantorovitch, A. Nikolov, A. Tokmakoff, X. Zhou, E. Motta, Position paper on realizing smart products: challenges for semantic web technologies, in: CEUR Workshop Proceedings, 2009.

[45] F. Calore, D. Lombardi, E. Mussi, P. Plebani, B. Pernici, Retrieving substitute services using semantic annotations: a foodshop case study, in: International Conference on Business Process Management, Springer, 2007.

[46] V. Espín, M.V. Hurtado, M. Noguera, Nutrition for elder care: a nutritional semantic recommender system for the elderly, Expert Systems (2016) (IF: 1.505 in 2017).

[47] Y.-L. Chi, T.-Y. Chen, W.-T. Tsai, A chronic disease dietary consultation system using OWL-based ontologies and semantic rules, Journal of Biomedical Informatics (2015) (IF: 2.882 in 2017).

[48] T. Pizzuti, G. Mirabelli, M.A. Sanz-Bobi, F. Goméz-Gonzaléz, Food track & trace ontology for helping the food traceability control, Journal of Food Engineering (2014).

[49] T. Pizzuti, G. Mirabelli, Ftto: an example of food ontology for traceability purpose, in: International Conference on Intelligent Data Acquisition and Advanced Computing Systems (IDAACS), IEEE, 2013.

[50] C.-J. Su, Y.-A. Chen, C.-W. Chih, Personalized ubiquitous diet plan service based on ontology and web services, International Journal of Information and Education Technology (2013).

[51] C.-J. Su, C.-Y. Chiang, M.-C. Chih, Ontological knowledge engine and health screening data enabled Ubiquitous Personalized Physical Fitness (UFIT), MDPI Sensors (2014) (IF: 3.031 in 2018).

[52] P. Tumnark, F.A.d. Conceição, J.P. Vilas-Boas, L. Oliveira, P. Cardoso, J. Cabral, N. Santibutr, Ontology-based personalized dietary recommendation for weightlifting, in: International Workshop on Computer Science in Sports, Atlantis Press, 2013.

[53] R.-C. Chen, Y.-D. Lin, C.-M. Tsai, H. Jiang, Constructing a diet recommendation system based on fuzzy rules and knapsack method, in: International Conference on Industrial, Engineering and Other Applications of Applied Intelligent Systems, Springer, 2013.

[54] P. Curiel, A. Lago, An infrastructure to enable lightweight context-awareness for mobile users, MDPI Sensors (2013) (IF: 3.031 in 2018).

[55] L. Miao, J. Chun, J.C. Yoshiyuki Higuchi, Ontology-based user preferences Bayesian model for personalized recommendation, Journal of Computational Information Systems (2013) (IF not found).

[56] N. Suksom, M. Buranarach, Y.M. Thein, T. Supnithi, P. Netisopakul, A knowledge-based framework for development of personalized food recommender system, in: International Conference on Knowledge, Information and Creativity Support Systems, 2010 (Conference ranking not found).

[57] U. Yasavur, R. Amini, C. Lisetti, N. Rishe, Ontology-based named entity recognizer for behavioral health, in: International FLAIRS Conference, 2013.

[58] G. Vadivu, S.W. Hopper, Semantic linking and querying of natural food, chemicals and diseases, International Journal of Computer Applications (2010) (IF: 3.12 in 2017).

[59] D.H. Fudholi, N. Maneerat, R. Varakulsiripunth, Y. Kato, Application of protégé, SWRL and SQWRL in fuzzy ontology-based menu recommendation, in: International Symposium on Intelligent Signal Processing and Communication Systems, IEEE, 2009 (Conference ranking not found).

[60] D.H. Fudholi, N. Maneerat, R. Varakulsiripunth, Ontology-based daily menu assistance system, in: International Conference on Electrical Engineering/Electronics, Computer, Telecommunications and Information Technology, IEEE, 2009.

[61] H. Gu, D. Wang, A content-aware fridge based on RFID in smart home for home-healthcare, in: International Conference on Advanced Communication Technology, IEEE, 2009 (Conference ranking not found).

[62] A. Sachinopoulou, J. Leppanen, H. Kaijanranta, J. Lahteenmaki, Ontology-based approach for managing personal health and wellness information, in: International Conference of the Engineering in Medicine and Biology Society, IEEE, 2007.

[63] H.-C. Li, W.-M. Ko, Automated food ontology construction mechanism for diabetes diet care, in: International Conference on Machine Learning and Cybernetics (ICMLC, B-Rank Conference), IEEE, 2007.

[64] Y.-H. Chen, T.-h. Huang, D.C. Hsu, J.Y.-j. Hsu, ColorCocktail: an Ontology-Based Recommender System, 2006.

[65] R. Ribeiro, F. Batista, J.P. Pardal, N.J. Mamede, H.S. Pinto, Cooking an ontology, in: International Conference on Artificial Intelligence: Methodology, Systems, and Applications, Springer, 2006 (Conference ranking not found).

[66] Open food facts project, https://world.openfoodfacts.org/who-we-are.

[67] S. Lohmann, V. Link, E. Marbach, S. Negru, WebVOWL: web-based visualization of ontologies, in: Knowledge Engineering and Knowledge Management, Springer, 2014.

[68] A. Gyrard, G. Atemezing, M. Serrano, PerfectO: An Online Toolkit for Improving Quality, Accessibility, and Classification of Domain-Based Ontologies, Springer, 2021.

[69] J. Seignalet, L'alimentation ou la troisième médecine, Editions du Rocher, 2012.

[70] L. Sophie, Les aliments qui guérissent, Édition Leduc, 2007.

[71] R. Burke, Knowledge-based recommender systems, in: Encyclopedia of Library and Information Systems, 2000.

[72] J. Wardle, J. Sarris, Clinical Naturopathy: an Evidence-Based Guide to Practice, Elsevier Health Sciences, 2014.

[73] M. Aćimović, Essential oils: inhalation aromatherapy – a comprehensive review, Journal of Agronomy, Technology and Engineering Management 4 (2) (2021) 547–557.

[74] D. Festy, Ma bible des huiles essentielles guide complet d'aromatherapie, Leduc, 2018.

[75] J. Kabat-Zinn, Au cœur de la tourmente, la pleine conscience, De Boeck Supérieur, 2016.

[76] J. Kabat-Zinn, Full Catastrophe Living, Revised Edition: How to Cope with Stress, Pain and Illness Using Mindfulness Meditation, Hachette, UK, 2013.

[77] R.J. Davidson, J. Kabat-Zinn, J. Schumacher, M. Rosenkranz, D. Muller, S.F. Santorelli, F. Urbanowski, A. Harrington, K. Bonus, J.F. Sheridan, Alterations in brain and immune function produced by mindfulness meditation, Psychosomatic Medicine (2003).

[78] J. Kabat-Zinn, R. Davidson, Z. Houshmand, L'esprit est son propre médecin: Le pouvoir de guérison de la méditation, Guy Saint-Jean Editeur, 2014.

[79] A. Gyrard, M. Serrano, G. Atemezing, Semantic web methodologies, best practices and ontology engineering applied to Internet of things, in: IEEE World Forum on Internet of Things, 2015.

[80] F. Morange-Majoux, Psycho-physiologie, 2eme edition, Dunod, 2017.

[81] ACCRA, Knowledge Engineering Framework for IoT Robotics Applied to Smart Healthcare and Emotional Well-Being, 2021.

SAREF4EHAW-compliant knowledge discovery and reasoning for IoT-based preventive health and well-being

9

IoT-based preventive health and well-being knowledge discovery and reasoning

Amelie Gyrard and Antonio Kung

Trialog, Paris, France

Contents

9.1 Introduction

Assisting people to stay independent at home and to decrease hospital costs but also social isolation for the (elderly) people is getting more attention. e-Health is a term used among others such as telehealth, m-Health, telemedicine, digital health, health IT [1]. The integration of heterogeneous technologies and devices require an interoperable solution to describe devices and data exchanged for preventive health and well-being.

Preventive health can be achieved by considering four main topics (see Fig. 9.1): (1) physical activity, (2) healthy diet, (3) emotional management with positive psychology [2] to manage our emotions to improve stress management, and (4) sleep management. The environment can affect health as well; it is called epigenetic [3], and has been demonstrated with twins: twins share the same DNA and a different environment lifestyle can change health. **Wellness** defines a healthy lifestyle, by taking into consideration mind, body, and spirit as a whole for an overall feeling of well-being. **Well-being** is the state of being comfortable, healthy, or happy.

Physical activities: Activities such as Yoga [4], Taichi [5], Qigong [6] benefits have been demonstrated. Tai Chi helps with depression, pulmonary disease, balance disorders, Parkinson's disease, cardiovascular health, osteoporosis, chronic pain, and cancer as it is scientifically proven by researchers at the Osher Center for Integrative Medicine at the Harvard Medical School [5]. Qigong improves well-being and reduces anxiety, stress, and depression [51]. Practicing sports will help reduce stress, enhance better sleep, etc. A sport is more and more recommended for a healthy lifestyle, to reduce stress, improve well-being, etc. More and more companies are creating new devices to support drills, such as smartwatches (e.g., Fitbit), etc. [7].

Affective Science (e.g., emotion): Martin Seligman created the Positive Psychology Center [2] at the University of Pennsylvania's Department of Psychology. He focuses on positive psychology and encourages well-being rather than reducing ill-being for better physical and mental health. Rosalind Picard [8] founded the Affective Computing Research Group at the MIT Media Lab and is the co-founder of the startups, Affectiva (facial and vocal emotion recognition) and Empatica (wearable to detect epilepsy crisis). Standford's research team[1] is also researching affective science. Daniel Goleman's research [9] focuses on emotional intelligence. There is also a new trend to improve well-being at work with "Happiness Manager," "Chief Happiness Officer," etc.

Healthy Diet: Seignalet et al. [10] emphasize that healthy food is considered as a third medicine as explained in his book.

Sleep and Relaxation: Benefits of meditation have already been proven with Davidson's affective neuroscience's research [11,12], along with Paul Eckman, Matthieu Ricard [13,14], and Allan Wallace [15,16].

---

[1] https://psychology.stanford.edu/research/department-areas/affective-science.

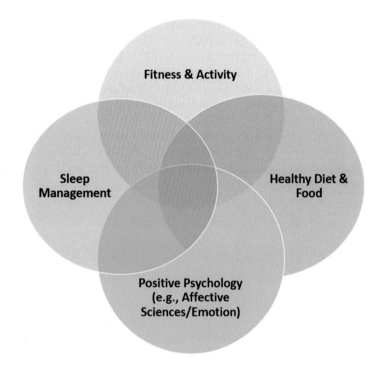

FIGURE 9.1

Healthy Lifestyle Venn Diagram: (1) physical activity, (2) healthy diet, (3) emotional management with positive psychology, and (4) sleep management.

To achieve preventive health, IoT technologies can be employed. **IoT for a healthcare survey** [17] provides several IoT-healthcare applications: (1) glucose level sensing, (2) ECG monitoring, (3) blood pressure monitoring, (4) body temperature monitoring, (5) oxygen saturation monitoring, (6) rehabilitation systems, (7) medication management, (8) wheelchair management, (9) imminent healthcare solutions, and (10) healthcare solutions using smartphones. Internet of things for Healthcare [17] does not cover enough fitness benefits.

We address the following **research questions (RQ)**:

- RQ1: What are the limitations of the existing ontology-based health IoT projects? Can we reuse the domain expertise designed in past projects?
- RQ2: What are the sensors relevant to the health domain? Are there standardized sensor dictionaries for the health domain? Are there ontology standards for the health domain?
- RQ3: What are the rules and reasoning mechanisms to interpret health sensor data to help developers faster design their IoT-based health applications?
- RQ4: How to prove the veracity of the reasoning engine?

The main **contributions** of this paper are:

- C1: A deep investigation of the ontology-based health projects shared though the LOV4IoT-Health ontology catalog; it addresses RQ1 in Section 9.2.1.
- C2: Alignment of the health sensor dictionary with the ETSI (European Telecommunications Standards Institute) SmartM2M SAREF ontology and its extensions: SAREF for eHealth ageing Well (SAREF4EHAW), and SAREF for Wearables (SAREF4WEAR); it addresses RQ2 in Section 9.3.1.
- C3: The reasoner retrieves knowledge to design health use cases; it addresses RQ3 in Section 9.3.4.
- C4: Provenance to keep track of the knowledge designed by domain experts is explicitly encoded in the ontology catalog and rule data sets to prove the veracity of the reasoning engine; it addresses RQ4 in Section 9.3.5.
- C5: **Compliant with Standardization:** Results are promoted within the ISO/IEC 21823-3 IoT semantic interoperability [18], and the AIOTI WG Standardization[2] Semantic Interoperability Expert Group [19,20] where the reasoner is taken as a baseline [21]. SAREF designers are also involved within AIOTI WG. Furthermore, semantic web technologies (RDF, RDFS, OWL, SPARQL) employed are supported by the W3C standard.

Structure of the paper: Related work on health ontology catalogs, standards such as ETSI Smart M2M SAREF4EHAW, and health knowledge graphs are described in Section 9.2. The sensor dictionary for health, the knowledge discovery, and its reasoner are described in Section 9.3. ETSI SmartM2M SAREF-compliant health scenarios are introduced in Section 9.4; the classification of the source of the knowledge to prove the veracity of our scenarios is included. Those results have impact in projects such as the ACCRA European project [22]. The ETSI SmartM2M SAREF limitations are summarized in Section 9.5. The paper concludes and envisions future work in Section 9.6.

9.2 Related work: ontology-based IoT project catalog for health

This section introduces related work on health ontology catalogs in Section 9.2.1, standards such as ISO or ETSI Smart M2M SAREF4EHAW ontology in Section 9.2.2, and health knowledge graphs in Section 9.2.3.

9.2.1 Ontology-based IoT project catalog for health with LOV4IoT-health

We collected and analyzed the state of the art to enrich our ontology-based IoT health project knowledge base (LOV4IoT-Health is summarized in Table 9.5). Other ontology catalogs such as BioPortal [23] and Linked Open Vocabularies (LOV) [24] are not focused on IoT-based health applications. LOV4IoT-Health is more focused on the ontologies and related scientific publications, sensors, and reasoning mechanisms employed as detailed in Table 9.5, which are not covered by other ontology catalogs. Our survey is the result of a continuous enrichment of the LOV4IoT ontology catalog [25] since 2012, dealing with more and more expertise and synonyms (e.g., emotion, affective science, health, well-being, fitness, etc.). We provide tools to support the reuse of the survey outcome (e.g., dump of ontology code, web services, and web-based ontology catalog) and release them for the AI4EU Knowledge Extraction

---

[2] https://aioti.eu/aioti-wg03-reports-on-iot-standards/.

for the Web of Things Challenge.[3] Meanwhile, we are aware of Systematic Literature Review (SLR) guidelines such as [26]. The survey on IoT-based health ontologies is also reused within the reasoning discovery explained hereafter; both are frequently updated (see Table 9.2 for URLs). Manual extraction and semiautomatic analysis [27,28] have been done to extract knowledge (see Section 9.3.3).

9.2.2 Standards: ISO and ETSI SmartM2M

We have investigated standards such as ISO 13606-5:2010 Health Informatics – Electronic Health Record communication standards, and ETSI SmartM2M SAREF4EHAW for e-Health/Aging-well ontology.

9.2.2.1 ETSI SmartM2M SAREF4EHAW for e-Health/Aging-well

Among all ontologies collected, analyzed, and summarized in Table 9.5 and Section 9.2.1, we selected the SAREF ontology since it is supported by the ETSI SmartM2M standard and the European Commission. Other ontologies are reused to extract knowledge from domain experts as explained in Section 9.3. The **ETSI Smart M2M SAREF4EHAW ontology** [29] aims to cover the following use cases: (1) elderly at home monitoring and support, (2) monitoring and support of healthy lifestyle for citizens, (3) Early Warning System (EWS) and cardiovascular accidents detection. The use cases are classified into the following categories: (1) daily activity monitoring, (2) integrated care for older adults under chronic conditions, (3) monitoring assisted persons outside the home and controlling risky situations, (4) emergency trigger, (5) exercise promotion for fall prevention and physical activeness, (6) cognitive simulation for mental decline prevention, (7) prevention of social isolation, (8) comfort and safety at home, and (9) support for transportation and mobility.

SAREF deliverables reviewed standards (IEEE, ETSI, SNOMED International, OneM2M), Alliances (AIOTI), IoT platforms, and European projects and initiatives, etc.

SAREF4EHAW investigated the following ontologies: (1) WSNs/measurement ontologies: OGC (Open Geospatial Consortium) Observations and Measurements (O&M), Sensor Model Language (SensorML), Semantic Sensor Web (SWE): W3C and OGC SOSA (Sensing, Observation, Sampling, and Actuation), and W3C SSN (Semantic Sensor Network). NASA QUDT (Quantities, Units, Dimensions, and Types). (2) e-Health/Ageing-well domain main ontologies: ISO/IEEE 11073 Personal Health Device (PHD) standards, ETSI SmartBAN Reference Data Model and associated modular ontologies, ETSI SmartM2M SAREF4EHAW, FHIR RDF (Resource Description Framework), FIESTA-IoT ontology to support the federation of testbeds, Bluetooth LE profiles for medical devices proposed by zontinua, MIMU-Wear (Magnetic and Inertial Measurement Units) ontology, and Active and Healthy Aging (AHA) platform wearables' device ontology. SAREF has been mapped with oneM2M base ontology in 2017.

SAREF ETSI TR 103 509 [29] defines 43 final ontological requirements and 59 additional service level assumptions of the e-Health/Aging-well domain (use cases included) are presented. For instance, the ontology will describe concepts to describe ECG devices.

---

[3] https://www.ai4eu.eu/ke4wot.

9.2.2.2 ISO 13606-5:2010 health informatics – electronic health record communication standards

ISO 13606-5:2010 Health Informatics – Electronic Health Record communication standard[4] defines an architecture for exchanging an Electronic Health Record (EHR) describing the patient's health status and to ease communication between EHR systems (e.g., clinicians applications, decision support systems).

9.2.3 Health knowledge graphs

We review health KGs and categorize them based on their interpolation with statistical and learning-based approaches in [30]. We explain the challenges in integrating heterogeneous models to maintain a PHKG utilizing various modalities. In [30], the "Summary of (Health) Knowledge Graph (KG) using or not Machine Learning (ML)" table provides references and summaries of the KG scientific articles. It also cites nonhealth KG such as Schema.org (designed by significant internet companies) to understand how KGs work. This literature analysis shows that there is no reasoner inferring abstraction from sensor data sets and later combining them with multidisciplinary external domain knowledge (e.g., health ontologies). Only two papers integrate KG and ML technologies for health: Shi et al. [31] and Rotmensch et al. [32]. Le Phuoc et al. [33] is the only work addressing IoT data, but not for the health use case. None of those KGs discusses the challenges mentioned above, and hence none is appropriate for use.

9.3 Knowledge discovery and reasoning for preventive health and well-being

The sensor dictionary for health is described in Section 9.3.1, ontology visualization in Section 9.3.2, knowledge extraction from ontologies and scientific publications in Section 9.3.3, and the health reasoner in Section 9.3.4.

9.3.1 ETSI SmartM2M SAREF-compliant semantic sensor health dictionary

Sensor Dictionary for Health and Well-Being: We built a sensor dictionary for the health domain compliant with standards such as ETSI SmartM2M SAREF. We designed a pattern to classify sensors for the health domain: for each sensor, we provide the produced measurements and the associated unit; we also deal with synonyms. Furthermore, we referenced for each sensor the source of knowledge using it (e.g., past projects referenced within the ontology-based IoT project catalog, see Section 9.2.1 and Table 9.5), and reasoning mechanisms to interpret health sensor data (see the rule discovery project in Section 9.3.4).

The ETSI SmartM2M SAREF-compliant sensor health dictionary is provided as a web service: http://linkedopenreasoning.appspot.com/saref/subclassOf/?sarefCoreClassName=Sensor&m3 ApplicationDomain=HealthM2MDevice&format=xml or can be hidden within Graphical User Interface (GUI) (see Section 9.3.4).

---

[4] http://www.en13606.org/information.html.

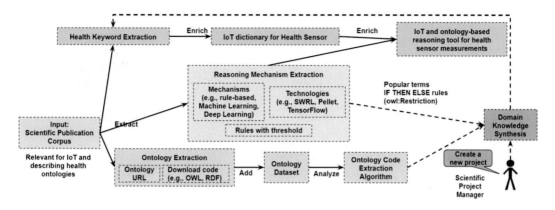

FIGURE 9.2

Knowledge extraction from health ontologies.

ETSI SmartM2M SAREF-Compliant Semantic Annotation: The sensor dictionary is compliant with the terms employed by SAREF when possible, as illustrated in Table 9.3. The limitations of ETSI SmartM2M SAREF are summarized in Section 9.5. Sensor health data sets (e.g., JSON, XML) follow the SenML format[5] to represent sensor measurement, its value, its unit, and its timestamp. Available demonstrators are providing code examples (see Table 9.2). A simple rule repository for the semantic annotation is already available such as if "t" or "temp" or "temperature" is used and located on the person; it probably is a body temperature and will be annotated following the dictionary mentioned above, which is implemented as an ontology.

9.3.2 Ontology visualization for preventive health and well being

To ease a quick discovery of the ontology content, the WebVOWL tool [34] helps to visualize ontologies available within LOV4IoT-Health (introduced in Section 9.2.1 and Table 9.5) and that are preselected if they can be loaded without errors. The set of issues while loading the ontology is described in [35] and shared to the WebVOWL team.

9.3.3 Semi-automatic knowledge extraction from preventive health and well being ontologies

We semiautomatically analyzed the ontologies with Natural Language Processing techniques (NLP): (1) ontology code and (2) scientific publications describing the ontologies.

From a set of publications on IoT health ontology-based projects, we extract knowledge such as which sensors are employed, which reasoning mechanisms are used to interpret sensor measurements, and is there an ontology available and reusable, etc. (as depicted in Fig. 9.2).

---

[5] https://tools.ietf.org/html/rfc8428.

Applicativ	Concept	SAREF	DemaCar	DemaCare	DemaCare	Yao2013	Paganelli	Manate2	OpenEHR	Registry	Mazuel20	Yao2013	Roose201	Jovic2011	Saref	Activag	Amies20
Healthcare	pressure, b	saref-cc	Not found	Not found	Not found	blood_pre	systolicblo	Not found	Not found	a_bloodty	Not found	bloodpres	Not found	Not found	Error	Error	systolicbl
Healthcare	pulse, hear	Not fou	Not found	Not found	heartrate	heart_fail	heartratefr	pulseread	Not found	patientdi	Not found	pulserate	Not found	Not found	Error	Error	restingpuls
Healthcare	GlucoseBl	Not fou	Not found	Not found	Not found	red_blood	glucosebl	Not found	Not found	Not found	Not found	Not found	Not found	Not found	Error	Error	glucosebl
Healthcare	SpO2 oxyg	Not fou	Not found	Not found	Not found	Not found	respirator	Not found	Not found	spo2	respiratory	Not found	Not found	Error	Error	respirator
Healthcare	Person Pat	patient	patient	patient	person	person	patient	patient	Not found	Person pa	patient	patient	Not found	Not found	Error	Error	Not found
Healthcare	Humidity	saref-cc	Not found	Not found	Not found	Not found	relativehu	Not found	Not found	Not found	Not found	humidity	Not found	Not found	Error	Error	Not found
Healthcare	Temperatu	saref-cc	Not found	Not found	skintempe	skin_chan	temperatu	Not found	Not found	Not found	temperat	temperat	Not found	Not found	Error	Error	Not found
Healthcare	skin condu	Not found	Not found	Not found	skincondu	skin_chan	Not found	Not found	Not found	Not found	Not found	Not found	Not found	Not found	Error	Error	Not found
Healthcare	ECG	Not fou	Not found	Not found	Not found	Not found	Not found	Not found	Not found	tachycard	ecg		Not found	Not found	Error	Error	Not found
Healthcare	EKG	Not fou	Not found	Not found	Not found	Not found	Not found	ekg	Not found	Not found	Not found	Not found	Not found	Not found	Error	Error	Not found
Healthcare	stress Data	Not fou	highstress	stressdat	Not found	Not found	Not found	Not found	Not found	Not found	Not found	stresslev	Not found	Not found	Error	Error	Not found
Healthcare	WheelChai	Not fou	Not found	Not found	Not found	wheelchai	Not found	wheelcha	Not found	Not found	Not found	Not found	Not found	Not found	Error	Error	Not found

FIGURE 9.3

Semiautomatic knowledge extraction from **health** ontologies (https://bit.ly/2EzzSGG).

9.3.3.1 Extracting specific terms from ontology code

We semiautomatically compare existing health ontologies with each other. The main goal is to extract a common pattern to later generate a unified and federated personalized health knowledge graph [30,36]. We focus on health sensors that can be employed for patient monitoring. Sensor-related terms are later included in the sensor dictionary (Section 9.3.1) to be later associated with reasoning mechanisms as mentioned in Section 9.3.4. As an example, we automatically retrieve key phrases (see the second column in Fig. 9.2) if they mention **health and sensor related keywords** such as: "person," "patient," "blood pressure," "temperature," "heart rate," "glucose," "sp02," "oxygen," "respiratory rate," "skin conductivity," "sensor," "device," "humidity," "ecg," "ekg," "activity." The entire document of health concept knowledge extraction is available.[6]

More sophisticated knowledge extraction methodologies that have been applied to other domain ontologies such as smart home, smart cities, weather, and transportation in [27,28], could be applied to health ontologies. Knowledge from scientific publications describing ontologies as well is used. Automatic rule extraction from the ontology code is available in Noura et al. [37], and semiautomatic knowledge extraction from ontologies in Fig. 9.3.

9.3.3.2 Extracting knowledge from scientific publications

To accelerate the investigation of ontology-based health projects, we collect scientific publications (see Fig. 9.2) with the set of criteria that we are interested in:

- Are ontology URLs available within the scientific article? (as described in Algorithm 2). Frequently, URLs are missing. Authors have been contacted to encourage them to follow FAIR principles to release ontology codes and we enriched the ontology code data set when receiving positive answers. LOV4IoT-Health collected 79 ontology-based projects; only 22 are sharing their ontologies online as of October 2021 (see Table 9.2 for demo URL). Other ontology-based projects are referenced in LOV4IoT-Health since they provide knowledge about sensors and reasoning mechanisms employed and described within their paper. Unfortunately, the ontologies cannot be automatically processed yet since they are not accessible.
- Are there UML diagrams describing the ontology architectures? We retrieve captions of figure and table captions.

---

[6] https://bit.ly/2EzzSGG.

- Which sensors are employed for the health domain? We defined a list of keywords to find them in the scientific paper and also within the ontology code. As an example, we automatically retrieve key phrases if they mention **health and sensor related keywords** such as: "heart beat," "heart rate," "body temperature," "blood glucose," "blood sugar level," etc.
- Are there reasoning mechanisms and already defined rules to interpret health sensor data to be employed by smart IoT-based health applications? We defined a list of keywords to find them in the scientific papers. As an example, we automatically retrieve key sentences if they mention **reasoning-related keywords** such as: "jena owl reasoner," "fact++," "hermit," "racer," "pellet," "rule," "reasoning," "logic-based," "infer," "OWL description logic," etc.
- Does the reference section provide more publications to investigate? We enrich our scientific publication data set accordingly (e.g., LOV4IoT-health ontology catalog). The methodology that updates the LOV4IoT ontology catalog is detailed in Algorithm 1.

We used the Apache OpenNLP[7] library for NLP. We split the PDF file with classes such as `SentenceDetectorME`, `SentenceModel` and use our list of specific keywords to automate our manual tasks.

Algorithm 1 Updating the Ontology Catalog with Additional **Health** Knowledge by Crawling the Web or Scientific Libraries.

1: **Procedure: MyProcedure–**

Input: Keyphrases (e.g., "smart health," "hear beat sensor," "health," "Electronic Health Records (EHR) Ontology," "wearable," "dementia," "diabetes," "Ambient Assisted-Living (AAL)," "remote monitoring for healthcare" + ("ontology" OR "reasoning" "rules"))

Output: list of scientific papers in PDF to analyze—health ontology catalog

FOR all papers

IF the paper talks about list of ontology-based health key phrases and sensors referenced in our health dictionary THEN add the paper to the list of papers to analyze

IF the paper has ontology code example or screenshot (OWL, RDF, RDFS) THEN save the ontology code (to be later automatically analyzed)

IF potentialURLOntology(String URL) THEN query the URL and save the ontology code

IF we access OWL/RDF/RDFS ontology code THEN save the ontology in a repository

Check URL related keywords such as "http:," "https://"
(potential URL for the ontology)

IF the paper talks about reasoning mechanisms (e.g., semantic web rule language THEN extract the paragraph, picture, table, etc.

IF the paper has rule patterns (IF THEN ELSE, ?X, ?Y ->?Z) THEN extract the paragraph, picture, table, etc.

---

[7] https://opennlp.apache.org/.

Algorithm 2 Finding an ontology URL in the scientific publication.

Procedure: MyProcedure–
Input: PDF file (e.g., scientific publication, deliverable)

Split the file into sentences

For each sentence

if sentence contains ".owl" OR starts with ("http://" OR "www." OR "https://") OR contains "#" at the end of the URL **then**

end if
Output: the list of sentences with URLs (list_URLs)

 Reasoning Discovery for the Healthcare Domain

How to deduce meaningful information from sensors producing data in the healthcare domain?

Sensor	Projects	
Skin Conductance Sensor Skin conductance is measured in units of siemens; formerly mhos).	Get project	Get rule
Stress Level Sensor E.g., a stress sensor based on Galvanic Skin Response (GSR), and controlled by ZigBee.	Get project	Get rule
Weight Sensor Weight Sensor (e.g. weighting scale form Nokia) to measure your weight in lb or kg. Useful for Health projects (asthma, bariatric surgery, etc.).	Get project	Get rule
Pulse Oxymeter, SpO2, Blood Oxygen Saturation Sensor,	Get project	Get rule

FIGURE 9.4

Subset of the Sensor-based Linked Open Reasoning (S-LOR) for the health domain retrieving projects and rules using specific sensors.

9.3.3.3 Usage of semiautomatic extraction within reasoning demonstrators

We semiautomatically extracted sensors and reasoning employed within ontology-based health projects to design the health sensor dictionary. The sensor dictionary is implemented as an ontology,[8] and is used within demonstrators such as SLOR-Health to easily retrieve sensors for the health domain and reasoning mechanisms for each sensor (see the list of health sensor on the left in Fig. 9.4). Such demonstrators are later used to design end-to-end scenarios.

---

[8] http://sensormeasurement.appspot.com/?p=m3.

9.3.4 Knowledge discovery and reasoning for preventive health and well-being (S-LOR health)

The sensor health dictionary described in Section 9.3.1 and depicted in Table 9.3 and Table 9.4 is employed within the health rule discovery [22] (see Table 9.2 for URL). Each sensor for the health domain can be automatically retrieved using SPARQL queries; the sensor dictionary is displayed in Fig. 9.4 and Fig. 9.5 (on the left part), and the content shown on the GUI is from Table 9.5 (on the right part).

9.3.5 Keeping track of provenance metadata

Provenance: To keep track of the source of knowledge, we add explicitly this information within the rule data set and LOV4IoT RDF data sets as shown in Listing 9.1, by using the W3C PROV-O ontology http://www.w3.org/ns/prov#. For instance, `prov:hadPrimarySource` keeps track of the source by providing the URL of the scientific publication mentioning the rule or reasoning mechanism.

```
1   <m3:Rule rdf:about="LowSPO2Hristoskova2014Health">
2     <rdfs:label xml:lang="en">LowSPO2, IF m3:SPO2 LESS THAN 90 m3:Percent THEN LowSPO2</rdfs:label>
3     <rdfs:comment xml:lang="en">Ontology−driven monitoring of patient's vital signs enabling personalized medical
          detection and alert [Hristoskova, Sakkalis et al. 2014]</rdfs:comment>
4     <prov:hadPrimarySource rdf:resource="https://www.ncbi.nlm.nih.gov/pmc/articles/PMC3926628/"/>
5     <m3:ruleUsingM2MDevice rdf:resource="&m3;PulseOxymeter"/>
6     <m3:fromM2MApplication rdf:resource="&lov4iot;Hristoskova2014Health"/>
7     <m3:hasUrlRule rdf:resource="&lorHealthSPO2;"/>
```

Listing 9.1: Adding provenance metadata using the W3C PROV ontology.

```
1    <dcat:Dataset rdf:about="http://linkedopenreasoning.appspot.com/dataset/rule−dataset">
2        <dc:title xml:lang="en">RDF distribution of the rule dataset</dc:title>
3        <dc:description xml:lang="en">RDF distribution of the rule dataset. </dc:description>
4        <versionInfo rdf:datatype="&xsd;decimal">5.4</versionInfo>
5        <dcterms:modified rdf:datatype="&xsd;date">2021−08−08</dcterms:modified>
6        <dcterms:issued rdf:datatype="&xsd;date">2013−01−01</dcterms:issued>
7        <rdfs:comment xml:lang="en">RDF distribution of the rule dataset.</rdfs:comment>
8        <vann:preferredNamespacePrefix>rule−dataset</vann:preferredNamespacePrefix>
9        <dc:creator xml:lang="en">Amelie Gyrard</dc:creator>
10       <vann:preferredNamespaceUri>http://linkedopenreasoning.appspot.com/dataset/rule−dataset</vann:
            preferredNamespaceUri>
11       <vs:term_status>Work in progress</vs:term_status>
12       <dcat:keyword>"reasoning", "rule",
13       "vocabulary","semantics","ontology",
14       "health", "sensor", "heart beat", "cholesterol", "blood glucose"</dcat:keyword>
15       <dcat:mediaType>"application/rdf+xml"</dcat:mediaType>
16       <dcat:downloadURL>http://linkedopenreasoning.appspot.com/dataset/rule−dataset</dcat:downloadURL>
17   </dcat:Dataset>
```

Listing 9.2: Describing the rule data set using the Data Catalog Vocabulary (DCAT) vocabulary.

Table 9.1 Ontology Name space.

Namespace prefix	Description name	Namespace URL
dcat	Data Catalog Vocabulary	http://www.w3.org/ns/dcat#
prov	Provenance Ontology	http://www.w3.org/ns/prov#
dc	Dublin Core (DC)	http://purl.org/dc/elements/1.1/
dcterms	Dublin Core Metadata Terms	http://purl.org/dc/terms/
vann	Vocabulary for Annotating Vocabulary Descriptions	http://purl.org/vocab/vann/
vs	Vocab Status ontology	http://www.w3.org/2003/06/sw-vocab-status/ns#
rdf	Resource Description Framework	http://www.w3.org/1999/02/22-rdf-syntax-ns#
rdfs	Resource Description Framework Schema (RDFS)	http://www.w3.org/2000/01/rdf-schema#
owl	Ontology Web Language	http://www.w3.org/2002/07/owl#
m3	Machine-to-Machine Measurement Sensor Dictionary	http://sensormeasurement.appspot.com/m3#
saref-core	Smart Applications REFerence ontology	https://saref.etsi.org/core/

FIGURE 9.5

S-LOR linked open reasoning for the health domain.

To describe the rule data set itself, we use the Data Catalog Vocabulary (DCAT) vocabulary http://www.w3.org/ns/dcat# as shown in Listing 9.2.

Table 9.1 reminds name spaces used: the first column for the prefix name, the second column for the name space description, and the third column for the ontology name space URL.

9.4 End-to-end knowledge-based health and well-being use cases

We retrieve knowledge expertise from existing projects by classifying sensors (sensor dictionary for health explained in Section 9.3.1), ontologies used to model data and applications (LOV4IoT ontology catalog), and the reasoning mechanisms used to interpret sensor data (S-LOR tool). We integrate the knowledge from several communities (health experts, IoT community designing ontologies, etc.). The collected knowledge can be automatically retrieved within online knowledge API demonstrators, as referenced in Table 9.2.

To summarize, we have defined a set of use cases using the SLOR reasoning discovery that extract knowledge from the LOV4IoT-Health ontology catalog. The knowledge acquired to infer meaningful information is implemented as rules. We simulated sensor data (e.g., data using the SenML format). We annotated the data to be compliant with ontologies (the file can be accessed though the demos referenced in Table 9.2). The semantic reasoner (based on the Apache Jena inference engine) is executed on the semantic sensor data set and the set of rules are all compliant with each other.

For instance, after executing the reasoning engine, a rule has been executed that deduces that a heart beat measurement, which is greater than 215 per minutes might be tachycardia, as illustrated in Fig. 9.8. A subset of rules example discovered from the SLOR rule discovery and employed in health scenarios is displayed in Table 9.6. We have more and more scenarios including various sensors (introduced in Table 9.3 and Table 9.4), as an example:

- **Blood pressure** to infer disorders such as hypertension (Fig. 9.6).
- **Blood glucose** to infer disorders such as hyperglycemia (Fig. 9.7).
- **Heart beat** to infer disorders such as tachycardia (Fig. 9.8).
- **Diet-related health data**: cholesterol, blood glucose level, magnesium, potassium, sodium, vitamin D, etc. (see Table 9.4).
- **Well-being food recommendation use case**: food recommended for depression, diabetes, cholesterol, anxiety, stress, fatigue, sleeping issues, headache, etc.
- **Other health data**: SPO2, frequency, skin conductance, etc. (see Table 9.3).
- **Activity data**: activity level data.
- **Air quality data**: NO2, PM10, O3, PM2.5, SO2, CO.
- More and more scenarios are being added within the same demos with different kind of sensor data (as referenced within Table 9.3 and Table 9.4) and more and more rules to deduce high level information.

The set of online demonstrators are summarized in Table 9.2. Demonstrators (more precisely the second column of Table 9.3) are classified according to: (1) physical activity, (2) healthy diet, (3) emotional management with positive psychology, and (4) sleep management; as shown in the Venn diagram depicted in Fig. 9.1.

Technologies employed in the implementation: Our demonstrators, referenced in Table 9.2, are implemented with the Apache Jena semantic web framework that can deal with RDF, RDFS, and OWL languages to implement the sensor dictionary. This dictionary is implemented within the M3 ontology which is refined for the health domain. Jena also provides an inference engine used for the reasoning process within scenarios. Java is used to develop REST web services with the JAX-RS library to hide the complexity of using semantic web technologies, and the Graphical User Interface (GUI) is implemented with Ajax, JQuery, JavaScript, HTML, and CSS.

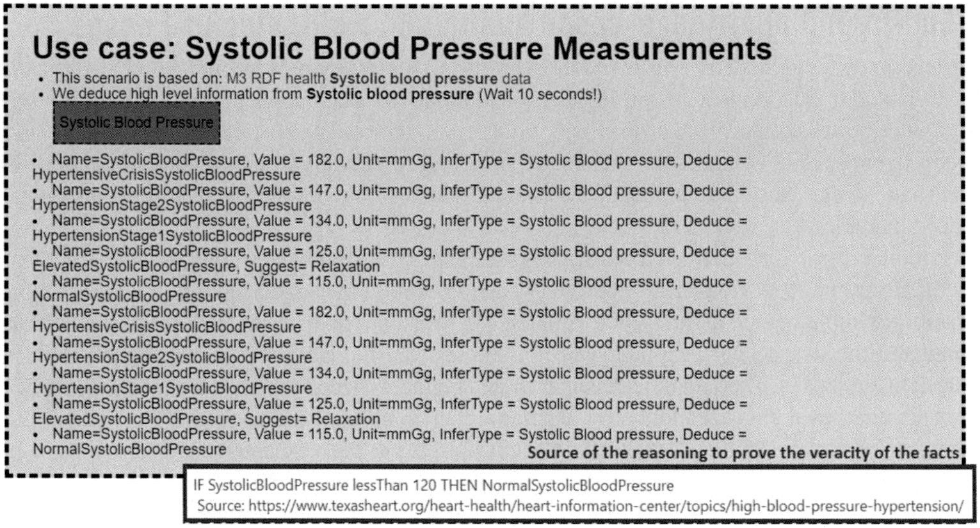

Use case: Systolic Blood Pressure Measurements

- This scenario is based on: M3 RDF health **Systolic blood pressure** data
- We deduce high level information from **Systolic blood pressure** (Wait 10 seconds!)

Systolic Blood Pressure

- Name=SystolicBloodPressure, Value = 182.0, Unit=mmGg, InferType = Systolic Blood pressure, Deduce = HypertensiveCrisisSystolicBloodPressure
- Name=SystolicBloodPressure, Value = 147.0, Unit=mmGg, InferType = Systolic Blood pressure, Deduce = HypertensionStage2SystolicBloodPressure
- Name=SystolicBloodPressure, Value = 134.0, Unit=mmGg, InferType = Systolic Blood pressure, Deduce = HypertensionStage1SystolicBloodPressure
- Name=SystolicBloodPressure, Value = 125.0, Unit=mmGg, InferType = Systolic Blood pressure, Deduce = ElevatedSystolicBloodPressure, Suggest= Relaxation
- Name=SystolicBloodPressure, Value = 115.0, Unit=mmGg, InferType = Systolic Blood pressure, Deduce = NormalSystolicBloodPressure
- Name=SystolicBloodPressure, Value = 182.0, Unit=mmGg, InferType = Systolic Blood pressure, Deduce = HypertensiveCrisisSystolicBloodPressure
- Name=SystolicBloodPressure, Value = 147.0, Unit=mmGg, InferType = Systolic Blood pressure, Deduce = HypertensionStage2SystolicBloodPressure
- Name=SystolicBloodPressure, Value = 134.0, Unit=mmGg, InferType = Systolic Blood pressure, Deduce = HypertensionStage1SystolicBloodPressure
- Name=SystolicBloodPressure, Value = 125.0, Unit=mmGg, InferType = Systolic Blood pressure, Deduce = ElevatedSystolicBloodPressure, Suggest= Relaxation
- Name=SystolicBloodPressure, Value = 115.0, Unit=mmGg, InferType = Systolic Blood pressure, Deduce = NormalSystolicBloodPressure

Source of the reasoning to prove the veracity of the facts

IF SystolicBloodPressure lessThan 120 THEN NormalSystolicBloodPressure
Source: https://www.texasheart.org/heart-health/heart-information-center/topics/high-blood-pressure-hypertension/

FIGURE 9.6

Inferring disease (e.g., hypertension) from blood pressure.

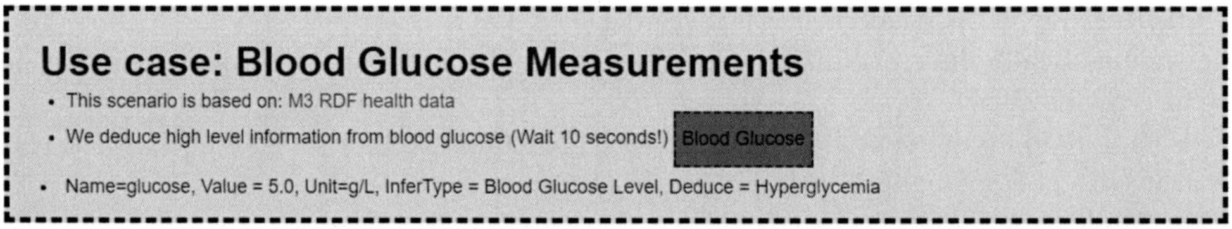

Use case: Blood Glucose Measurements

- This scenario is based on: M3 RDF health data
- We deduce high level information from blood glucose (Wait 10 seconds!)
- Name=glucose, Value = 5.0, Unit=g/L, InferType = Blood Glucose Level, Deduce = Hyperglycemia

FIGURE 9.7

Inferring disease (e.g., hyperglycemia) from blood glucose.

Use case: Heartbeat Measurements

- This scenario is based on: M3 RDF health heartbeat data
- We deduce high level information from heartbeat (Wait 10 seconds!) Heart beat

- Name=HeartBeat, Value = 38.0, Unit=beatperminute, InferType = Heart beat, Deduce = Bradycardia
- Name=HeartBeat, Value = 38.0, Unit=beatperminute, InferType = Heart beat, Deduce = VerySlowHeartBeat
- Name=HeartBeat, Value = 100.0, Unit=beatperminute, InferType = Heart beat, Deduce = NormalHeartBeat
- Name=HeartBeat, Value = 105.0, Unit=beatperminute, InferType = Heart beat, Deduce = FastHeartBeat, Suggest= Rest
- Name=HeartBeat, Value = 55.0, Unit=beatperminute, InferType = Heart beat, Deduce = SlowHeartBeat
- Name=HeartBeat, Value = 215.0, Unit=beatperminute, InferType = Heart beat, Deduce = VeryFastHeartBeat, Suggest= Rest
- Name=HeartBeat, Value = 215.0, Unit=beatperminute, InferType = Heart beat, Deduce = Tachycardia

Rule: IF m3:HeartBeat greaterThan 150 m3:BeatPerMinute THEN Tachycardia
Source: Ontology-driven monitoring of patient's vital signs enabling personalized medical detection [Hristoskova 2014]

Source of the reasoning to prove the veracity of the facts

FIGURE 9.8

Inferring disease (e.g., tachycardia) from heartbeat.

Table 9.2 Set of online demonstrators: ontology catalogs, rule discovery, and full scenarios.

Tool name	Tool URL
LOV4IoT-Health Ontology-based IoT Project Catalog	http://lov4iot.appspot.com/?p=lov4iot-health
LOV4IoT-Health Web Service and Dumps (for Developers)	http://lov4iot.appspot.com/?p=queryHealthOntologiesWS
S-LOR Health Rule Discovery	http://linkedopenreasoning.appspot.com/?p=slor-health
M3-Health Full Scenarios	http://sensormeasurement.appspot.com/?p=health
M3-Health (Naturopathy) Full Scenarios	http://sensormeasurement.appspot.com/?p=naturopathy

ACCRA European–Japan project: Some of those scenarios such as heartbeat are also designed for the ACCRA European-funded project[9] that can be embedded in more sophisticated robot GUIs to automatically alert the physicians when needed. More explanations are published within [22].

9.5 Key contributions and lessons learned

We designed a sensor dictionary and reasoner compliant with those standards: SAREF (SAREF-Core, SAREF4EHAW, and SAREF4WEAR), IEC 61360 Common Data Dictionary, and W3C SOSA/SSN employed within our semantic datasets, SPARQL queries, and rules. We have analyzed the following limitations of SAREF specifications which highlights the need of our unified dictionary explained in Section 9.3.1.

- **Measurement types**: `saref:Property`, `m3:MeasurementType` and `sosa:ObservableProperty` are similarly designed.
- **Inconsistency, lack of unification, or duplication** are found for naming such as `saref-core:TemperatureSensor` and `saref4agri:Thermometer`, it demonstrates the complexity to search for the right terms and handle synonyms. We also have to deal with missing concepts and handle cross-domains (e.g., diet is dependent to health; there is no SAREF for food).
- **Sensor data values**: `saref:hasValue`, `m3:hasValue`, and `sosa:hasSimpleResult` are similarly designed; all can be used within our semantic data sets, SPARQL queries, and rules.
- **Units**: `saref:isMeasuredIn` and `m3:hasUnit` are similarly designed; all can be used within our semantic data sets, SPARQL queries, and rules. Ontologies' main goal is to explicitly describe the data, descriptions such as `saref:TemperatureUnit` does not exactly remove ambiguities regarding the unit used such as Celsius or Fahrenheit, which might lead to mistakes with automatic reasoning.
- **Unifying sensor metadata**: We structure sensor data in Table 9.3 and Table 9.4: (1) sensor name, (2) the produced measurement, (3) the associated unit (e.g., `Hertz` to be more explicit than `saref4envi:FrequencyUnit`). There is a need of domain experts to verify synonyms (e.g., heart beat, heart rate). Each row of the table (sensor, measurement, unit) is implemented within the M3 ontology designed and maintained since 2012, (see also the M3-lite[10] refined for the FIESTA-IoT

[9] https://www.accra-project.org/en/sample-page/.
[10] https://github.com/fiesta-iot/ontology/blob/master/m3-lite.owl.

Table 9.3 Subset of the sensor dictionary: M3 ontology extended for emotional well-being and preventive health. See the SLOR-Health online demo for a more exhaustive sensor dictionary (frequently updated).

Sensor, measurement name and unit	M3 scenarios	SAREF-Core, SAREF4Health SAREF4Wear	Other names and standards, e.g., IEC 61360 Common Data Dictionary
m3:PulseOxymeter m3:SPO2, xsd:GramPerLiter	✓ M3-Health	✗	–
m3:SkinConductanceSensor m3:SkinConductance, m3:Siemens	✓ M3-Health (Emotion)	✗	SC (Galvanic skin response) GSR, electrodermal activity, EDA
m3:HeartBeatSensor m3:HeartBeat, m3:BeatPerMinute	✓ M3-Health (Emotion)	✗	HeartRate, Breath rate, Respiration rate, Pulse wave, Breathing rate bpm
m3:FrequencySensor m3:Frequency, m3:Hertz	✓ M3-Health	✗ s4envi:FrequencyUnit	alpha, beta, theta, delta, gamma wave
m3:SystolicBloodPressureSensor m3:SystolicBloodPressure, m3:mmGg	✓ M3-Health	✗	blood volume pressure blood volume pulse (BVP)
m3:ActivityLevelSensor m3:ActivityLevel xsd:int	✓ M3-Health (Fitness)	✗ s4ehaw:hasActivity (owl:ObjectProperty)	–
m3:Pedometer m3:NumberStep, xsd:int	✓ M3-Health (Fitness)	✗	–
m3:PollenLevelSensor m3:PollenLevel, xsd:int	✓ M3-Health	✗	–
m3:SnoringLevelSensor m3:SnoringLevel, m3:Decibel	✓ M3-Health	✗	–
m3:BodyThermometer m3:BodyTemperature m3:DegreeCelsius m3:DegreeFahrenheit	✓ M3-Health	✗	–
m3:MagneticFieldSensor m3:MagneticField, m3:Tesla	✓ M3-Health	✗	–
m3:MagneticFluxDensitySensor m3:MagneticFluxDensity, m3:Gauss	✓ M3-Health	✗	–
m3:NeuronNumberSensor m3:NeuronNumber, xsd:int	✓ M3-Health	✗	–
m3:HemoglobinSensor m3:Hemoglobin, m3:GramPerLiter	✓ M3-Health	✗	Hemoglobin sensor [40]

H2020 European-funded project running from 2015 to 2018 [38]). SAREF is supported by ETSI M2M since 2015.

- **Provenance metadata**: SAREF does not keep track the provenance of the information. For each sensor, we reference sources such as scientific publications or project deliverables referenced on the LOV4IoT ontology catalog project (Section 9.2.1 and Table 9.5). Similarly, the S-LOR project (see Section 9.3.4) suggests reasoning mechanisms (e.g., rules) for specific sensor type. It is not provided by SAREF.
- **Interlinking ontologies**: our proposed solution, compared to SAREF, is to link and unify existing ontologies to achieve semantic interoperability. We added links such as `rdfs:subclassOf`, `owl:equivalentClass`, and `rdfs:seeAlso`. For each heath sensor, we explicitly add a link such as `<rdfs:subClassOf rdf:resource="&saref-core;Sensor"/>` in the sensor dictionary. SAREF or

Table 9.4 Subset of the sensor dictionary: M3 ontology extended for emotional well-being and preventive health, focused on diet-related data. See the SLOR-Health online demo for a more exhaustive sensor dictionary (frequently updated).

Sensor, measurement name and unit	M3 scenarios	SAREF-Core, SAREF4Health SAREF4Wear	Other names and standards, e.g., IEC 61360 Common Data Dictionary
m3:CholesterolSensor, m3:Cholesterol m3:GramPerLiter, m3:MmolPerLiter	✓ M3-Health (Diet)	✗	–
m3:Glucometer m3:BloodGlucose, xsd:Percent	✓ M3-Health (Diet)	✗	–
m3:Magnesium, m3:MagnesiumSensor m3:GramPerLiter	✓ M3-Health (Diet)	✗	Magnesium sensor [41]
m3:PotassiumSensor, m3:Potassium m3:GramPerLiter	✓ M3-Health (Diet)	✗	Potassium wearable sensor [42]
m3:SodiumSensor, m3:Sodium m3:GramPerLiter	✓ M3-Health (Diet)	✓ saref4watr:Sodium	Sodium wearable sensor [42]
m3:VitaminDSensor, m3:VitaminD m3:GramPerLiter	✓ M3-Health (Diet)	✗	Vitamin D sensor [43]
m3:ZincSensor, m3:Zinc m3:GramPerLiter	✓ M3-Health (Diet)	✗	[43]

W3C SOSA do not consider the IEC 61360 – Common Data Dictionary standard[11] that we address in Table 9.3.

- **IoT Alliance**: Our work is taken as a baseline in AIOTI[12] (Alliance for the Internet of Things Innovation) WG03 Standards – IoT semantic interoperability expert group within white papers [19, 21].

9.6 Conclusion and future work

Monitoring remotely patient's vital signals can assist (elderly) people to stay independent at home and reduce health care costs. Designing preventive health applications requires cross-domain knowledge acquired from heterogeneous communities (e.g., health, affective science, fitness, diet, sleep, well-being, IoT, Ambient Assisted Living, etc.). Health applications will use more wearables to monitor patient's vital signs. Integrating machine interpretable knowledge implemented within ontologies from various domains is challenging. A solution is needed to reuse expertise from domain experts when they are not available. Our experience and expertise is shared within standards (e.g., editors of the ISO/IEC 21823-3 IoT semantic interoperability).

---

[11] https://cdd.iec.ch/cdd/iec61360/iec61360.nsf/TreeFrameset?OpenFrameSet.
[12] https://aioti.eu/.

Short-term challenges: LOV4IoT is relevant for the IoT community. The results are encouraging to update the data set with additional domains and ontologies. LOV4IoT leads to the AIOTI (The Alliance for the Internet of Things Innovation) IoT ontology landscape survey form[13] and analysis result,[14] executed by the WG03 Standards – Semantic Interoperability Expert Group. It aims to help industrial practitioners and nonexperts to answer those questions: Which ontologies are relevant in a certain domain? Where to find them? How to choose the most suitable? Who is maintaining and taking care of their evolution?

Mid-term challenges: Automatic knowledge extraction from ontologies and scientific publication describing the ontology purpose is challenging, as highlighted in our AI4EU Knowledge Extraction for the Web of Things (KE4WoT) Challenge. The challenge encourages to reuse the expertise designed by domain experts and make the domain knowledge usable, interoperable, and integrated by machines. We released the set of ontologies, as dumps, web services, tutorials, and make them available for the challenge.

Long-term challenges: To improve the veracity and the evaluation of the recommender, involving physicians such as psychologists, fitness coaches, nutritionists would enhance our health applications, by proving more the facts. We investigated psycho-physiology [39] research field to prove such facts. Collecting emotion ontologies has been initiated within the ACCRA project (robots for healthy ageing that provide well-being applications) [22].

Acknowledgments

This work has partially received funding from the European Union's Horizon 2020 research and innovation program under project grant agreement ACCRA No. 738251, StandICT.eu 2023 No. 951972 (open call), and AI4EU No. 825619 (challenge open call). We would like to thank project partners for their valuable comments. The opinions expressed are those of the authors and do not reflect those of the sponsors.

Appendix 9.A IoT-based ontologies for health

The ontologies are collected and analyzed to extract knowledge to: (1) build a sensor dictionary for health (see Section 9.3.1), (2) extract rules from ontologies or scientific publications (see Section 9.3.4), and (3) simulate full health scenarios by providing provenance of the information (see Section 9.4).

We summarize a subset of ontologies collected in Table 9.5 to convey the purpose of the designed ontologies:

- Ambient Assisted-Living (AAL) ontologies in Section 9.A.1,
- Disease-related ontologies such as cardiology and diabetes in Section 9.A.2,
- Electronic Health Records (EHR) ontologies in Section 9.A.3,
- Wearable ontologies in Section 9.A.4,
- Health Ontologies designed for EU Projects in Section 9.A.5, and
- Other health-related ontologies in Section 9.A.6.

---

[13] https://ec.europa.eu/eusurvey/runner/OntologyLandscapeTemplate.
[14] https://bit.ly/3fRpQUU.

Table 9.5 Subset of ontology-based health IoT projects and reasoning mechanisms employed.

Authors	Year	Project	OA	Reasoning	Sensors
BioPortal [23]	2021 2009	Biomedical ontology catalog but IoT ontologies for smart health not found	✓	✗	✗
LOV [24] Linked Open Vocabularies	2021 2015	Ontology Catalog designed by the Semantic Web Community	✓	✗	✗
ACTIVAGE [76]	2018	ACTIVAGE Ontologies	✓	✓ Data Analytics	✓
SAREF4EHAW [29] Moreira et al. [81,82]	2020 2018	SAREF for e-Health and Ageing Well	✓	✗	✗
SAREF4WEAR [83]	2020	SAREF for Wearables	✓	✗	✗
Nachabe et al. [75,84]	2015 2014	**Wearable** – WBAN for mobile application for **sport** exercises	✓	✓ SWRL	
Villalonga et al. [74] MIMU-Wear	2017	**Wearable** – Activity recognition	✓	Rules for wearable replacement	
Enshaeifar, Barnaghi et al. [65]	2018	FHIR4TIHM Data model based on HL7 and not ontology	✓	✓ Data analytics and ML	✓ temperature humidity, blood pressure, pulse
Aloulou et al. [50]	2013	Ambient Assisted Living (AAL)	✓	✓ Euler, FOL rules reasoning engine	✓
Lemlouma et al. [53] Roose, Abdelaziz	2013	Elderly Dependency in Smart Homes	✓	✓ SWRL, Jess rule engine	✓ Shower, Proximity, presence, vibrator, motion, mattress sensor
Yao, Akhil Kumar et al. [59,60]	2013 2009	Context Awareness in Healthcare heart failure	✓	✓ Drools, SWRL Pellet, Jess	✗
Brandt, Lukkien Liu et al. [46,47]	2013	ContoExam ontology for Remote Patient Monitoring (RPM)	✓	✗	✓ ECG
Paganelli, Giuli et al. [44,45]	2011 2007	Ontology for health monitoring and patient chronic condition in home	✓	✓ rule engine	✓ heart rate body temperature
Jovic et al. [58]	2011	Heart failure ontology	✓	✓ Pellet, SWRL	✓ Echocardiography
Zhao, Samwald et al. [78,79]	2010	Chinese Medicine as Linked Open Data	✓	✗	✗
Mazuel et al. [54] INSERM, France	2009	SNOMED to align 3 ontologies: hypertension, pneumology, surgery resuscitation	✓	✗	✗
Lafti et al. [48] – HIT Canada	2007	Telehealth Smart Home Elderly in Loss of Cognitive Autonomy	✓	✓ SWRL, OWL restrictions	✓ motion, temp, presence, light, blood pressure, actuator, fall detector
Seneviratne et al. [63]	2021	Personal Health KG for Diet	✗	✓	
Li et al. [66]	2020	Medical KG from EMRs – Parkinson's	✗	✓ ML	
Chatterjee et al. [49]	2021	Healthy Lifestyle Management – Obesity	✗	✓ Hermit, SWRL	✓
Otto et al. [1]	2020	Ontology for telemedicine terms	✗	✗	✗
Lyons et al. [77]	2021	SORBET (Sensor Ontology for Reusable Biometric Expressions and Transformations)	✗	✗	✗
Adel et al. [67]	2020	Ontology for EHR	✗	✗	✗
Rhayem et al. [51]	2017	HealthIoT Ontology patient monitoring Diagnosis	✗	✓ Drools inference engine, SWRL rules	temperature, heart rate, cholesterol, blood pressure
Sherimon et al. [61]	2016	OntoDiabetic ontology Process ontology (clinical guidelines)	✗	✓ OWL2 rules forward chaining inference	✗
Gai et al. [71]	2015	EHR for diagnosis error prevention	✗	✓ EPAA	✗
Mukasine et al. [62]	2014	ontology for glucose, insuline, diabete, diet	✗	✓	✗

Legend: Ontology Availability (OA), when the code is available, the ontologies are classified on the top. Then the ontology-based projects are classified by year of publications. First-Order Logic (FOL), Error Prevention Adjustment Algorithm (EPAA), Machine Learning (ML).

9.A.1 Ambient assisted-living (AAL)/ remote monitoring for health ontologies using IoT technologies

Ontology-based context model for health monitoring (Paganelli et al. [44,45]) and handling patient chronic conditions are addressed in a home-based care scenario. A patient personal domain ontology and a rule-based reasoning approach analyze patient data such as body temperature and heart rate to detect patient abnormal conditions and raise alarm when threshold values are reached.

ContoExam ontology for Remote Patient Monitoring (RPM) (Brandt, Liu et al. [46,47]) is used in an e-Health application to increase quality of life by curing remotely and to improve decision making to ease professional decisions. An epileptic seizure scenario is provided.

Telehealth Smart Home (TSH) ontology (Lafti et al. [48]) is designed for elderly in loss of cognitive autonomy (e.g., intellectual deficiency because of Alzheimer or similar disease, physical-deficiency due to age-related disease, visual deficiency, auditive deficiency). The TSH ontology is used to understand if the patient is safe or he is at risk, and is employed by the Bayesian network that recognizes the patient's activity and understands its life habits. Telehealth Smart Home (TSH) ontology comprises seven subontologies: (1) Person and medical history ontology describes the person who needs care and his medical history, or the person taking care and his duties, (2) Behavior ontology to understand life habits and critical physiological parameters, (3) Equipment smart home ontology describes furniture equipment, the household equipment, and the technical equipment to ensure the patient safety, (4) Task smart home ontology, (5) Software smart home ontology, (6) Habitat smart home ontology describes where the patient lives such as rooms, doors, windows, and (7) Event/decision smart home ontology detects a critical situation or a change of habits.

UiA e-Health ontology/UiAeHo (Chatterjee et al. [49]) is used to annotate personal, physiological, behavioral, and contextual data from heterogeneous health and wellness data (sensor, questionnaire, and interview). The ontology is employed within an e-Coach system, a rule-based decision support system (DSS) to predict the probability for health risk and provide a lifestyle recommendation generation plan against adverse behavioral risks. The ontology is integrated with W3C SSN and SNOMED-CT. The ontology is applied to an obesity use case but could be extended to other lifestyle diseases. Five experts have been consulted with a research background in ICT, e-Health, nursing, and nutrition for simulating activity and nutrition data. Obesity-related information and guidelines were obtained from the World Health Organization (WHO), the National Institute for Health and Care Excellence (NICE), and the Norwegian Dietary Guidelines.

Aloulou et al. [50] achieved Ambient Assisted Living (AAL) deployment experience in Singapore within three rooms of a nursing home with eight patients and two caregivers. Three services to assist patients, based on semantic Plug&Play mechanisms and sensors, are designed: (1) wandering at night without going back to sleep, (2) showering for a long period of time, and (3) leaving the wash-room tap on after washing hands.

HealthIoT Ontology for patient monitoring (Rhayem et al. [51]) is integrated within the IoT Medicare system for diagnosis and decision making for doctors to assist patients. Sensors employed are temperature, heart rate, cholesterol, blood pressure, and blood glucose.

Cognitive Semantic Sensor Network ontology (CoSSN) ontology (Zgheib et al. [52]). The semantic IoT healthcare application development framework provides a simple software API usable in various application domains and to alleviate the tasks of software developers. The framework is based on the OSGi Java framework and the Kura platform from the Eclipse foundation to integrate MQTT communication protocol and facilities to add IoT devices.

Lemloula, Laborie, Roose et al. [53] focus on Activities of Daily Living (ADL) on French AGGIR elderly people (Autonomy Gerontology Iso-Resources Group). Sensors such as temperature, video, sound, presence, etc. are used to detect activities of daily living such as hygiene, toilet use, eating, resting, and dressing.

9.A.2 Disease-related ontologies

Mazuel et al. [54] align three ontologies with SNOMED v3.5: a pneumology ontology (OntoPneumo based on CIM-10), arterial hypertension (OntoHTA based on SNOMED-CT), and surgery resuscitation ontology (OntoReaChir based on CIM-10).

9.A.2.1 Cardiology-related ontologies

Colicchio et al. [55] annotate patients' care context cardiology data as represented in clinical notes and spoken communications during outpatient visits.

Patient's Vital Signs Monitoring (Hristoskova et al. [56,57]) is an Ambient Intelligence (AmI) framework that provides real-time monitoring of patients diagnosed with Congestive Heart Failure (CHF). The ontology-based reasoning enables personalized medical detection and alert (e.g., rules for SP02, heart rate, and blood pressure).

Heart Failure (HF) ontology (Jovic et al. [58]) is designed for the EU FP6 project HEARTFAID. The expert system for patient related warnings, suggestions, and/or decisions includes 200 rules implemented in SWRL with Pellet that are in a form similar to the presented rule for systolic heart failure. Height groups of rules including: diagnosis, alternative diagnosis, severity assessment, prognosis, medication prescription, and medication related warnings, and acute decompensation detection. This HF ontology has been developed mainly by technical people by reading medical literature, primarily HF guidelines published by the European Society of Cardiology.

Hospital ontology (Heart failure from the HEARTFAID team and clinical ontology) from ConFlexFlow (Clinical cONtext based flexible workFlow) (Kumar, Yao et al. [59,60]) is used for Clinical Decision Support Systems (CDSS). ConFlexFlow supports clinical workflows to follow standards such as HL7 and clinical guidelines. ConFlexFlow integrates medical knowledge in the form of rules (using SWRL, Drools, and Jess), 18 SWRL rules for describing heart failure procedural knowledge (detection, diagnosis, and treatment of chronic heart failure, systolic heart failure, hypertensive heart failure, etc.). It focuses on the diagnosis and treatment of heart failure patients based on context information. Experiments are done with 30 patients, 25 hospital personnel, 40 assets, and 40 locations as test cases. Pellet is employed to validate the model with logical consistency, concept satisfaction, and classification. Key measures of quality (KPIs) are: number of treatment errors because of drug interactions (or allergies), number of diagnosis errors, number of cases of treatment not covered by patient's insurance, number of treatment failures for lack of available resources, complication rate per patient, patient satisfaction, etc.

9.A.2.2 Diabetes and diet-related ontologies

OntoDiabetic (Sherimon et al. [61]) is an ontology-based reasoning to recommend the suitable treatment for diabetic patients by considering the current medical status. Patients in the studies have three main complications: cardiovascular disease (CVD), diabetic nephropathy, and hypertension. The reasoner (forward chaining inference) reasons by processing the semantic profile as input with the clinical guidelines defined within the process ontology stored knowledge to infer risk scores and treatment suggestions. OntoDiabetic system computes the score and predicts the risk of diabetic patients due to smoking, alcohol, physical activity, sexual, and cardiovascular disease that mainly affects diabetes.

Mukasine et al. [62] design an ontology-based knowledge base to manage diabetic patient. The ontology supports data sharing and can generate recommendations for the diabetic patient such as medications and diets in the remote supervision.

Personal Health KG for diet (that includes an ontology) (Seneviratne et al. [63]) considers user's temporal personal health data (and mentioned our Personalized Health KG past work [36]) to personalize dietary recommendations for diabetic patients. Food recommendations are provided to answer such questions: (1) What should I eat for breakfast? (2) What foods can I eat if I have a dairy allergy? (3) What can I substitute for food Y? (patient's taste preferences).

A more comprehensive review of food-related ontologies are available in our parallel work [64] that encourages a healthy diet. Hereafter, an overview of research was published in 2021.

9.A.2.3 Dementia models, ontologies or KGs: Parkinson's, Alzheimer's, etc.

Technology Integrated Health Management (TIHM), FHIR4TIHM, (Enshaeifar, Barnaghi et al. [65]) supports 700 patients with dementia, healthcare practitioners, and patient's caregivers to improve their quality of life: (1) learn daily patterns, (2) detect agitated/irritated patients, and (3) detect Urinary Tract Infections (UTIs). TIHM is deployed in the Chertsey Hospital (Surrey, England). FHIR4TIHM is a model, based on HL7 and not an ontology, that uses IoT technologies (e.g., 25 sensors/apps per home). Security and privacy issues are addressed.

Medical KG from large-scale Electronic Medical Records (EMRs) (Li et al. [66]) infers possible diseases and recommend medical orders. Eight steps are needed to build the medical KG: data preparation, entity recognition, entity normalization, relation extraction, property calculation, graph cleaning, related-entity ranking, and graph embedding. The data set contains 16,217,270 deidentified clinical visit data of 3,767,198 patients from Southwest Hospital in China. The KG contains 22,508 entities and 579,094 quadruplets (instead of usual triplets). The ontology term is not explicitly mentioned. Use cases for Parkinson's disease and lung cancer are provided. The International Classification of Diseases (ICD-9) standard is used to map disease, diagnosis, and surgery terms. Two experiments are performed: (1) Bi- LSTM network, and (2) graph embedding to a neural network (Bi- LSTM network) task to predict medicine prescription by diseases using the MIMIC3 data set.

Telehealth Smart Home (TSH) ontology (Lafti et al. [48]) (mentioned previously) is designed for elderly in the loss of cognitive autonomy (e.g., intellectual deficiency because of Alzheimer's or a similar disease, physical-deficiency due to an age-related disease).

9.A.3 Electronic Health Records (EHR) ontologies

Adel et al. [67] cover health standards such as IHTSDO, DICOM, CDISC, IHE, HL7, CEN/ISO 13606, and openEHR. Adel et al. exploit the semantic web technologies to support EHRs by designing a unified ontology-based framework to deal with data integration between heterogeneous systems. It is aimed to build a more realistic, applicable, accurate, medically acceptable, reliable, and global EHR interoperable environment. The fuzzy ontology-based semantic interoperability framework for distributed EHR systems [68] is designed to help physicians query patient data from distributed locations using near-natural language queries. The EHR MIMIC-III intensive care unit data set includes 100 patients.

Ontology and HL7 Reference Information Model (RIM)-based middleware (Plastiras et al. [69,70]) provides interoperability between Personal Health Record (PHR) and Electronic Health Record (EHR) systems.

Ontology-based EHR Error Prevention Model (OEHR-EPM) (Gai et al. [71]) is designed to alert physicians and assist them with medical diagnosis using the Error Prevention Adjustment Algorithm (EPAA) algorithm. A scenario assists physicians to avoid misdiagnose by distinguishing the gastritis from pancreatitis since they have similar symptoms that can cause a misdiagnose with a serious consequence, such as surgery and death.

OpenEHR project [72] provides an Electronic Health Record (EHR) ontology.

9.A.4 Wearable ontologies

Hodges et al. [73] manually align ontologies having concepts such as disease, symptom, anatomy, device, and physical property.

MIMU-Wear (Villalonga et al. [74]), is an extensible ontology that describes wearable sensor platforms consisting of mainstream magnetic and inertial measurement units (MIMUs), measurement properties, and the characteristics of wearable sensor platforms including their on-body location.

Generic Ontology for Wireless Body Area Networks (WBANs) (Nachabe et al. [75]) is designed for the Android m-Health mobile application to calculate burned calories and trajectory of a runner while doing his running exercise.

9.A.5 Ontologies from European projects: ACTIVAGE and HEARTFAID

H2020 ACTIVAGE Semantic Interoperability Layer (SIL) ontology (Kalamaras et al. [76]) is used by real-time data analytics within the Data Lake infrastructure. The ACTIVAGE ontology is based on existing IoT ontologies to combine and extend them: W3C SSN, ETSI SmartM2M SAREF, oneM2M, IoT-Lite, and OpenIoT. ACTIVAGE is a project that supports large-scale IoT applications for health assistance for older people. A smart home use case eases decision making (e.g., understand daily activities, facilitate clinicians in monitoring patient's health, and detect anomalies). The mobility use case monitors and assists the older person while moving in a city, providing information and alerts when needed.

FP6 HEARTFAID project produced the Heart Failure (HF) ontology (Jovic et al. [58]) and the Hospital ontology (Yao et al. [59,60]), as mentioned above.

9.A.6 Other ontologies

SORBET (Sensor Ontology for Reusable Biometric Expressions and Transformations) (Lyons et al. [77]) designed for the Medidata Sensor Cloud product. It is based on wearable technologies to collect sensor data from the patient's daily lives to be used in clinical trials.

Ontology for telemedicine terms (Otto et al. [1]) is designed to remove ambiguities of telemedicine terms: (1) the definition of relevant terms, (2) their interrelations, and (3) a description of specific application types of telemedicine.

Chinese Medicine (CM) Linked Open Data (Zhao, Samwald et al. [78,79]) is published on the web.

Ontology for Diagnostic classification (Bertaud-Gounot et al. [80]), applied to diseases such as spondyloarthritis (SpA), is experimented with 30 real patient cases. The HermiT reasoner API is employed for ontology validation, consistency checking, and taxonomy classification.

Table 9.6 Subset of rules example discovered from the SLOR rule discovery and employed in health scenarios.

Sensor measurement type	Rules	Source
Air Quality Index (AQI)	Good AQI US (0–50 AQI) Moderate AQI US (51–100 AQI) Unhealthy sensitive group AQI US (101–150 AQI) Unhealthy AQI US (151–200 AQI) Very unhealthy AQI US (201–300 AQI) Hazardous AQI US (301–500 AQI)	Air quality web site [WR21]
Pollen Index	Low pollen level (0–2.4) Low medium pollen level (2.5–4.8) Medium pollen level (4.9–7.2) Medium high pollen level (7.3–9.6) High pollen level (9.7–12)	Pollen web site [WR10]
Outside Humidity	Dry humidity (30%–40%) Normal humidity (40%–70%) Very moist humidity (80%–100%) Very dry humidity (0%–30%) Moist humidity (70%–80 %)	Staroch et al. [85]
Inside Humidity	Low humidity (< 50%) High humidity (51%–69%) High humidity (> 70%)	Yacchirema et al. [86]
Activity (Minutes active, Sedentary minutes, Minutes lightly active, Number of steps)	Sedentary person (< 5000 steps count) Mildly active person (5000–7499 steps count) Moderately active person (7500–9999 count) Active person (10,000–12,499 steps count) Highly active person (>= 12,500 steps count)	Yacchirema et al. [86]
Snoring	Normal snoring level (< 40 dB) Mild snoring level (40-50 dB) Moderate snoring level (50-60 dB) Severe snoring level (>= 60 dB)	Yacchirema et al. [86]
Sleep (Minutes REM sleep, Minutes Light sleep, Minutes Deep sleep, number minutes active, minutes asleep, minutes awaken, Number of awakenings, Time in bed)	Not found yet	Yacchirema et al. [86] Laxminarayan [87] Angelidou [88] Mueller et al. 2011 [89] PhD Sleep Activity Ontology

References

[1] L. Otto, L. Harst, P. Timpel, B. Wollschlaeger, P. Richter, H. Schlieter, Defining and delimitating telemedicine and related terms – an ontology-based classification, Studies in Health Technology and Informatics (2020).

[2] M. Seligman, Flourish: A Visionary New Understanding of Happiness and Well-Being, Simon and Schuster, 2012.

[3] V. Urman, La révolution épigénétique: votre mode de vie compte plus que votre hérédité, Albin Michel, 2018.

[4] M. Wei, J.E. Groves, The Harvard Medical School Guide to Yoga: 8 Weeks to Strength, Awareness, and Flexibility, Da Capo Lifelong Books, 2017.

[5] P.M. Wayne, M. Fuerst, The Harvard Medical School Guide to Tai Chi: 12 Weeks to a Healthy Body, Strong Heart, and Sharp Mind, Shambhala Publications, 2013.

[6] F. Wang, J.K. Man, E.-K.O. Lee, T. Wu, H. Benson, G.L. Fricchione, W. Wang, A. Yeung, The effects of qigong on anxiety, depression, and psychological well-being: a systematic review and meta-analysis, Evidence-Based Complementary and Alternative Medicine (2013).

[7] C.E. King, M. Sarrafzadeh, A survey of smartwatches in remote health monitoring, Journal of Healthcare Informatics Research (2018).

[8] R.W. Picard, Affective Computing, MIT Press, 2000.

[9] D. Goleman, R.E. Boyatzis, A. McKee, Primal Leadership: Unleashing the Power of Emotional Intelligence, Harvard Business Press, 2013.

[10] J. Seignalet, L'alimentation ou la 3e médecine, FX de Guibert ed, 2001.

[11] R.J. Davidson, J. Kabat-Zinn, J. Schumacher, M. Rosenkranz, D. Muller, S.F. Santorelli, F. Urbanowski, A. Harrington, K. Bonus, J.F. Sheridan, Alterations in brain and immune function produced by mindfulness meditation, Psychosomatic Medicine 65 (4) (2003) 564–570.

[12] R.J. Davidson, A. Lutz, Buddha's brain: neuroplasticity and meditation [in the spotlight], IEEE Signal Processing Magazine 25 (1) (2008) 176–174.

[13] P. Ekman, R.J. Davidson, M. Ricard, B. Alan Wallace, Buddhist and psychological perspectives on emotions and well-being, Current Directions in Psychological Science (2005).

[14] M. Ricard, L'art de la Méditation, NiL, 2011.

[15] G. Desbordes, L.T. Negi, T.W. Pace, B.A. Wallace, C.L. Raison, E.L. Schwartz, Effects of mindful-attention and compassion meditation training on amygdala response to emotional stimuli in an ordinary, non-meditative state, Frontiers in Human Neuroscience 6 (2012) 292.

[16] E.L. Rosenberg, A.P. Zanesco, B.G. King, S.R. Aichele, T.L. Jacobs, D.A. Bridwell, K.A. MacLean, P.R. Shaver, E. Ferrer, B.K. Sahdra, et al., Intensive meditation training influences emotional responses to suffering, Emotion 15 (6) (2015) 775.

[17] S.R. Islam, D. Kwak, M.H. Kabir, M. Hossain, K.-S. Kwak, The Internet of things for health care: a comprehensive survey, IEEE Access 3 (2015) 678–708.

[18] ISO, ISO/IEC 21823 Semantic Interoperability: Interoperability for Internet of Things Systems Part 3 Semantic Interoperability, 2020.

[19] M. Bauer, H. Baqa, S. Bilbao, A. Corchero, L. Daniele, I. Esnaola, I. Fernandez, O. Franberg, R. Garcia-Castro, M. Girod-Genet, P. Guillemin, A. Gyrard, C.E. Kaed, A. Kung, J. Lee, M. Lefrancois, W. Li, D. Raggett, M. Wetterwald, Semantic IoT Solutions – A Developer Perspective (Semantic Interoperability White Paper Part I), 2019.

[20] M. Bauer, H. Baqa, S. Bilbao, A. Corchero, L. Daniele, I. Esnaola, I. Fernandez, O. Franberg, R. Garcia-Castro, M. Girod-Genet, P. Guillemin, A. Gyrard, C.E. Kaed, A. Kung, J. Lee, M. Lefrançois, W. Li, D. Raggett, M. Wetterwald, Towards semantic interoperability standards based on ontologies (Semantic Interoperability White Paper Part II), 2019.

[21] P. Murdock, L. Bassbouss, M. Bauer, M.B. Alaya, R. Bhowmik, P. Brett, R.N. Chakraborty, M. Dadas, J. Davies, W. Diab, et al., Semantic Interoperability for the Web of Things, 2016.

[22] Amelie Gyrard, Kasia Tabeau, et al., Knowledge Engineering Framework for IoT Robotics Applied to Smart Healthcare and Emotional Well-Being, 2021.

[23] N.F. Noy, N.H. Shah, P.L. Whetzel, B. Dai, M. Dorf, N. Griffith, C. Jonquet, D.L. Rubin, M.-A. Storey, C.G. Chute, et al., Bioportal: ontologies and integrated data resources at the click of a mouse, Nucleic Acids Research 37 (suppl_2) (2009) W170–W173.

[24] P.-Y. Vandenbussche, G.A. Atemezing, M. Poveda-Villalón, B. Vatant, Linked Open Vocabularies (LOV): a gateway to reusable semantic vocabularies on the Web, Semantic Web Journal (2016).

[25] A. Gyrard, C. Bonnet, K. Boudaoud, M. Serrano, LOV4IoT: a second life for ontology-based domain knowledge to build semantic web of things applications, in: IEEE International Conference on Future Internet of Things and Cloud, 2016.

[26] B. Kitchenham, R. Pretorius, D. Budgen, O.P. Brereton, M. Turner, M. Niazi, S. Linkman, Systematic Literature Reviews in Software Engineering – A Tertiary Study, Information and Software Technology, 2010.

[27] M. Noura, A. Gyrard, S. Heil, M. Gaedke, Automatic Knowledge Extraction to build Semantic, Web of Things Applications, 2019.

[28] M. Noura, A. Gyrard, S. Heil, M. Gaedke, Concept extraction from the web of things knowledge bases, in: International Conference WWW/Internet 2018, Elsevier, 2018, Outstanding Paper Award.

[29] ETSI TS 103 410-8 V1.1.1 (2020-07) SmartM2M; Extension to SAREF; Part 8: eHealth/Ageing-well Domain, 2020.

[30] A. Gyrard, U. Jaimini, M. Gaur, S. Shekarpour, K. Thirunarayan, A. Sheth, Reasoning over personalized healthcare knowledge graph: a case study of patients with allergies and symptoms, in: Semantic Models in IoT and e-Health Applications, Elsevier, 2022.

[31] L. Shi, S. Li, X. Yang, J. Qi, G. Pan, B. Zhou, Semantic health knowledge graph: semantic integration of heterogeneous medical knowledge and services, BioMed Research International (2017).

[32] M. Rotmensch, Y. Halpern, A. Tlimat, S. Horng, D. Sontag, Learning a health knowledge graph from electronic medical records, Scientific Reports (2017).

[33] D. Le-Phuoc, H.N.M. Quoc, H.N. Quoc, T.T. Nhat, M. Hauswirth, The graph of things: a step towards the live knowledge graph of connected things, Journal of Web Semantics (2016).

[34] S. Lohmann, V. Link, E. Marbach, S. Negru, WebVOWL: web-based visualization of ontologies, in: Knowledge Engineering and Knowledge Management, Springer, 2014.

[35] A. Gyrard, G. Atemezing, M. Serrano, PerfectO: An Online Toolkit for Improving Quality, Accessibility, and Classification of Domain-Based Ontologies, Springer, 2021.

[36] A. Gyrard, M. Gaur, K. Thirunarayan, A. Sheth, S. Shekarpour, Personalized health knowledge graph, in: 1st Workshop on Contextualized Knowledge Graph (CKG) Co-Located with International Semantic Web Conference (ISWC), 8–12 October 2018, Monterrey, USA, 2018.

[37] M. Noura, Y. Wang, S. Heil, M. Gaedke, OntoSpect: IoT ontology inspection by concept extraction and natural language generation, in: ICWE, 2021.

[38] R. Agarwal, D.G. Fernandez, T. Elsaleh, A. Gyrard, J. Lanza, L. Sanchez, N. Georgantas, V. Issarny, Unified IoT ontology to enable interoperability and federation of testbeds, in: IEEE World Forum on Internet of Things, 2016.

[39] F. Morange-Majoux, Psycho-physiologie, 2eme edition, Dunod, 2017.

[40] S.K. Biswas, S. Chatterjee, S. Bandyopadhyay, S. Kar, N.K. Som, S. Saha, S. Chakraborty, Smartphone-enabled paper-based hemoglobin sensor for extreme point-of-care diagnostics, ACS Sensors (2021).

[41] M. Algarra, C.M. Jiménez-Herrera, J.C. Esteves da Silva, Recent applications of magnesium chemical sensors in biological samples, Critical Reviews in Analytical Chemistry (2015).

[42] P. Pirovano, M. Dorrian, A. Shinde, A. Donohoe, A.J. Brady, N.M. Moyna, G. Wallace, D. Diamond, M. McCaul, A wearable sensor for the detection of sodium and potassium in human sweat during exercise, Talanta (2020).

[43] S. Kia, S. Bahar, S. Bohlooli, A novel electrochemical sensor based on plastic antibodies for vitamin d3 detection in real samples, IEEE Sensors Journal (2019).

[44] F. Paganelli, D. Giuli, An ontology-based system for context-aware and configurable services to support home-based continuous care, IEEE Transactions on Information Technology in Biomedicine (2010).

[45] F. Paganelli, D. Giuli, An ontology-based context model for home health monitoring and alerting in chronic patient care networks, in: 21st International Conference on Advanced Information Networking and Applications Workshops, 2007, AINAW'07, Vol. 2, IEEE, 2007, pp. 838–845.

[46] P. Brandt, T. Basten, S. Stuijk, V. Bui, P. de Clercq, L.F. Pires, M. van Sinderen, Semantic interoperability in sensor applications making sense of sensor data, in: 2013 IEEE Symposium on Computational Intelligence in Healthcare and e-Health (CICARE), IEEE, 2013, pp. 34–41.

[47] H. Liu, Improving semantic interoperability in remote patient monitoring, Master's thesis, 2013.

[48] F. Latfi, B. Lefebvre, C. Descheneaux, Ontology-based management of the telehealth smart home, dedicated to elderly in loss of cognitive autonomy.

[49] A. Chatterjee, A. Prinz, M. Gerdes, S. Martinez, An automatic ontology-based approach to support logical representation of observable and measurable data for healthy lifestyle management: proof-of-concept study, Journal of Medical Internet Research 23 (4) (2021) e24656.

[50] H. Aloulou, M. Mokhtari, T. Tiberghien, J. Biswas, P. Yap, An adaptable and flexible framework for assistive living of cognitively impaired people, IEEE Journal of Biomedical and Health Informatics (2013).

[51] A. Rhayem, M.B.A. Mhiri, M.B. Salah, F. Gargouri, Ontology-based system for patient monitoring with connected objects, Procedia Computer Science 112 (2017) 683–692.

[52] R. Zgheib, Semom, a semantic middleware for iot healthcare applications, Ph.D. thesis, Université Paul Sabatier-Toulouse III, 2017.

[53] T. Lemlouma, S. Laborie, P. Roose, Toward a context-aware and automatic evaluation of elderly dependency in smart homes and cities, in: IEEE International Symposium on a World of Wireless, Mobile and Multimedia Networks (WoWMoM), IEEE, 2013.

[54] L. Mazuel, J. Charlet, et al., Alignement entre des ontologies de domaine et la snomed: trois études de cas, in: Actes de la 20ème conférence Ingénierie des Connaissances – IC2009, 2009.

[55] T.K. Colicchio, P.I. Dissanayake, J.J. Cimino, Formal representation of patients' care context data: the path to improving the electronic health record, Journal of the American Medical Informatics Association (2020).

[56] A. Hristoskova, V. Sakkalis, G. Zacharioudakis, M. Tsiknakis, F. De Turck, Ontology-driven monitoring of patient's vital signs enabling personalized medical detection and alert, Sensors 14 (1) (2014) 1598–1628.

[57] A. Hristoskova, W. Boffé, T. Tourwé, F. De Turck, Integrated semantic and event-based reasoning for emergency response applications, in: International Conference on Geo-Information for Disaster Management (Gi4DM 2012), 2012.

[58] A. Jovic, D. Gamberger, G. Krstacic, Heart failure ontology, Bio-algorithms and Med-systems 7 (2011).

[59] W. Yao, A. Kumar, Conflexflow: integrating flexible clinical pathways into clinical decision support systems using context and rules, Decision Support Systems 55 (2) (2013) 499–515.

[60] W. Yao, C. Chu, A. Kumar, Z. Li, Using ontology to support context awareness in healthcare, in: Proceedings of the 19th Workshop on Information Technologies and Systems, Dec 14–15, 2009, Phoenix, AZ, USA, 2009.

[61] P. Sherimon, R. Krishnan, OntoDiabetic: an ontology-based clinical decision support system for diabetic patients, Arabian Journal for Science and Engineering (2016).

[62] A. Mukasine, Ontology-based personalized system to support patients at home, Master's thesis, Universitetet i Agder, University of Agder, 2014.

[63] Oshani Seneviratne, Jonathan Harris, Ching-Hua Chen, Deborah L. McGuinness, Personal health knowledge graph for clinically relevant diet recommendations, arXiv preprint, arXiv:2110.10131, 2021.

[64] A. Gyrard, K. Boudadoud, A naturopathy knowledge graph and recommendation system to boost the immune system, in: Semantic Models in IoT and e-Health Applications, Elsevier, 2022.

[65] S. Enshaeifar, P. Barnaghi, S. Skillman, A. Markides, T. Elsaleh, S.T. Acton, R. Nilforooshan, H. Rostill, The Internet of Things for Dementia Care, IEEE Internet Computing, 2018.

[66] L. Li, P. Wang, J. Yan, Y. Wang, S. Li, J. Jiang, Z. Sun, B. Tang, T.-H. Chang, S. Wang, et al., Real-world data medical knowledge graph: construction and applications, Artificial Intelligence in Medicine (2020).

[67] E. Adel, S. El-Sappagh, S. Barakat, M. Elmogy, A semantic interoperability framework for distributed electronic health record based on fuzzy ontology, International Journal of Medical Engineering and Informatics (2020).

[68] E. Adel, S. El-Sappagh, S. Barakat, J.-W. Hu, M. Elmogy, An extended semantic interoperability model for distributed electronic health record based on fuzzy ontology semantics, Electronics (2021).

[69] P. Plastiras, D.M. O'sullivan, Combining ontologies and open standards to derive a middle layer information model for interoperability of personal and electronic health records, Journal of Medical Systems (2017).

[70] P. Plastiras, D. O'Sullivan, P. Weller, An ontology-driven information model for interoperability of personal and electronic health records, 2014.

[71] K. Gai, M. Qiu, L.-C. Chen, M. Liu, Electronic health record error prevention approach using ontology in big data, in: IEEE International Conference on High Performance Computing and Communications, IEEE International Symposium on Cyberspace Safety and Security, and IEEE International Conference on Embedded Software and Systems, IEEE, 2015.

[72] Roman, Electronic health record (ehr) ontology documentation (1912), http://trajano.us.es/~isabel/EHR/.

[73] Jack Hodges, Mareike Kritzler, Florian Michahelles, Stefan Lueder, Erik Wilde, Ontology alignment for wearable devices and bioinformatics in professional health care, Citeseer.

[74] C. Villalonga, H. Pomares, I. Rojas, O. Banos, MIMU-wear: ontology-based sensor selection for real-world wearable activity recognition, Neurocomputing (2017).

[75] L. Nachabe, M. Girod-Genet, B. El Hassan, F. Aro, Applying ontology to WBAN for mobile application in the context of sport exercises, in: International Conference on Body Area Networks, Institute for Computer Sciences, Social-Informatics and Telecommunications Engineering (ICST), 2014.

[76] I. Kalamaras, N. Kaklanis, K. Votis, D. Tzovaras, Towards big data analytics in large-scale federations of semantically heterogeneous IoT platforms, in: IFIP International Conference on Artificial Intelligence Applications and Innovations, Springer, 2018.

[77] R. Lyons, G.R. Low, C.B. Congdon, M. Ceruolo, M. Ballesteros, S. Cambria, P. DePetrillo, Towards an extensible ontology for streaming sensor data for clinical trials, in: Proceedings of the 12th ACM Conference on Bioinformatics, Computational Biology, and Health Informatics, 2021, pp. 1–6.

[78] J. Zhao, et al., Publishing Chinese medicine knowledge as linked data on the web, Chinese Medicine 5 (1) (2010) 1–12.

[79] M. Samwald, M. Dumontier, J. Zhao, J.S. Luciano, M.S. Marshall, K. Cheung, Integrating findings of traditional medicine with modern pharmaceutical research: the potential role of linked open data, Chinese Medicine (2010).

[80] V. Bertaud-Gounot, R. Duvauferrier, A. Burgun, Ontology and medical diagnosis, in: Informatics for Health and Social Care, 2012.

[81] J. Moreira, L.F. Pires, M. Van Sinderen, L. Daniele, M. Girod-Genet, Saref4health: towards iot standard-based ontology-driven cardiac e-health systems, Applied Ontology (2020).

[82] J. Moreira, L.F. Pires, M. van Sinderen, L. Daniele, Saref4health: Iot standard-based ontology-driven healthcare systems, in: FOIS, 2018.

[83] ETSI TS 103 410-9 V1.1.1 (2020-07): "SmartM2M; Extension to SAREF; Part 9: Wearables Domain", 2020.

[84] L. Nachabe, M. Girod-Genet, B. El Hassan, Unified data model for wireless sensor network, IEEE Sensors Journal 3 (2015) (IF: 3.076 in 2020).

[85] P. Staroch, A weather ontology for predictive control in smart homes, Master's thesis, 2013.

[86] D.C. Yacchirema, D. Sarabia-Jácome, C.E. Palau, M. Esteve, A smart system for sleep monitoring by integrating iot with big data analytics, IEEE Access (2018).

[87] P. Laxminarayan, Exploratory analysis of human sleep data, Ph.D. thesis, Worcester Polytechnic Institute, 2004.

[88] R. Angelidou, Development of a portable system for collecting and processing bio-signals and sounds to support the diagnosis of sleep apnea, Master's thesis, 2015.

[89] R.S. Mueller, Ontology-driven Data Integration for Clinical Sleep Research, Ph.D. thesis, Case Western Reserve University, 2011.

Reasoning over personalized healthcare knowledge graph: a case study of patients with allergies and symptoms

10

Amelie Gyrard[a], **Utkarshani Jaimini**[a,c], **Manas Gaur**[a,c], **Saeedeh Shekharpour**[b],
Krishnaprasad Thirunarayan[a], **and Amit Sheth**[a,c]

[a]*Ohio Center of Excellence in Knowledge-enabled Computing (Kno.e.sis), Wright State University, Dayton, OH, United States*
[b]*University of Dayton, Dayton, OH, United States*
[c]*University of South Carolina, Columbia, SC, United States*

Contents

10.1 Introduction

Recent studies have shown the need for a variety of effective solutions to maintain a healthy lifestyle (e.g., Apple Health Kit [WR2]) and are the basis for emerging applications such as digital personalized health coach applications that track and interpret health and well-being data [1]. Other related areas contributing to these applications include Wireless Body Area Networks (WBANs) [2], medical Internet of Things (mIoT), m-Health, e-Health, and Ambient Assisted Living (AAL) [WR3]. These applications utilize: (1) inexpensive IoT devices and wearables to collect raw data (e.g., Fitbit to measure activity and sleep patterns), (2) a knowledge graph (e.g., obesity ontology [WR4]) to structure and abstract the data, and (3) a reasoning mechanism to deduce insights and recommendations (e.g., using environmental data with a rule-based inference engine to provide high-level abstractions). The World Health Organization (WHO) [WR1] estimates 235 million people had asthma in the year 2018. In cases such as asthma, a multifactorial disease, a lot of different data types are relevant. However, the state-of-the-art lacks appropriate exploitation of contextual and personalized data and their integration with background medical knowledge bases for monitoring user health.

Context-awareness refers to the use of external data that can impact the user's situation. For instance, IoT devices can be used to monitor the surrounding environment. Interpretation of fine-grained IoT data using a background model for abstraction provides contextual awareness to clinicians. Clinical protocols should take these into account to determine a patient's condition [WR5]. For example, each patient reacts differently when exposed to environmental factors (e.g., air pollution or different types of pollen).

Personalization adjusts the treatment according to the patient's vulnerability and severity. Augmented Personalized Healthcare (APH), a strategy for patient empowerment [3] consists of self-monitoring, Self-Appraisal, self-management, intervention, and disease progression tracking and prediction. Self-monitoring is defined as continuous monitoring of patient health data using IoT devices, wearables, environmental sensors, etc. Self-appraisal is the ability of the patient to evaluate the relevance of the data and observations within the context of the patient's health objectives. Self-management is the decision and behavioral changes a patient engage in to impact one's health objectives. Intervention is the change in the patient's care protocol prescribed by the clinician. Disease progression tracking and prediction is the longitudinal collection of Patient Generated Health Data (PGHD) and environmental data, which can facilitate monitoring of patient's disease progress, predict a significant change in their health status, and take remedial actions. Patient's multimodal data is obtained from clinical documents (demographic information, clinician's observations, lab tests, data collected during clinical visits), PGHD including sensor and social data (see Fig. 10.1, e.g., involving more than 25 types of data for each pediatric asthma patient in [4]). The PHKG understands treatments based on the patient's vulnerabilities, triggers, and symptoms. Personalization, in conjunction with predictive analytics, is needed to realize self-management and intervention [5]. For instance, the dosage of long-term controller medication prescribed for an asthmatic patient to control the symptoms is tailored to the person's asthma severity, the potential for environmental triggers, and the history [6].

The personalized healthcare knowledge graph [WR6] is a representation of all the relevant medical knowledge and personal data for a patient. A knowledge graph is comprised of an interlinked set of ontologies and data sets. PHKG formalizes medical information in terms of relevant relationships between entities. For instance, an asthma KG can describe causes, symptoms, and treatments for asthma, and PHKG can be the subgraph containing the causes, symptoms, and treatments that are tailored and applicable to a given patient. PHKG can support the development of innovative applications such as digital personalized health coach application that can keep patients informed, help manage their chronic condition, and empower the clinicians to make effective decisions on health-related issues or receive timely alerts as needed through continuous monitoring.

Societal Challenges (SC) and **Technical Challenges (TC)** for building a PHKG are investigated as follows:

- **SC1**: What recommendations can be suggested by health applications to assist patients?
- **SC2**: How can personalized health coach applications help clinicians?
- **SC3**: Web sites such as airnow.gov and pollen.com provide visualization of environmental factor's data sets according to their quality (e.g., low or high), which is mainly used by humans to under-

stand the current environmental condition. How can machines automatically interpret and get this information? We need to provide the range (e.g., between X and Y then it is considered HIGH) to the machine to understand the data either with rule-based reasoning (used in this chapter) or machine learning techniques (kept as future work).

- **SC4**: Can a machine be trained and allowed to diagnose a disease such as asthma?
- **SC5**: What environmental conditions (e.g., pollen level) impact an individual patient and trigger asthma symptoms (e.g., cough)?
- **TC1**: How to build a PHKG? How are Google Health KG or IBM Watson KG made? Information provided in tutorials such as [7] does not offer concrete steps to construct a KG.
- **TC2**: How to deduce meaningful information from a data set or actionable information from PHKG?
- **TC3**: How to maintain data privacy and security of patient data? Our IRB deals with this critical issue, including anonymization (information that identifies a patient remains in the DCH'SHIPAA compliant scope and is not associated with the kHealth-Asthma managed data). Additional exploration is discussed in [8] and is not part of this chapter.

Addressing these challenges has to lead to our main **Innovations (I)** as discussed below:

- **I1**: The PHKG for Asthma, extending our position paper [9].
- **I2**: A set of predefined queries provided as a tutorial that uses the kAO ontology and query the enriched semantic data sets.
- **I3**: A PHKG rule-based inference engine applied to asthma disease, which interlinks domain knowledge.
- **I4**: A set of interoperable rules following the linked data and linked vocabularies trends.

To design the PHKG prototype, we set out with the following **Assumptions (A)**:

- **A1**: All scientific papers mentioning an ontology and rules should provide a reproducible result, which means that the code or a demonstrator is available online and reproduction of the experiments should be quick and easy. Moreover, according to ontology engineering best practices (e.g., FAIR principles), ontologies are expected to be shared online.
- **A2**: Each IoT data type (e.g., air temperature, pollen) can be annotated with a concept from the kAO ontology, which is interlinked with other relevant ontologies to enrich raw data sets.
- **A3**: All data sets are already converted to Sensor Measurement Lists (SenML) format [WR7] supported by Cisco and Ericsson, a format to unify sensor measurements. The automatic semantic annotator supporting heterogeneous data formats (CSV, various JSON formats, etc.) is considered as future work.

Use Case: We demonstrate the use of PHKG with the kHealth-Asthma project [WR8,WR35] developed at Kno.e.sis Research Center in collaboration with Dayton Children's Hospital. kHealth-Asthma is a framework for continuous monitoring of the patient's health signals, indoor and outdoor environmental data. It generates notifications as needed to assist both the patient and the clinicians [10]. The kHealth-Asthma data set integrates data from three different sources: (1) clinicians provided anonymized extracts (throughout kHealth-Asthma, a patient is only identified by an id, patient identity is only known to the relevant clinical personnel at DCH for HIPAA compliance), (2) *Environmental data* collected using IoT devices (e.g., Foobot) and by Web Services (for outside humidity, outdoor temperature, air quality [WR9], and pollen index and type [WR10]), and (3) *Personal health signals* recorded using mobile application (e.g., a short questionnaire) and IoT devices (e.g., Fitbit, peak flow meter) to provide data on sleep, activity, and heart rate, etc.

The remainder of this chapter is organized as follows. First, we discuss the related work limitations in Section 10.2. Then we introduce the PHKG reasoner in Section 10.3. The implementation, results, and evaluation are explained in Section 10.4. Discussions are provided in Section 10.5. Finally, we draw our conclusion in Section 10.6.

10.2 Related work

We review the state-of-the-art healthcare KGs in this section, and categorize them based on their interpolation with statistical and learning-based approaches.

Knowledge Vault and **Google Knowledge Graph** can be considered as pioneering works for the 'knowledge graph" key phrase (aka semantic web technologies and philosophy) [11]. Knowledge graphs quality (e.g., error detection) and evaluation are surveyed in [12]. The survey does not provide references regarding Healthcare KG but compares Machine Learning (ML)-based KGs. Major search engines Bing, Google, and Yahoo created **Schema.org** [13], a set of vocabularies to annotate web pages. Annotations are embedded in websites to structure data on the web. The first application of Schema.org is Google's Rich Snippets. Schema.org supports RDFa, JSON-LD, and microdata part of HTML5. Schema.org demonstrates the wide deployment and usage of semantic web technologies.

Personal Health KG for diet (that includes an ontology) (Seneviratne et al. [14]) considers user's temporal personal health data (and mentioned our personalized health KG past work [9]) to personalize dietary recommendations for diabetic patients. Food recommendations are provided to answer such questions: (1) What should I eat for breakfast? (2) What foods can I eat if I have a dairy allergy? (3) What can I substitute for food Y? (patient's taste preferences).

Medical KG from large-scale Electronic Medical Records (EMRs) (Li et al. [15]) infers possible diseases and recommend medical orders. Eight steps are needed to build the medical KG: data preparation, entity recognition, entity normalization, relation extraction, property calculation, graph cleaning, related-entity ranking, and graph embedding. The data set contains 16,217,270 deidentified clinical visit data of 3,767,198 patients from Southwest Hospital in China. The KG contains 22,508 entities and 579,094 quadruplets (instead of usual triplets). Use cases for Parkinson's disease and lung cancer are provided. International Classification of Diseases (ICD-9) standard is used to map disease, diagnosis, and surgery terms. Two experiments are performed: (1) Bi- LSTM network and (2) graph embedding to a neural network (Bi- LSTM network) task to predict medicine prescription by diseases using the MIMIC3 data set.

Depression KG [16] is a disease-centric KG applied to major depressive disorder, and it addresses several challenges: (1) Heterogeneity of data sets, (2) text processing, (3) incompleteness, inconsistency, and incorrectness of data sets, and (4) expressive, representation of medical knowledge. Depression KG, utilizing rule-based reasoning over the KG, can help psychiatric doctors with no KG expertise. A **health KG to analyze EMRs** is designed by Rotmensch et al. [17] to process high-quality knowledge bases automatically (e.g., ICD-9 and UMLS) linking 156 diseases and 491 symptoms directly from EMRs. Medical concepts are extracted from 273,174 patient records. Three probabilistic models are used to construct the KG: Logistic regression, Naive Bayes and a Bayesian network using noisy-OR gates. The authors use the manually curated Google Health KG [WR36] to demonstrate that their work provides better precision and recall. Shi et al. [18] designed a **pneumonia KG**, and perform a contextual inference pruning algorithm on knowledge graphs. **Active Semantic Electronic Medical Record (ASEMR)** [19] is a rule-based application that semantically annotates EMRs that aims to (1) reduce medical errors, (2) improves physician efficiency, (3) improves patient safety and satisfaction in medical practice, (4) improves quality of billing records leading to a better payment, and (5) makes it easier to capture and analyze health outcome measures.

Traditional Chinese Medicine (TCM) Health preservation KG [WR11] integrates a visualization functionality to browse the KG [20]. The KG (designed in Chinese) has relationships between symptoms, syndromes, diseases, prescriptions, and treatments. The comparison of this work with the literature survey is missing and lacks a clear explanation of the personalization recommendation. Another **Traditional Chinese Medicine (TCM) KG** contains Electronic Medical Records (EMRs) in Chinese that are automatically processed and analyzed [21]. The KG focuses on symptoms and symptom-related entity extraction. Data related to symptoms, diseases, and medicines were collected from eight mainstream healthcare websites: Familydoctor, PCbaby, fh21, JIANKE, 120 ask, 39Health, 99Health, and QQYY. The Linked Open Data set (LOD) (in Chinese) uses UMLS and is shared on CKAN [WR12]. The work combines machine learning and knowledge graphs for the medical domains. A Conditional Random Field (CRF) model extracts symptoms from EMRs. A decision tree classifier is employed with seven labels: department, TCM, western medicine, symptom, disease, examination, and others. A framework for automated extraction method for TCM medical KG construction is designed in [22]. The KG construction framework uses an ontology model and deep learning technique (Recurrent Neural Network). The framework comprises of four main modules: (1) A medical ontology constructor explicitly adds knowledge (i.e., metadata) from unstructured clinical texts by utilizing NLP techniques (e.g., named entity recognition and text classification), (2) a knowledge element generator generates triples, (3) a structured knowledge data set generator extracts medical words to build the medical ontology, and (4) a graph model constructor uses deep learning algorithms to build the KG. The evaluation of the framework is done with a data set, which comprises of 886 patient cases for a hypertension scenario.

The need to combine KG and ML is introduced in [23] with a set of challenges of dealing with knowledge such as (1) incompleteness, (2) implicitness, (3) heterogeneity, and (4) different models. However, no healthcare use case is mentioned. Reviews on statistical models that can be trained on KGs is introduced in [24]. Statistical models of graphs can be combined with text-based information extraction methods for automatically constructing KGs from the Web, as demonstrated within Google's Knowledge Vault.

The section below, explains the challenges in integrating heterogeneous models to maintain a PHKG utilizing various modalities supported by kHealth for an asthma patient. Table 10.1 provides references and abstract summaries of the KG scientific articles (discussed above). It also cites nonhealthcare KG such as Schema.org (designed by significant internet companies) to understand how KGs work. This literature analysis shows that there is no reasoner inferring abstraction from sensor data sets and later combining them with multidisciplinary external domain knowledge (e.g., health and weather ontologies). Only two papers integrate KG and ML technologies for healthcare: Shi et al. [18] and Rotmensch et al. [17]. Le Phuoc et al. [25] is the only work addressing IoT data, but not for the healthcare use case. None of those KGs discusses the challenges mentioned above, and hence none is appropriate for use by for kHealth-Asthma project.

Table 10.1 Summary of (Healthcare) Knowledge Graph (KG) using or not Machine Learning (ML).

Authors	Year	Subject	KG	ML	Health KG
Seneviratne et al. [14]	2021	Health KG and Diet	✓	✗	✓
Li et al. [15]	2020	Medical KG from EMRs	✓	✓	✓
Shi et al. [18]	2017	Health KG	✓	✓	✓
Yu et al. [20]	2017	KG TCM Visualization	✓	✗	✓
Ruan et al. [21]	2017	KG TCM	✓	✗	✓
Weng et al. [22]	2017	KG TCM	✓	✗	✓
Wilcke et al. [23]	2017	KG and ML	✓	✓	✗
Rotmensch et al. [17]	2017	Health KG and EMR	✓	✓	✓
Huang et al. [16]	2017	Depression KG	✓	✗	✓
Paulheim et al. [12]	2016	Survey KG	✓	✗	✗
Guha et al. [13]	2016	Google KG, Schema.org	✓	✗	✗
Nickel [24], Murphy, Tresp,	2016	KG and ML, Google KG	✓	✓	✗
Dong, Gabrilovich et al. [11]	2014	Knowledge Vault, Google KG			
Le-Phuoc Graph of Things [25]	2016	KG for IoT	✓	✗	✗
Ramaswami et al. [WR36]	2015	Google Health KG	✓	✗	✗
Sheth et al. [19]	2007	Ontology and rule-based system for EMRs – Active Semantic Electronic Medical Record (ASEMR)	✓	✗	✓

Legend: Traditional Chinese Medicine (TCM).

10.3 A reasoner for personalized health knowledge graph

PHKG Architecture: We explain the method to build the PHKG in terms of (1) its architecture (introduced in [18]), (2) the use cases considered, (3) the medical data sets obtained from the Linked Open Data (LOD) cloud [26], (4) the deductive reasoning mechanism to get high-level information from IoT data sets, and (5) an online ontology catalog tool to reuse and share the domain knowledge. The architecture designed to build our PHKG is introduced in Fig. 10.1. PHKG uses heterogeneous knowledge sources: (1) IoT data provided by sensors, (2) medical data sets from Kno.e.sis Alchemy API [WR31] that allows access to SNOMED-CT [WR13], UMLS [WR14], and ICD-10 [WR15], (3) ontology catalogs to reuse models (e.g., asthma ontology), and (4) a set of unified deductive rules to interpret data.

Data workflow is highlighted in Fig. 10.2. Data comes from data sets provided by the kHealth project to infer higher-level knowledge using our KG-based reasoner. Each kHealth data set (e.g., asthma) must be semantically annotated according to an ontology (e.g., the asthma data set will be designed according to the kAO ontology [WR37]). It is a required step to later execute the kHealth reasoner to infer new triples as explained below. To infer meaningful information from data sets, we designed a set of rules. Those rules are mainly extracted from scientific publications when available or from web services having the domain expertise required to get abstractions from the data. The main novelty of our kHealth reasoner is the combination of different domain knowledge through IF THEN ELSE rules.

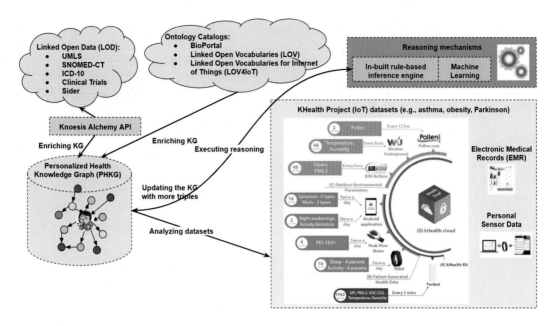

FIGURE 10.1

Personalized Health Knowledge Graph (PHKG) architecture.

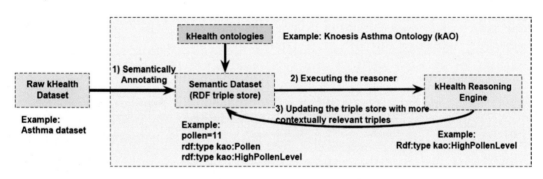

FIGURE 10.2

Enriching health and IoT data sets with rules: Overview of the reasoner workflow.

The ELSE part enables us to deduce meaningful abstraction from IoT measurements. Once the rules are executed, the IoT data is linked to specific domain ontologies (e.g., those referenced in Table 10.2) or data sets.

A detailed implementation of the kHealth reasoner is given in Fig. 10.3, which demonstrates the data enrichment with a simple scenario: as input data, a raw value (e.g., pollen level 10) is given; as an output, a recommendation is given "do not got outside" since the pollen is considered HIGH. Our kHealth reasoner comprises several components as follows:

Table 10.2 Relevant knowledge (implemented as ontologies or data sets) for the kHealth projects to build the PHKG.

Topic	Knowledge expertise	Ontology Catalog/ Data set Catalog	Ontology URL
Asthma	Not found yet	BioPortal, AberOWL	Asthma Ontology (AO) [WR24]
Asthma	KHealth Project introduced above	LOV4IoT	Kno.e.sis Asthma Ontology (kAO) [WR16]
Person	Schema.org [13]	–	[WR32]
Sensors	W3C SOSA/SSN	LOV, LOV4IoT	URL [WR33]
Smart home and weather	Staroch et al. 2013 [28]	LOV4IoT, LOV	Staroch's ontology [WR22]
Smart home and weather	Kofler et al. 2011 [29]	LOV4IoT, LOV	Kofler's ontology [WR23]
Food	SMART PRODUCTS	LOV4IoT	URL [WR28b]
Health	SNOMED-CT	BioPortal	URL [WR13]
Health	ICD-10	BioPortal	URL [WR15]
Health	UMLS	BioPortal	URL [WR14]
Health	Clinical trials	BioPortal	URL [WR27]
Health	RxNorm	BioPortal	URL [WR25]
Health	SIDER	–	URL [WR26] [WR26b]
Health	MEDDRA	BioPortal	URL [WR27]
Health	Diseasome (disease)	–	Data set URL [WR40]
Health	DrugBank [WR39]	BioPortal	XSD Schema URL [WR40]
Patient, Health	Hristoskova et al. 2014 [30]	LOV4IoT	Not yet
Air pollution	Oprea et al. 2009 [31]	LOV4IoT	Not yet

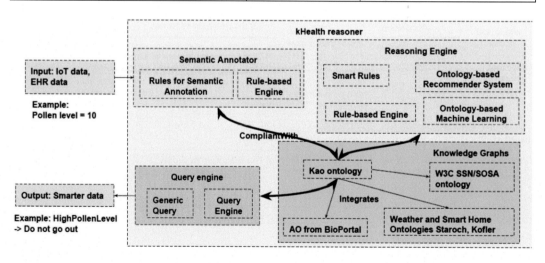

FIGURE 10.3

End-to-end data workflow to automatically enrich and query smarter data.

- **Semantic Annotator** automatically annotates heterogeneous data according to the kAO ontology.
- **Reasoning Engine** infers abstraction from semantic data. It comprises a set of interoperable rules compliant with the kAO ontology.
- **Knowledge Graph (KG)** reuses health knowledge expertise from reputed and reliable ontology catalogs such as BioPortal, LOV, LOV4IoT, and AberOWL (as depicted in Table 10.2). Our Kno.e.sis Asthma Ontology (kAO) [WR16] incorporates parts of several domain-specific ontologies:
  - Asthma Ontology (AO) [WR24] from BioPortal to reuse relevant concepts.
  - W3C SOSA/SSN [WR33] to semantically annotate sensor observations (e.g., the *peak flow meter* is a subclass of the sosa:Sensor class).
  - Weather to deduce meaningful information from weather data sets (e.g., outside temperature and humidity).
  - Smart home ontologies [27,28] to interpret inside temperature and humidity data.
  - Schema.org [WR32] to describe people.
  - RXNORM ontology [WR25] has been used to get additional information about asthma medications (e.g., Ventolin, Abdulterol) (see Appendix 10.B).
  - SMART PRODUCTS ontologies [WR28b] that describe food, recipes, and food allergies (e.g., nut scenario) (see Appendix 10.B).
  - More ontologies are planned to be integrated (e.g., SNOMED-CT [WR13], MEDDRA [WR30]).
- **Query engine** provides a generic request to query smart data. The first column in Table 10.3 contains the type of data to retrieve (e.g., pollen). A generic SPARQL query is illustrated in Listing 10.6.

The kAO ontology (Fig. 10.4) has been manually designed for the needs of the kHealth asthma project in collaboration with the Dayton Children's Hospital and evaluated with semantic web quality tools.

The integration of the domain knowledge provides several challenges: (1) *understanding the context* is done by formalizing domain knowledge, which is tailored to achieve a specific application, and (2) *knowledge implementation is not shared online*: a significant debate about reproducible results (e.g., FAIR principles) within research papers [WR29]. We have to reimplement the explicit knowledge, which is time-consuming and costly, and does not follow the primary goal of ontologies (sharing and reusing knowledge).

10.4 Implementation, results, and evaluation

Our rule-based reasoner has been implemented (explained in Section 10.4.1) to interpret health and environmental data sets, enriched with domain knowledge expertise available within the PHKG (summarized in Table 10.2). The PHKG is evaluated according to semantic web best practices (Section 10.4.2).

10.4.1 Implementation

Technologies: The current implementation was done in the Java 7 language, and using the Jena[1] semantic web framework. Jena is an open-source Java framework (supported by Apache) to develop semantic

---

[1] https://jena.apache.org/.

Table 10.3 Relevant rules for health projects (e.g., asthma, obesity, sleeping disorders).

Sensor measurement type	Rules	Source
Air Quality Index (AQI)	Good AQI US (0–50 AQI) Moderate AQI US (51–100 AQI) Unhealthy Sensitive Group AQI US (101–150 AQI) Unhealthy AQI US (151–200 AQI) Very Unhealthy AQI US (201–300 AQI) Hazardous AQI US (301–500 AQI)	Air quality web site [WR21]
Pollen Index	Low Pollen Level (0–2.4) Low MediumPollen Level (2.5–4.8) Medium Pollen Level (4.9–7.2) Medium High Pollen Level (7.3–9.6) High Pollen Level (9.7–12)	Pollen web site [WR10]
Outside Humidity	DryHumidity (30%–40%) NormalHumidity (40%–70%) VeryMoistHumidity (80%–100%) VeryDryHumidity (0%–30%) MoistHumidity (70%–80%)	Staroch et al. [28]
Inside Temperature	AboveRoomTemperature (25–30 °C) AboveRoomTemperature (>25 °C) BelowRoomTemperature (10–20 °C) BelowRoomTemperature (<20 °C) RoomTemperature (20–25 °C) Frost (0 °C) Frost (0–(-25.0) °C) ExtremeFrost (>-25.0 °C) ExtremeHeat (>37 °C) Heat (>30 °C) AboveZeroTemperature (>0 °C)	Staroch et al. [28] Kofler et al. [29] Staroch et al. 2013 [28] Kofler et al. 2011 [29] Staroch et al. [28], Kofler et al. [29] Staroch et al. [28] Kofler et al. [29] Kofler et al. [29] Kofler et al. [29] Staroch et al. [28], Kofler et al. [29] Kofler et al. [29]
Inside Humidity	Low Humidity (<50%) High Humidity (51–69%) High Humidity (>70%)	Yacchirema et al. [32]
Activity (Minutes active, Sedentary minutes, Minutes lightly active, Number of steps)	Sedentary Person (< 5000 steps Count) Mild Active Person (5000–7499 steps Count) Moderate Active Person (7500–9999 steps Count) Active Person (10,000–12,499 steps Count) Highly Active Person (>= 12,500 steps Count)	Yacchirema et al. [32]
Snoring	Normal Snoring Level (< 40 dB) Mild Snoring Level (40–50 dB) Moderate Snoring Level (50–60 dB) Severe Snoring Level (>= 60 dB)	Yacchirema et al. [32]
Peak Flow Meter	Asthma Green Zone Asthma Yellow Zone Asthma Red Zone	American Lung Association Peak Flow rate [WR20]
Sleep (Minutes REM Sleep, Minutes Light sleep, Minutes Deep sleep, number minutes active, minutes asleep, minutes awaken, Number of Awakenings, Time in Bed)	Not found yet	Yacchirema et al. [32] Laxminarayan [33] Angelidou [34] Mueller et al. 2011 [35] PhD Sleep Activity Ontology
CO_2	Not found yet	Oprea et al. [31]

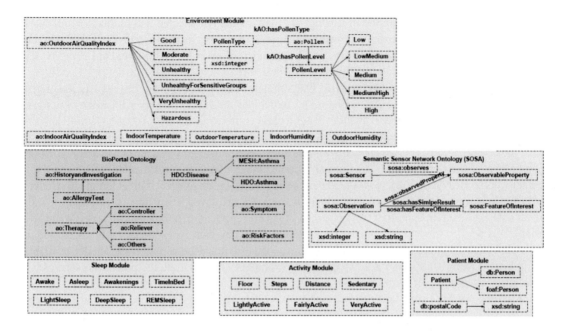

FIGURE 10.4

Overview of the kAO ontology.

web applications. The Jena framework provides an inference engine (rule-based reasoning) to deduce meaningful knowledge from semantic data sets. AndroJena, a light version of the Jena framework, compatible with Android devices, also provides the query engine and the inference engine, which can be considered for further extensions. Semantic languages, supported by W3C, such as RDF, RDFS, and OWL are employed to develop the KG made of ontologies (e.g., kAO ontology) and used to create the automatic semantic annotation component (see Fig. 10.3) to unify data sets. W3C SPARQL language is used to query the PHKG.

Input data example: Data sets are provided in JSON. A simple pollen data JSON example is depicted in Listing 10.1. Listings are provided within the Appendix section.

We use the SenML/XML language to unify data in this chapter since it is a format used by companies such as CISCO and W3C Web of Things standards. An example of a simple SenML data sample to describe pollen is depicted in Listing 10.2.

Semantic annotator: It automatically annotates SenML/XML data according to the kAO ontology. For instance, pollen will be explicitly annotated as kAO:PollenLevel. Pollen is described as a sosa:observedProperty, that demonstrates the usage of the W3C SSN/SOSA ontology. A rule example to automatically annotate the data compliant with kAO is depicted in Listing 10.3. The execution of the rules will explicitly add the new triple ?measurementUri rdf:type kAO:PollenLevel.

Semantic rule-based inference engine: The rules are compliant with the kAO ontology. The Jena inference engine (also previously used for the automatic semantic annotation) is used to infer high-level abstractions by executing a set of "common sense" rules. A rule example of deducing the high-level abstraction `HighPollen` is illustrated in Listing 10.4. We manually collected and unified a set of rules (as represented in Table 10.3) extracted from guidelines or scientific publications, which are based on various rule languages and semantic reasoners. The grammar of our rules is illustrated in Listing 10.10, which has been inspired by the Jena rule syntax and structure.

Ontology-based recommender system: Our ontology-based recommender system is illustrated in Listing 10.5. The property `hasRecommendation` combines the kAO ontology with external domain knowledge. In this example, the `HighPollenLevel` class is linked to the `DoNotGoOut` class. `HighPollenLevel` must be a class since it is used to annotate the data semantically and to provide the explicit `rdf:type HighPollenLevel`, which is processable and understandable by machines. `DoNotGoOut` could be defined as a class or instance. The current recommendation system implementation is being done manually and considered as future work to semiautomatically enrich our KG with external knowledge.

Generic query engine: The generic SPARQL query is displayed in Listing 10.6. The query retrieves data classified in Table 10.3 (first column). A generic application template is provided where the variable `?semanticAnnotationTypeUri` can be replaced by any data types already referenced within kAO (e.g., pollen, humidity). `?semanticAnnotationTypeUri` must be replaced by an IRI available within kAO. Listing 10.8 shows a SPARQL query example to demonstrate the powerful usage of the knowledge-based reasoner with a patient having allergies to nuts. An allergy to nuts is the perfect example since we can query the food taxonomy [WR28b], which explicitly mentions and references differences types of nuts (e.g., peanuts, macadamia nuts). Further, with an ASK SPARQL query, we can illustrate the potential of our reasoner compared to usual keyword-based approach (e.g., butternut squash is not a nut).

Output data: A SPARQL result (Listing 10.7) is returned by the Jena framework when the generic SPARQL query is executed. The result can be easily parsed (by any developers, even with no semantic web expertise) and returned through web services. Listing 10.7 is a simple output example providing:

- *SemanticAnnotationType* is the explicit information added by the semantic annotation component with a set of rules to deal with synonyms (precipitation = rainfall), descriptions (t=temp=temperature), typo issues, etc. It is also compliant with the kAO ontology.
- *Deduce* is the result returned once the inference engine is executed with a set of rules that provide higher abstractions.
- *Suggest* is the result provided by the generic SPARQL query executed when rules have been interlinked with external domain ontologies and data sets.

The implementation use case for asthma, obesity, and sleep disorders: To implement the rule data set, we investigated the literature in-depth to acquire knowledge from domain experts (air quality, pollen, outside environment, inside environment, etc.). The data set of rules is summarized in Table 10.3.

The current implementation is tested with *various use cases* to interpret different data types (as depicted in Fig. 10.5 and Fig. 10.6) to demonstrate the genericity of the reasoner: (1) pollen, (2) air quality, (3) inside temperature, (4) outside humidity, (5) heart rate (6) steps count, (7) food caloric level (8) snoring level, (9) body mass index, (10) peak expiratory flow, and (11) sleep disorder breathing.

We are integrating more and more scenarios in Table 10.3. Progressively, we will obtain a generic application comprising the components (ontology, semantic annotator, rule-based engine) compatible with each other.

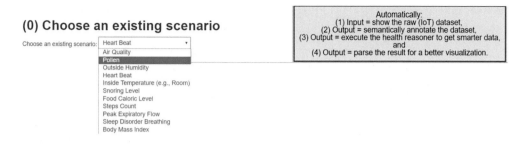

FIGURE 10.5

Scenarios supported by the health reasoner (pollen data, air quality, outside humidity, heart beat, inside room temperature, snoring level, food caloric level, steps count, body mass index, sleep disorder breathing).

FIGURE 10.6

The health reasoner enriches raw data with high level abstractions and suggestions.

10.4.2 KAO ontology evaluation

To develop the ontology, we follow FAIR principles, criteria, and best practices recognized by the semantic web community: (1) ontology methodology developments [36,37], (2) the International Semantic Web Conference (ISWC) Resource Track [WR17] encourages FAIR principles for ontology submissions, and (3) the PerfectO methodology [38,39] [WR18].

Ontology methodologies recommend defining the *competency questions*. The primary purpose of designing the ontology in our project is to infer high-level abstractions from sensor data. The inference engine executes a set of rules (Table 10.3) compliant with the kAO ontology.

ISWC Resource Track guidelines: The kAO ontology has been evaluated following the semantic web best practices preconized by the ISWC Resource Track guidelines, which provide the following criteria: (1) impact, (2) reusability, (3) design and technical quality, and (4) availability.

Table 10.4 Table from SemWeb Best Practices for Dummies with PerfectO [WR19] (Part I).

Rule number	Description	Difficulty	Status
Rule 1	Finding a good ontology name	*	DONE (kao.owl)
Rule 2	Finding a good ontology name space	**	DONE (http://purl.org/kAO#)
Rule 3	Sharing your ontology online	**	DONE
Rule 4	Adding ontology metadata	**	DONE
Rule 5	Adding rdfs:label, rdfs:comment, dc:description for each concept and property	*	DONE
Rule 6	All classes should start with an uppercase and properties with a lowercase.	*	DONE
Rule 7	Submitting your ontology to ontology catalogs	**	ONGOING
Rule 8	Reusing and linking ontologies	***	DONE (see above all ontologies reused)
Rule 9	Dereferenceable URI copy paste the namespace URL of your ontology in a web browser to get the code	**	DONE
Rule 10	Checking syntax validator	*	DONE
Rule 11	Adding ontology documentation	*	ONGOING
Rule 12	Adding ontology visualization	*	ONGOING
Rule 13	Improving ontology design	***	ONGOING
Rule 14	Improving dereferencing URI and content negotiation	***	ONGOING
Rule 15	Protege ontology editor	***	DONE (can be loaded under Jena and Protege)

Legend: DONE means that the kao ontology is integrated with the tools mentioned. ONGOING means additional efforts are required (e.g., incompatibility issues to fix).

- *Impact and reusability*: The kAO ontology has been exploited in various scenarios (as demonstrated above). Automatic documentation can be provided with the LODE tool; the kAO documentation is available [WR16].
- *Design and technical quality*: The kAO ontology is being validated with a set of validation tools designed by the semantic web community. The design of the ontology is available online [WR36] as a graph visualization with WebVOWL. We have improved the ontology with the Oops tools, which automatically detects common pitfalls and provides recommendations to fix them. Oops employed with our kAO ontology can be tested online. The Vapour tool integrated with the kAO ontology has been used to check dereferencing URI and content negotiation. Finally, TripleChecker checks that the use of existing ontologies has been correctly used within our kAO ontology.
- *Availability*: The ontology is published at a persistent PURL URL. PURL is highly encouraged to define the ontology URL, so that the ontology can be hosted on different servers, and PURL offers URL redirections.

Best practices with PerfectO: We have also followed the PerfectO methodology [38], which synthesizes a set of additional best practices and eases their achievements with the step-by-step tutorial [WR38] that helps improve the ontology. The set of fifteen best practices suggested by PerfectO is summarized in Table 10.4. The first column indicates the rule number, the second column describes the rule, the third column provides the difficulty level to follow the rule from 1 to 3, and the fourth column indicates if the rules are supported or not with two options DONE or ONGOING. For instance, rule 3 encourages to share the ontology online, which has been done: http://purl.org/kao.

10.5 Discussions and extensions for future work

Digital Phenotype Score discussion and it's use case. The kHealth team developed a set of scores for the asthma pediatric patients [4]: (1) symptom score, (2) rescue score, (3) controller compliance score, (4) activity score, (5) awakening score, (6) asthma control test score, and (7) digital phenotype score.

Technical limitations of the reasoner. Several extensions are possible to improve this work: (1) Enriching the health reasoner with additional scenarios, diseases, etc., (2) extending the semantic annotator to support heterogeneous data formats, and (3) reasoner reliability (consistency and completeness), and proving suggestions (e.g., by keeping the provenance of the facts).

Health reasoner enriched with additional scenarios: Measurement types that are supported are shown in Table 10.3. The first column illustrates measurement types for the health projects: peak flow, air quality, pollen index, outside humidity, inside temperature, sleep (REM, light sleep, deep sleep, number of minutes active), CO2, activity (minutes active, sedentary minutes, minutes lightly active, number of steps), outside temperature, inside humidity, ozone, VOC, and PM 2.5. If the kHealth projects aggregate more data, apparently the new data type must be supported to deduce abstractions when the reasoner is executed. The second column illustrates the high-level abstraction (e.g., high pollen level) inferred by the inference engine to interpret simple values (most of the time just a number). The third column indicates knowledge provenance (e.g., scientific publications, website) to infer high-level abstractions.

Semantic annotator extension to support heterogeneous data formats: Our assumption is that raw data sets were in the SenML/XML format. Our automatic semantic annotator needs to be extended to support raw data sets and heterogeneous formats (CSV, JSON, Google Sheets, etc.) and heterogeneous ways to describe data and metadata. Integrating solutions such as SPARQL-Generate [WR34] can help achieve this extension.

Reasoner reliability (consistency and completeness): The inference engine can be enriched with additional rules to cover more data sets. However, consistency and completeness must be maintained. We currently perform a manual checking [WR28a]. When several knowledge sources are defining the same measurement, we exploit the range designed by experts providing fine-grained rules. For instance, one source of knowledge is defining eight rules to interpret data, while a second source of knowledge only three rules, then we will ignore the second source of knowledge. The refinement of the recommendation system is considered as future work.

10.6 Conclusion

The PHKG will assist patients and clinicians, as a long-term vision, in automatically interpreting a very broad variety of data (EHR and IoT) that can be relevant to a multifactorial disease such as asthma for developing a personalized digital health application.

PHKG is integrated within an architecture and a methodology, which is supported by a prototype and a tutorial. The architecture is comprised of three main components: (1) a semantic annotator to unify heterogeneous data set sources and formats according to the kAO ontology, and (2) a rule-based reasoner to automatically infer abstractions from data sets, and (3) a generic query engine to retrieve smarter data according to the data type (e.g., pollen level). The three components are compliant with the kAO ontology. The ontology-based reasoner for personalized healthcare is an innovative solution to integrate various domains and knowledge.

Short-term challenges: The rule-based PHKG is designed for the asthma disease but could be extended to predict diagnosis, provide more recommendations, and applied to other conditions (e.g., obesity, dementia, epilepsy, and Parkinson's).

Mid-term challenges: Working on efficiency and scalability of the proposed approach. Additional knowledge can be integrated with natural language processing and machine learning/deep learning techniques by semiautomatically extracting knowledge from scientific publications, or any relevant sources.

Long-term challenges: The methodology to build the personalized knowledge graph could be generalized to be applied to other domains such as agriculture with smart irrigation to deal with different crop types, robotics in smart homes for aging people, smart energy, etc. For instance, Google is designing a research agenda on personal knowledge graphs [40], and adopted the ideas/built upon our personalized health knowledge graph work [9].

Acknowledgments

Thanks to the kHealth team for fruitful discussions and feedback, and Amanuel Alombo for the Kno.e.sis Alchemy API design. This work is partially funded by kHealth NIH 1 R01 HD087132-01 and Hazards SEES NSF Award EAR 1520870. The opinions expressed are those of the authors and do not reflect those of the sponsors.

Appendix 10.A Listings: code example

```
1   "_index": "currentpollenbymonitoringsites",
2        "_type": "pollen",
3        "_id": "AV4bMcjbj1AKCoZ−TYub",
4        "_score": 1,
5        "_source": {
6          "pin": {
7            "location": "39.006735,−84.623182"
8          },
9          "timestamp": "2017−08−25T21:00:01.111Z",
10            "Pollen_Index": 10.4,
11           "WunderGround": null} }
```

Listing 10.1: Pollen data set example in JSON.

```
1  <zone name="Environment">
2     <senml bn="urn:body:uuid:c68ad78b−09eb−4303−ae3c−d5d23149ee96">
3        <e n="pollen" t="0" u="X" v="10.4"/>
4     </senml>
5  </zone>
```

Listing 10.2: Pollen data set example in SenML/XML.

```
1  [PollenLevel: (?measurementUri rdf:type kAO:PollenLevel)
2              <−
3              (?measurementUri sosa:observedProperty "pollen")
4              (?sensor m3:produces ?measurementUri)
5              (?sensor m3:observes m3:Environment)
6  ]
```

Listing 10.3: Pollen data set semantically annotated to be compliant with the KAO ontology.

```
1  [HighPollenLevel:
2              (?measurement rdf:type kAO:PollenLevel)
3              (?measurement sosa:hasSimpleResult ?v)
4              greaterThan(?v,9.7)
5              lessThan(?v,12)
6           −>
7         (?measurement rdf:type kAO:HighPollenLevel)
8  ]
```

Listing 10.4: Rule example to deduce high-level abstraction from the pollen data set.

```
1  <owl:ObjectProperty rdf:ID="hasRecommendation">
2          <rdfs:label xml:lang="en">hasRecommendation</rdfs:label>
3          <rdfs:comment xml:lang="en">used for the kHealth recommender system (e.g., high pollen level −> do not go out)</
              rdfs:comment>
4  </owl:ObjectProperty>
5
6  <owl:Class rdf:ID="HighPollenLevel">
7          <rdfs:label xml:lang="en">High Pollen Level</rdfs:label>
8          <rdfs:comment xml:lang="en"></rdfs:comment>
9          <rdfs:subClassOf rdf:resource="#PollenLevel"/>
10 <kAO:hasRecommendation rdf:resource="http://purl.org/kAO#DoNotGoOut"/>
11     </owl:Class>
12
13 <owl:Class rdf:ID="DoNotGoOut">
14         <rdfs:label xml:lang="en">Do Not Go Out</rdfs:label>
15         <rdfs:comment xml:lang="en">Do not go out when pollen is high.</rdfs:comment>
16 </owl:Class>
```

Listing 10.5: Linking cross-domain knowledge to provide suggestions.

```
 1  PREFIX kAO: <http://purl.org/kAO#>
 2  PREFIX sosa: <http://www.w3.org/ns/sosa/>
 3  PREFIX qudt: <http://qudt.org/schema/qudt#>
 4  PREFIX rdfs: <http://www.w3.org/2000/01/rdf−schema#>
 5  PREFIX rdf: <http://www.w3.org/1999/02/22−rdf−syntax−ns#>
 6  PREFIX dc: <http://purl.org/dc/elements/1.1/>
 7
 8  SELECT DISTINCT ?name ?value ?unit ?semanticAnnotationType ?deduce ?suggest ?suggest_comment ?deduceUri ?
        suggestUri WHERE{
 9          ?measurement sosa:observedProperty ?name.
10          ?measurement sosa:hasSimpleResult ?value.
11          ?measurement sosa:resultTime ?time.
12          ?measurement qudt:unit ?unit.
13          ?measurement rdf:type ?semanticAnnotationTypeUri.
14
15          OPTIONAL {
16                  ?semanticAnnotationTypeUri rdfs:label ?semanticAnnotationType.
17                  FILTER(LANGMATCHES(LANG(?semanticAnnotationType), "en"))}
18
19          OPTIONAL {
20                  ?measurement rdf:type ?deduceUri .
21                  ?deduceUri rdfs:label ?deduce.
22                  FILTER(LANGMATCHES(LANG(?deduce), "en"))
23                  FILTER(str(?deduceUri) != str(sosa:ObservableProperty) )
24                  FILTER(str(?deduceUri) != str(?semanticAnnotationTypeUri) )
25
26                  OPTIONAL{
27
28  ?deduceUri kAO:hasRecommendation ?suggestUri . # e.g high pollen −> do not go out
29                          ?suggestUri rdfs:label ?suggest.
30                          FILTER(LANGMATCHES(LANG(?suggest), "en"))
31                          OPTIONAL{
32                                  ?suggestUri rdfs:comment ?suggest_comment. #dc:description
33                                  FILTER(LANGMATCHES(LANG(?suggest_comment), "en"))
34                          }
35                  }
36          }}
```

Listing 10.6: Generic SPARQL query example to retrieve data semantic annotated with the kao ontology and enriched once the reasoner is executed.

```
1   <?xml version="1.0"?>
2   <sparql xmlns="http://www.w3.org/2005/sparql-results#">
3     <head>
4       <variable name="name"/>
5       <variable name="value"/>
6       <variable name="unit"/>
7       <variable name="semanticAnnotationType"/>
8       <variable name="deduce"/>
9       <variable name="suggest"/>
10      <variable name="suggest_comment"/>
11      <variable name="deduceUri"/>
12      <variable name="suggestUri"/>
13    </head>
14    <results>
15      <result>
16        <binding name="name">
17          <literal datatype="http://www.w3.org/2001/XMLSchema#string">pollen</literal>
18        </binding>
19        <binding name="value">
20          <literal datatype="http://www.w3.org/2001/XMLSchema#decimal">10.4</literal>
21        </binding>
22        <binding name="unit">
23          <literal datatype="http://www.w3.org/2001/XMLSchema#string">X</literal>
24        </binding>
25        <binding name="semanticAnnotationType">
26          <literal xml:lang="en">Pollen Level</literal>
27        </binding>
28        <binding name="deduce">
29          <literal xml:lang="en">High Pollen Level</literal>
30        </binding>
31        <binding name="suggest">
32          <literal xml:lang="en">Do Not Go Out</literal>
33        </binding>
34        <binding name="suggest_comment">
35          <literal xml:lang="en">Do Not Go Out when pollen is high.</literal>
36        </binding>
37        <binding name="deduceUri">
38          <uri>http://purl.org/kAO#HighPollenLevel</uri>
39        </binding>
40        <binding name="suggestUri">
41          <uri>http://purl.org/kAO#DoNotGoOut</uri>
42        </binding>
43      </result>
44    </results>
```

Listing 10.7: XML result returned as a suggestion.

```
 1  PREFIX rdfs: <http://www.w3.org/2000/01/rdf−schema#>
 2  PREFIX kao: <http://purl.org/kao#>
 3  PREFIX schema: <https://schema.org/>
 4  PREFIX rdf: <http://www.w3.org/1999/02/22−rdf−syntax−ns#>
 5
 6  SELECT DISTINCT ?patientURI ?patientLabel ?patientComment ?riskFactorLabel WHERE{
 7          ?patientURI rdfs:label ?patientLabel.
 8          ?patientURI rdfs:comment ?patientComment.
 9
10          ?patientURI kao:hasRiskFactor ?riskFactorURI. # to consider allergies
11          #?riskFactorURI can be replaced by <http://kmi.open.ac.uk/projects/smartproducts/ontologies/food.owl#Nuts>.
12          #?riskFactorURI can be replaced by <http://kmi.open.ac.uk/projects/smartproducts/ontologies/food.owl#Gluten>.
13          #?riskFactorURI rdfs:label ?riskFactorLabel.
14
15          ?patientURI rdf:type schema:Person
16  }
17
18  # RDF example:
19  # <owl:NamedIndividual rdf:about="http://purl.org/kao#Patient_scenario_test_allergy_nut">
20  # <rdfs:label xml:lang="en">Patient scenario test has allergies (e.g., nuts).</rdfs:label>
21  # <rdfs:comment xml:lang="en">Patient scenario test has allergies (e.g., nuts).</rdfs:comment>
22  # <rdf:type rdf:resource="https://schema.org/Person"/>
23  # <kao:hasRiskFactor rdf:resource="http://kmi.open.ac.uk/projects/smartproducts/ontologies/food.owl#Nuts"/>
24  # <dcterms:modified rdf:datatype="http://www.w3.org/2001/XMLSchema#date">2019−01−16</dcterms:modified>
25  # <dcterms:issued rdf:datatype="http://www.w3.org/2001/XMLSchema#date">2019−01−16</dcterms:issued>
26  # </owl:NamedIndividual>
```

Listing 10.8: A patient has risk factors (nuts, pollen, etc.): SPARQL query and semantic annotation example.

```
 1  PREFIX rdfs: <http://www.w3.org/2000/01/rdf−schema#>
 2  PREFIX food_taxonomy: <http://kmi.open.ac.uk/projects/smartproducts/ontologies/food_taxonomy.owl>
 3
 4  ASK WHERE{
 5          <http://kmi.open.ac.uk/projects/smartproducts/ontologies/food.owl#ButternutSquash> rdfs:subClassOf <http://kmi.
                open.ac.uk/projects/smartproducts/ontologies/food.owl#Nuts>.
 6          #?nutURI rdfs:label ?nutLabel.
 7  }
 8
 9  # RDF example:
10  # <!−− http://kmi.open.ac.uk/projects/smartproducts/ontologies/food.owl#ButternutSquash −−>
11  #
12  # <owl:Class rdf:about="http://kmi.open.ac.uk/projects/smartproducts/ontologies/food.owl#ButternutSquash">
13  # <rdfs:label>Butternut Squash</rdfs:label>
14  # <rdfs:subClassOf rdf:resource="http://kmi.open.ac.uk/projects/smartproducts/ontologies/food.owl#WinterSquash"/>
15  # </owl:Class>
16
17  # returns false
```

Listing 10.9: A keyword-based approach will deduce that butternut squash might make me a nut, but not a knowledge graph-based approach. This ASK SPARQL query example checks if butternut squash is a nut and returns false.

```
 1  phkg−rule := [ruleName : analyze−phkg−data]
 2
 3  analyze−phkg−data := term, ... term −> inferred−data // forward rule
 4
 5  inferred−data := term or [ base−rule]
 6
 7  term := (node, node, node) //triple pattern
 8          or functionComparison node, ... node) // e.g., greaterThan
 9
10  node := phkg_type
11          or uri−ref // e.g., http://example.com
12          or prefix−localname // e.g., rdf:type
13          or <uri−def> // e.g., <mySchema:myUri>
14          or ?varname // variable
15          or 'a literal' // a plain string literal
16          or 'lex'^^typeURI // a typed literal, xsd:* type anmes supported
17          or number // e.g., 42 or 25.5
18
19  phkg_type := measurement // e.g., kao:Temperature
20              or unit // e.g., kao:DegreeCelcius
21              or domain // e.g., kao:Health
```

Listing 10.10: Syntax of rules.

Appendix 10.B **Tutorials: SPARQL queries and End-to-End Scenarios**

Hereafter, a set of tutorials developed using the PHKG (e.g., kao ontology). A set of SPARQL queries are provided and illustrated with the asthma, obesity, or food-related diseases (Figs. 10.7–10.11).

> ∨ ▦ khealth.tutorialSparqlQuery.asthma
> 　> ▯ Tutorial01_GetPatientAge_DONE.java
> 　> ▯ Tutorial02A_GetPatientSymptom_DONE.java
> 　> ▯ Tutorial02B_GetPatientSymptomCount_DONE.java
> 　> ▯ Tutorial02C_GetPatient_SpecificSymptomCannotTalkInFullSentence_DONE.java
> 　> ▯ Tutorial02D_GetPatient_SpecificSymptomChestTightness_DONE.java
> 　> ▯ Tutorial02E_GetPatient_SpecificSymptomAirwayInflammation_DONE.java
> 　> ▯ Tutorial02F_GetPatient_SpecificSymptomCough_DONE.java
> 　> ▯ Tutorial02G_GetPatient_SpecificSymptomNoseOpenWide_DONE.java
> 　> ▯ Tutorial02H_GetPatient_SpecificSymptomShortnessBreath_DONE.java
> 　> ▯ Tutorial02I_GetPatient_SpecificSymptomWheeze_DONE.java
> 　> ▯ Tutorial03A_GetAllPatients_hasRiskFactorAllergy_DONE.java
> 　> ▯ Tutorial03B_GetSpecificPatient_hasRiskFactorAllergy_DONE.java
> 　> ▯ Tutorial03C_GetSpecificPatient_hasRiskFactorAllergyOutdoorAirQualityIndex_DONE.java
> 　> ▯ Tutorial03D_GetAllPatients_hasRiskFactorAllergyPollenLevel_DONE.java
> 　> ▯ Tutorial04A_GetPatientTherapy_DONE.java
> 　> ▯ Tutorial04B_GetAllPatientTherapyInhalerControllerMedicationSymbicort_DONE.java
> 　> ▯ Tutorial04C_GetAllPatientTherapyInhalerRescueMedicationProairVentolin_DONE.java
> 　> ▯ Tutorial05_GetPatientWokeUpLastNight_DONE.java

FIGURE 10.7

Example of SPARQL query tutorials for asthma disease to get patient information.

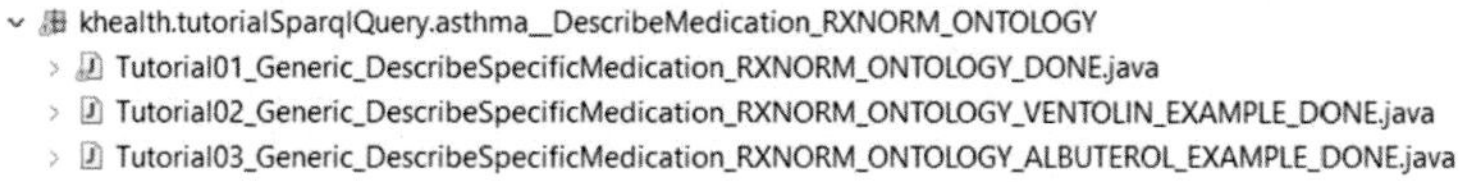

> ∨ ▦ khealth.tutorialSparqlQuery.asthma__DescribeMedication_RXNORM_ONTOLOGY
> 　> ▯ Tutorial01_Generic_DescribeSpecificMedication_RXNORM_ONTOLOGY_DONE.java
> 　> ▯ Tutorial02_Generic_DescribeSpecificMedication_RXNORM_ONTOLOGY_VENTOLIN_EXAMPLE_DONE.java
> 　> ▯ Tutorial03_Generic_DescribeSpecificMedication_RXNORM_ONTOLOGY_ALBUTEROL_EXAMPLE_DONE.java

FIGURE 10.8

Example of DESCRIBE SPARQL query tutorials to retrieve more information.

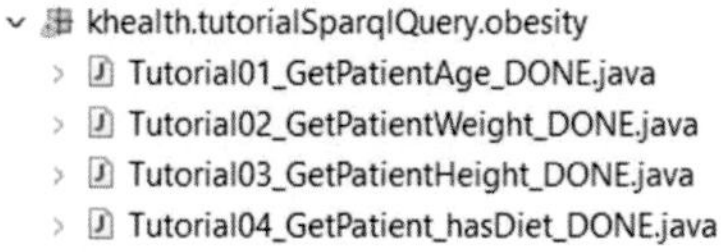

> ∨ ▦ khealth.tutorialSparqlQuery.obesity
> 　> ▯ Tutorial01_GetPatientAge_DONE.java
> 　> ▯ Tutorial02_GetPatientWeight_DONE.java
> 　> ▯ Tutorial03_GetPatientHeight_DONE.java
> 　> ▯ Tutorial04_GetPatient_hasDiet_DONE.java

FIGURE 10.9

Example of SPARQL query tutorials for the obesity disease.

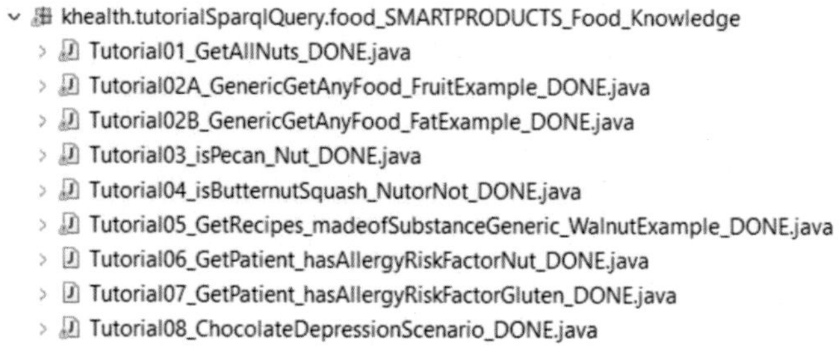

- ⌄ 🏢 khealth.tutorialSparqlQuery.food_SMARTPRODUCTS_Food_Knowledge
  - › 🗎 Tutorial01_GetAllNuts_DONE.java
  - › 🗎 Tutorial02A_GenericGetAnyFood_FruitExample_DONE.java
  - › 🗎 Tutorial02B_GenericGetAnyFood_FatExample_DONE.java
  - › 🗎 Tutorial03_isPecan_Nut_DONE.java
  - › 🗎 Tutorial04_isButternutSquash_NutorNot_DONE.java
  - › 🗎 Tutorial05_GetRecipes_madeofSubstanceGeneric_WalnutExample_DONE.java
  - › 🗎 Tutorial06_GetPatient_hasAllergyRiskFactorNut_DONE.java
  - › 🗎 Tutorial07_GetPatient_hasAllergyRiskFactorGluten_DONE.java
  - › 🗎 Tutorial08_ChocolateDepressionScenario_DONE.java

FIGURE 10.10

Example of SPARQL query tutorials for food-related diseases (e.g., allergy to nuts). It also demonstrates the potential of reusing food-related knowledge bases (e.g., SMART PRODUCTS project).

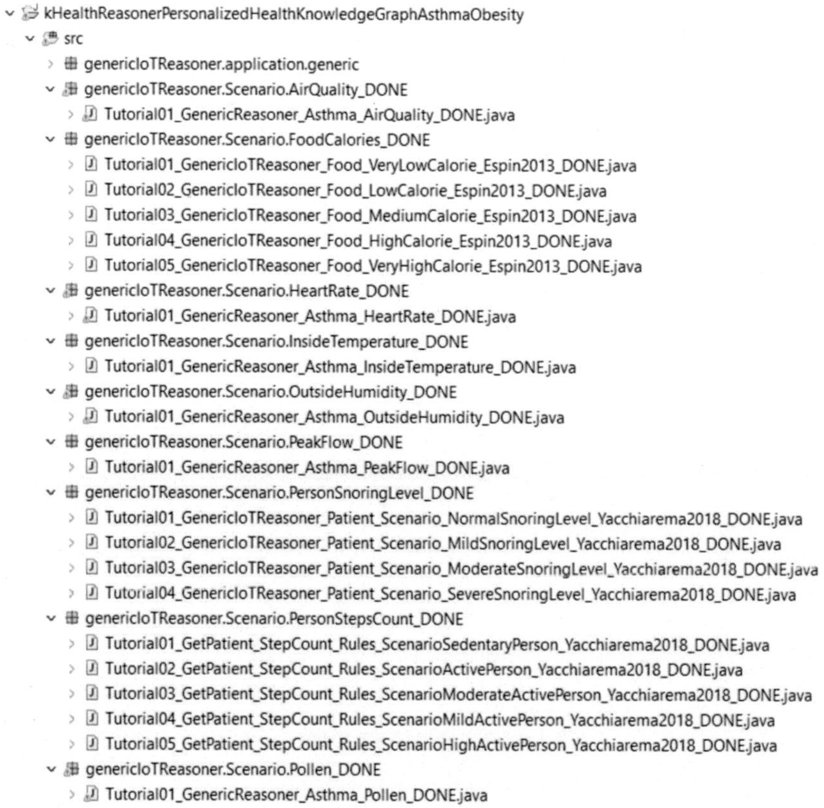

- ⌄ 🗂 kHealthReasonerPersonalizedHealthKnowledgeGraphAsthmaObesity
  - ⌄ 📁 src
    - › ⊞ genericIoTReasoner.application.generic
    - ⌄ 🏢 genericIoTReasoner.Scenario.AirQuality_DONE
      - › 🗎 Tutorial01_GenericReasoner_Asthma_AirQuality_DONE.java
    - ⌄ ⊞ genericIoTReasoner.Scenario.FoodCalories_DONE
      - › 🗎 Tutorial01_GenericIoTReasoner_Food_VeryLowCalorie_Espin2013_DONE.java
      - › 🗎 Tutorial02_GenericIoTReasoner_Food_LowCalorie_Espin2013_DONE.java
      - › 🗎 Tutorial03_GenericIoTReasoner_Food_MediumCalorie_Espin2013_DONE.java
      - › 🗎 Tutorial04_GenericIoTReasoner_Food_HighCalorie_Espin2013_DONE.java
      - › 🗎 Tutorial05_GenericIoTReasoner_Food_VeryHighCalorie_Espin2013_DONE.java
    - ⌄ ⊞ genericIoTReasoner.Scenario.HeartRate_DONE
      - › 🗎 Tutorial01_GenericReasoner_Asthma_HeartRate_DONE.java
    - ⌄ ⊞ genericIoTReasoner.Scenario.InsideTemperature_DONE
      - › 🗎 Tutorial01_GenericReasoner_Asthma_InsideTemperature_DONE.java
    - ⌄ ⊞ genericIoTReasoner.Scenario.OutsideHumidity_DONE
      - › 🗎 Tutorial01_GenericReasoner_Asthma_OutsideHumidity_DONE.java
    - ⌄ ⊞ genericIoTReasoner.Scenario.PeakFlow_DONE
      - › 🗎 Tutorial01_GenericReasoner_Asthma_PeakFlow_DONE.java
    - ⌄ ⊞ genericIoTReasoner.Scenario.PersonSnoringLevel_DONE
      - › 🗎 Tutorial01_GenericIoTReasoner_Patient_Scenario_NormalSnoringLevel_Yacchiarema2018_DONE.java
      - › 🗎 Tutorial02_GenericIoTReasoner_Patient_Scenario_MildSnoringLevel_Yacchiarema2018_DONE.java
      - › 🗎 Tutorial03_GenericIoTReasoner_Patient_Scenario_ModerateSnoringLevel_Yacchiarema2018_DONE.java
      - › 🗎 Tutorial04_GenericIoTReasoner_Patient_Scenario_SevereSnoringLevel_Yacchiarema2018_DONE.java
    - ⌄ ⊞ genericIoTReasoner.Scenario.PersonStepsCount_DONE
      - › 🗎 Tutorial01_GetPatient_StepCount_Rules_ScenarioSedentaryPerson_Yacchiarema2018_DONE.java
      - › 🗎 Tutorial02_GetPatient_StepCount_Rules_ScenarioActivePerson_Yacchiarema2018_DONE.java
      - › 🗎 Tutorial03_GetPatient_StepCount_Rules_ScenarioModerateActivePerson_Yacchiarema2018_DONE.java
      - › 🗎 Tutorial04_GetPatient_StepCount_Rules_ScenarioMildActivePerson_Yacchiarema2018_DONE.java
      - › 🗎 Tutorial05_GetPatient_StepCount_Rules_ScenarioHighActivePerson_Yacchiarema2018_DONE.java
    - ⌄ ⊞ genericIoTReasoner.Scenario.Pollen_DONE
      - › 🗎 Tutorial01_GenericReasoner_Asthma_Pollen_DONE.java

FIGURE 10.11

End-to-end scenarios to semantically annotate IoT data sets (e.g., air quality, pollen, humidity, temperature) and enrich data sets with abstractions by executing the reasoning engine (e.g., HighPollenLevel).

References

[1] D.V. Dimitrov, Medical internet of things and big data in healthcare, Healthcare Informatics Research (2016).

[2] R. Negra, I. Jemili, A. Belghith, Wireless body area networks: applications and technologies, Procedia Computer Science (2016).

[3] A. Sheth, U. Jaimini, H.Y. Yip, How will the internet of things enable augmented personalized health?, IEEE Intelligent Systems (2018).

[4] U. Jaimini, K. Thirunarayan, M. Kalra, R. Venkataraman, D. Kadariya, A. Sheth, "how is my child's asthma?" digital phenotype and actionable insights for pediatric asthma, JMIR Pediatrics and Parenting (2018).

[5] A. Sheth, U. Jaimini, K. Thirunarayan, T. Banerjee, Augmented personalized health: how smart data with iots and ai is about to change healthcare, in: IEEE International Forum on Research and Technologies for Society and Industry (RTSI), IEEE, 2017.

[6] L. Agertoft, S. Pedersen, Effects of long-term treatment with an inhaled corticosteroid on growth and pulmonary function in asthmatic children, Elsevier Respiratory Medicine Journal (1994).

[7] A. Bordes, E. Gabrilovich, Constructing and mining web-scale knowledge graphs: KDD 2014 tutorial, in: International Conference on Knowledge Discovery and Data Mining (SIGKDD), ACM, 2014.

[8] S. Sharma, K. Chen, A. Sheth, Towards practical privacy-preserving analytics for iot and cloud based healthcare systems, 2018.

[9] A. Gyrard, M. Gaur, K. Thirunarayan, A. Sheth, S. Shekarpour, Personalized health knowledge graph, in: Workshop on Contextualized Knowledge Graph (CKG) Co-Located with International Semantic Web Conference (ISWC), 2018.

[10] A. Sheth, P. Anantharam, K. Thirunarayan, khealth: proactive personalized actionable information for better healthcare, in: Workshop on Personal Data Analytics in the Internet of Things (PDA@ IOT), collocated at VLDB, 2014.

[11] X. Dong, E. Gabrilovich, G. Heitz, W. Horn, N. Lao, K. Murphy, T. Strohmann, S. Sun, W. Zhang, Knowledge vault: a web-scale approach to probabilistic knowledge fusion, in: ACM International Conference on Knowledge Discovery and Data Mining (SIGKDD), ACM, 2014.

[12] H. Paulheim, Knowledge graph refinement: a survey of approaches and evaluation methods, Semantic Web Journal 8 (3) (2017) 489–508.

[13] R.V. Guha, D. Brickley, S. Macbeth, Schema.org: evolution of structured data on the web, Communications of the ACM (2016).

[14] Oshani Seneviratne, Jonathan Harris, Ching-Hua Chen, Deborah L. McGuinness, Personal health knowledge graph for clinically relevant diet recommendations, arXiv preprint, arXiv:2110.10131, 2021.

[15] L. Li, P. Wang, J. Yan, Y. Wang, S. Li, J. Jiang, Z. Sun, B. Tang, T.-H. Chang, S. Wang, et al., Real-world data medical knowledge graph: construction and applications, Artificial Intelligence in Medicine (2020).

[16] Z. Huang, J. Yang, F. van Harmelen, Q. Hu, Constructing knowledge graphs of depression, in: ICHIS, 2017.

[17] M. Rotmensch, Y. Halpern, A. Tlimat, S. Horng, D. Sontag, Learning a health knowledge graph from electronic medical records, Scientific Reports (2017).

[18] L. Shi, L. Li, X. Yang, J. Qi, G. Pan, B. Zhou, Semantic health knowledge graph: semantic integration of heterogeneous medical knowledge and services, BioMed Research International (2017).

[19] A. Sheth, S. Agrawal, J. Lathem, N. Oldham, H. Wingate, P. Yadav, K. Gallagher, Active semantic electronic medical records, in: The Semantic Web, Springer, 2007.

[20] T. Yu, J. Li, Q. Yu, Y. Tian, X. Shun, L. Xu, L. Zhu, H. Gao, Knowledge graph for TCM health preservation: design, construction, and applications, Elsevier Artificial Intelligence in Medicine Journal (2017).

[21] T. Ruan, M. Wang, J. Sun, T. Wang, L. Zeng, Y. Yin, J. Gao, An automatic approach for constructing a knowledge base of symptoms in Chinese, Journal of Biomedical Semantics (2017).

[22] H. Weng, Z. Liu, S. Yan, M. Fan, A. Ou, D. Chen, T. Hao, A framework for automated knowledge graph construction towards traditional Chinese medicine, in: International Conference on Health Information Science, Springer, 2017.

[23] X. Wilcke, P. Bloem, V. De Boer, The knowledge graph as the default data model for learning on heterogeneous knowledge, Data Science (2017).

[24] M. Nickel, K. Murphy, V. Tresp, E. Gabrilovich, A review of relational machine learning for knowledge graphs, Proceedings of the IEEE (2016).

[25] D. Le-Phuoc, H.N.M. Quoc, H.N. Quoc, T.T. Nhat, M. Hauswirth, The graph of things: a step towards the live knowledge graph of connected things, Journal of Web Semantics (2016).

[26] S. Auer, V. Bryl, S. Tramp, Linked Open Data – Creating Knowledge Out of Interlinked Data: Results of the LOD2 Project, vol. 8661, Springer, 2014.

[27] M.J. Kofler, C. Reinisch, W. Kastner, A semantic representation of energy-related information in future smart homes, Energy and Buildings (2012).

[28] P. Staroch, A weather ontology for predictive control in smart homes, Master's thesis, 2013.

[29] M. Kofler, C. Reinisch, W. Kastner, An intelligent knowledge representation of smart home energy parameters, in: Proceedings of the World Renewable Energy Congress (WREC), 2011.

[30] A. Hristoskova, V. Sakkalis, G. Zacharioudakis, M. Tsiknakis, F. De Turck, Ontology-driven monitoring of patient's vital signs enabling personalized medical detection and alert, Sensors (2014).

[31] M.M. Oprea, Air_pollution_onto: an ontology for air pollution analysis and control, in: IFIP International Conference on Artificial Intelligence Applications and Innovations, Springer, 2009.

[32] D.C. Yacchirema, D. Sarabia-Jácome, C.E. Palau, M. Esteve, A smart system for sleep monitoring by integrating iot with big data analytics, IEEE Access (2018).

[33] P. Laxminarayan, Exploratory analysis of human sleep data, Ph.D. thesis, Worcester Polytechnic Institute, 2004.

[34] R. Angelidou, Development of a portable system for collecting and processing bio-signals and sounds to support the diagnosis of sleep apnea, Master's thesis, 2015.

[35] R.S. Mueller, Ontology-driven Data Integration for Clinical Sleep Research, Ph.D. thesis, Case Western Reserve University, 2011.

[36] M.C. Suarez-Figueroa, A. Gomez-Perez, M. Fernandez-Lopez, The NeOn methodology for ontology engineering, in: Ontology Engineering in a Networked World, Springer, 2012.

[37] N.F. Noy, D.L. McGuinness, et al., Ontology development 101: a guide to creating your first ontology, 2001.

[38] A. Gyrard, G. Atemezing, M. Serrano, PerfectO: an online toolkit for improving quality, accessibility, and classification of domain-based ontologies, Springer, 2021.

[39] A. Gyrard, M. Serrano, G. Atemezing, Semantic web methodologies, best practices and ontology engineering applied to internet of things, in: IEEE World Forum on Internet of Things, 2015.

[40] K. Balog, T. Kenter, Personal knowledge graphs: a research agenda, in: Proceedings of the 2019 ACM SIGIR International Conference on Theory of Information Retrieval, 2019, pp. 217–220.

Web references

[WR1] World Health Organization Asthma, https://www.who.int/respiratory/asthma/en/.

[WR2] Apple Health Kit, https://www.apple.com/ios/health/.

[WR3] Ambient Assisted Living (AAL), https://goo.gl/rLLYCC.

[WR4] Obesity Ontology Picture, https://goo.gl/JWyze4.

[WR5] Patient of the future, https://www.technologyreview.com/s/426968/the-patient-of-the-future/.

[WR6] Knoesis KG, http://wiki.knoesis.org/index.php/KnoesisKnowledgeGraph.

[WR7] SENML language, https://tools.ietf.org/html/draft-ietf-core-senml-08.

[WR8] kHealth project, https://goo.gl/quEfjH.

[WR9] Air quality, https://airnow.gov/index.cfm?action=aqibasics.aqi.

[WR10] Pollen, https://www.pollen.com.

[WR11] Traditional Chinese Medicine KG, http://www.tcmkb.cn/kg/.

[WR12] Symptoms data set, http://datahub.io/dataset/symptoms-in-chinese.

[WR13] SNOMED-CT, http://bioportal.bioontology.org/ontologies/SNOMEDCT.

[WR14] UMLS, https://www.nlm.nih.gov/research/umls/.

[WR15] ICD10, https://bioportal.bioontology.org/ontologies/ICD10.

[WR16] kAO ontology, http://wiki.knoesis.org/index.php/KHealthAsthmaOntology.

[WR17] ISWC Resource Track, http://iswc2018.semanticweb.org/call-for-resources-track-papers/.

[WR18] Semantic Web Best Practices, http://perfectsemanticweb.appspot.com/.

[WR19] Semantic Web Best Practices Documentation, http://perfectsemanticweb.appspot.com/documentation/
SemanticWebBestPracticesForDummies.pdf.

[WR20] Peak flow meter, http://www.lung.org/assets/documents/asthma/peak-flow-meter.pdf.

[WR21] Air quality index, http://aqicn.org.

[WR22] Smart home and weather ontology, http://paul.staroch.name/thesis/SmartHomeWeather.owl.

[WR23] Weather Ontology, https://www.auto.tuwien.ac.at/downloads/thinkhome/ontology/WeatherOntology.
owl.

[WR24] Asthma Ontology, https://bioportal.bioontology.org/ontologies/AO.

[WR25] RXNORMS, https://bioportal.bioontology.org/ontologies/RXNORM.

[WR26] SIDER, http://sideeffects.embl.de.

[WR26b] SIDER ontology, http://download.bio2rdf.org/#/current/sider/.

[WR27] Clinical Trials, https://bioportal.bioontology.org/ontologies/CTO.

[WR28a] Completude and completeness of rules, http://sensormeasurement.appspot.com/documentation/
NomenclatureSensorData.pdf.

[WR28b] Smart Products Project (food ontologies), http://projects.kmi.open.ac.uk/smartproducts/ontology.html.

[WR29] Code should be released for conference publications, https://goo.gl/7nBJnk.

[WR30] MEDDRA, https://www.meddra.org/.

[WR31] Kno.e.sis Alchemy API, http://wiki.knoesis.org/index.php/Knoesis_Alchemy_of_Healthcare.

[WR32] Schema.org Person https://schema.org/Person.

[WR33] W3C SOSA SSN ontology, https://www.w3.org/TR/vocab-ssn/.

[WR34] SPARQL-Generate, https://ci.mines-stetienne.fr/sparql-generate/.

[WR35] Slides "kHealth: Semantic Multi-sensory Mobile Approach to Personalized Asthma Care" https://goo.
gl/iPwNLw.

[WR36] Google Health, KG Blog (2015), https://www.blog.google/products/search/health-info-knowledge-
graph/.

[WR37] kao ontology namespace and code, http://purl.org/kao.

[WR38] Slides step-by-step tutorial to improve the ontology quality, dissemination, reuse, etc. Semantic Web
Best Practices, https://goo.gl/Rg4cGr.

[WR39] DrugBank data set, https://www.drugbank.ca/releases/latest.

[WR40] Diseasome data set, https://old.datahub.io/dataset/fu-berlin-diseasome.

Integrated context-aware ontology for MNCH decision support

11

Patience U. Usip[a], **Moses E. Ekpenyong**[a,b], **Funebi F. Ijebu**[a], and **Kommomo J. Usang**[a]

[a]Department of Computer Science, University of Uyo, Uyo, Nigeria
[b]Centre for Research and Development, University of Uyo, Uyo, Nigeria

Contents

11.1 Introduction

Maternal, neonatal, and child mortality reduction is a key target of the third Sustainable Development Goal (SDG), which aims at improving health and general well-being of people and nations [1]. As the duration for actualizing this goal gradually winds down, assessment of progress reveals uneven momentum among the committee of nations. While developed nations have achieved substantial progress on this goal, the efforts of developing countries are adjudged to be grossly insufficient [2]. In low- and medium-countries such as Nigeria, the level of implementation has slowed down tremendously, coupled with the poor health system and the unavailability of automated services to drive real-time clinical data and research on maternal, neonatal, and child health (MNCH). Furthermore, the over dependence on manual information and lack of technological innovation in the health sector has hindered the implementation of intelligent analytics, sufficient for mining the highly unstructured data generated by health facilities to support informed decision making and policy formulation. Information is highly required for decision-making processes, but intractable to acquire from unstructured data, except the use of intelligent reasoning processes are applied [3]. An enormous amount of data is constantly generated at health facilities manually and some trundle out to the web daily with available technologies [4]. Though much of these data in computer readable form are unstructured, they contain valuable information with potential impact when processed. The information derived from such data are useful for decision making, policy formulation, and more. As researchers continue to work

Semantic Models in IoT and eHealth Applications. https://doi.org/10.1016/B978-0-32-391773-5.00017-0

toward developing reasoning and data mining techniques for information extraction from unstructured data, more areas needing such attention and applied knowledge continue to emerge [5]. Methods proposed in the literature and results from these methods have achieved different data extraction tasks from unstructured text, as there are great potentials in filtering unstructured data into a structured form. Nevertheless, the obvious time-consuming nature of manually filtering unstructured data could lower trust and accuracy—as individuals are bound to express themselves differently thereby exhibiting different reasoning patterns and varying semantic protocols that produce a diverse representation of ideas. In domain-specific contexts typical of a health system, most patients' information for instance, are documented in a natural language, and which recording pattern permits a mixture of free, short- or long-hand, textual form, but can be clustered around certain attributes or fields such as patient name, age, sex, location, symptoms, doctors' notes/diagnose, and more. The key to an accurate documentation of unstructured patients records therefore rest on the following activities synonymous with the objectives of this study, including: (1) converting the textual materials into a machine-readable format, (2) resolving constraints surrounding the extraction of unique records, (3) indexing the textual data on recognized parts-of-speech, (4) creating a sound representation for establishing relation(s) between key elements of the text/data. (5) supporting decisions using an extracted knowledge graph that generates pointers to the key attributes of the system. As the health sector advances toward the Internet of Things (IoT), the most flourishing technology, with healthcare introducing a myriad of real-time applications, the major benefits derived from this technology include:

1. Simultaneous reporting and monitoring—including smart medical/healthcare services in real-time.
2. End-to-end connectivity and affordability—automating healthcare and patient healthcare workflow through healthcare mobility solutions, enabling interoperability, machine-to-machine communication, information exchange, and enhanced data transport for a more cost-effective healthcare service delivery.
3. Intelligent data crowdsourcing and analytics—managing a huge amount of data (structured and unstructured) requires an intelligent system to manage data collected in real-time through IoT-enabled mobile devices. This will reduce the collection of raw data and drive vital healthcare analytics and data-driven insights, for informed decision making.
4. Embedded service tracking and alerts—real-time tracking and alerts in life-threatening situations can safeguard patients in critical condition. With constant notifications and real-time alerts, proper/early monitoring, analysis, and diagnosis of patients' condition is achieved.
5. Remote medical assistance—connecting to a doctor due to distance and lack of knowledge/information is made possible by the IoT-enabled mobility solutions. With this solution, proper medical assistance will be a button away and patients can take medical prescriptions at home through healthcare delivery chains connected to patients through IoT devices.

IoT are not without challenges, given the numerous components embedded in this technology and the principle that these components must synchronize to produce the desired benefits. Hence, top on the list of challenges of IoT in healthcare include:

1. Data security and privacy—most IoT-enabled mobile devices capture data in real-time; hence, they lack adherence to data protocols and standards. Significant ambiguity regarding data ownership and regulation also exists. As such, data stored within IoT-enabled devices are prone to data thefts and may compromise personal health information.

2. Multiple devices and protocols integration—integrating multiple device types can cause hindrance in the implementation of IoT in the healthcare sector, because some device manufacturers are yet to reach a consensus regarding communication protocol and standards.
3. Data overload and accuracy—the nonuniformity of data and communication protocols presented by multidevice types stiffens the aggregate of data for vital insights and analysis. This in turn affects decision-making processes in the long run.
4. Cost-costs constitute one of the greatest challenges when developing an IoT app development for healthcare solutions. However, the return on investment from the developed system overshadows the costs.

The purpose of this research is to create a context-aware ontology that generates location-based representations from unstructured/semistructured data for supporting informed healthcare decisions and Internet of Health Things (IoHT). The contributions of this research to the growing field of medicine/healthcare include:

1. Availability of semi-structured MNCH data sets. Recently, we have excavated patients' clinical records directly from patients' files to build a semistructured MNCH data set. These data sets, which are meant to advance future research progress in the field, are available in a comma separated value (CSV) file and is currently being refined to serve as domain-specific data sets for the sub-Saharan African region. To ensure consistency of the records, the data sets were normalized and assigned unique record sequence numbers.
2. Integrated ontology-based framework. An ontology-based framework is proposed in this chapter to drive a knowledge base that generates spatiotemporal events for enhancing decision support systems. The knowledge extraction process is preposition enabled and useful for representing ge-olocation information.
3. Knowledge base for decision support systems. Creating suitable ontologies would drive intelligent analytic systems in a data-driven approach and serve the unmet needs of the health sector. The benefits of this solution are numerous as a more precise Smart medical assistant, sufficient for driving semantic processes of IoH services is certain.

The remainder of this chapter is structured as follows. Section 11.2 considers related works in the field and outlines the necessary gaps to fill. Section 11.3 presents the PeSONT framework and discusses the tools and components of the system. Section 11.4 documents the PeSONT, highlighting the concepts, sub-concepts, objects, and data properties, including cross references among the concepts. Section 11.5 discusses the designed PeSONT in relation to state-of-art. Section 11.6 concludes on the research and offer future research perspectives.

11.2 Related works

Natural language processing (NLP) systems are generally inaccurate at resolving the numerous challenges that trail natural languages. Natural languages systems pose severe problems for machine learning and intelligent systems and can sometimes lead to a misclassification of related-concepts or features of the system. These misclassifications render decision support systems ineffective, as a seemingly high rate of false positive or negative alarms are produced. Some major challenges inherent in natural languages include: (1) contextual words, phrases, and homonyms, (2) synonyms, (3) irony and sarcasm, (4) ambiguity (lexical, semantic, syntactic), (5) text and speech errors, (6) colloquialisms and slang, (7) domain-specific language, (8) low-resource languages, and (9) dearth in research and development.

In [3], sentences containing prepositions could either be spatial, geospatial, or nonspatial. While nonspatial prepositions do not describe or point to a location, spatial prepositions identify locations that are mostly within proximity (i.e., not geographically distinct). Geospatial prepositions on the other hand describe locations that are geographically distinguishable from another. Related research works [6–9] have focused on geospatial identification and extraction from text.

Place description is a conventional recurrence in conversations involving place recommendation and person direction in the absence of a compass or a navigational map. A place description provides locational information in terms of spatial features and the spatial relations between them. When place descriptions are verbally performed, automatic extraction of spatial features might be more difficult due to non-satisfaction of locative expression requirements. However, when such place description is present in natural language text, the location can easily be extracted because of the unavoidable prepositional inclusion in the written description. This inclusion of a proposition before location naming and description is referred to as locative expression [10]. Although spatial descriptor identification is easy for any fluent language speaker, several computational algorithms are still inefficient in this regard.

Kordjamshidi [6] proposed the pipeline joint learning approach for spatial sense identification using a triple of located object, spatial relation, and reference location. From the reference location, an object's position is identified and extracted as either spatial or geospatial content using the spatial relation.

On information extraction from plain text, Adnan and Akbar [11] opines that supervised learning, deep learning, and transfer learning techniques are the most suitable techniques to apply. An interesting clause in utilizing these methods is that the data set for information extraction has to be large for the efficient visualization. To perform similar information extraction operations on small data sets, the named entity recognition technique has been identified to be effective. Named entity recognition is a process where entities are identified and semantically classified into precharacterized classes or groups [11]. The corpus-based extraction performed in Hou et al. [12] corroborates Adnan and Akbar [11] but adopts a graph-based approach to data extraction for automatic domain knowledge construction. Their method, called GRAONTO, utilized a domain corpus consisting of documents with text in the natural language for information terms classification. With an intension to eliminate the manual time-consuming procedures of ontology design by knowledge engineers and other researchers, Markov clustering and random walk terms weighting approaches were adopted for concept extraction. Ontologies showed relations between terms or entities, hence the gSpan algorithm was used for relation extraction through subgraph mining.

Similarly, [13] presented a semiautomatic method for domain ontology extraction from Wikipedia. The similarity in both works lies in their direction of automatic ontology development even though different domain ontologies were considered.

To improve and run an effective healthcare delivery system supported by technology, a patient-clinic path mapping is useful. Such support system will enable patients to digitally visualize and consider paths to a choice health facility. Mapping patient location to a health facilities location would aid the identification of medical facilities and promote health equity among the populace. Furthermore, resources and healthcare personnel can be effectively managed [14]. To efficiently represent MNCH information and create a path link to a health facility location for semantic search, an ontology is required.

To provide a solution to the patient-clinic path mapping limitation, [17] highlighted the lack of georeferenced information and a comprehensive public health facility database for sub-Saharan Africa. They proposed a spatial inventory of public health facilities in the region. Their database is reported to have populated a collection of health facilities in over 50 countries in the region, including governmental and nongovernmental owned. In developing the multimethod geocoded inventory of health facilities in sub-Saharan Africa, [17] consulted the Ministries of Health websites including related data warehousing portals. Hu et al. [18] presented a modified random walk algorithm for location-based service delivery to users. They implemented an ontology-based design using current context information to determine the user's preferred location. In a similar study, [19] utilized spatiotemporal information from travelers' photos to discern decision about a traveler. Context awareness has also been of concern in the design of location-based ontologies. In Roussaki et al. [20], the use of ontologies for a locational object representation exhibits prominent advantages in integrated pervasive environments, resulting in the ever-improving application of GIS technologies in mobile environments for location identification and tour guides. Such technologies have been very useful for time management during location identification, and for providing new entrants into a city, personalized information about landmarks and venues for events.

To enable smart healthcare delivery services, there is need for a formal representation of clinical data ranging from clinical resources to patients' health records, including location information. IoHT devices capture heterogeneous data, which would certainly affect the quality of ontologies designed. Mishra and Jain [21–23] conclude that ontologies should be semantically analyzed by evaluation to ensure the design, structure, and incorporated concepts and their relations are efficient for reasoning. They proposed the use of QueryOnto for ontology verification and validation. Tiwari and Abraham [24] designed a smart healthcare ontology (SHCO) for healthcare information captured with IoT devices. They however adopted an integrated approach for the evaluation of SHCO and utilized evaluation tools like Themis and Test-Driven Development (TDD) Onto for verification of the test cases while Protégé and object-oriented programming (OOP) were used for validation of the modeled knowledge in the ontology.

Panchal and his colleagues [25] designed an ontology for Public Higher Education (AISHE-Onto) by using semantic web technologies OWL/RDF and SPARQL queries have been applied to perform reasoning with the proposed ontology.

In Lee et al. [26], an activity ontology that focused on determining the shortest path between an outdoor or indoor location and an indoor destination of interest was presented. The design connected a road/outdoor network model with an indoor topological network model to produce a 3-dimensional GIS-based topological model whose data comprised university indoor activity locations that can be shared, managed, and queried semantically.

11.3 Preposition-enabled spatial ontology: PeSONT

Since clinical notes were targeted at providing improved MNCH information, the proposed PeSONT provides a general domain ontology for health services where MNCH resides. An ontology is constructed using OWL (the Web Ontology Language), which classifies the extracted data and formally

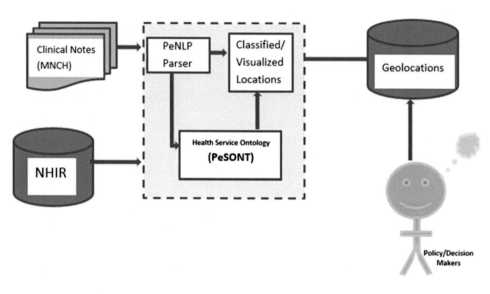

FIGURE 11.1

PeSONT framework.

specifies the concepts and their relations in the medical domain based on the data collection form for hospitals and clinics. The use of an open-standard format such as OWL allows this framework to support and improve spatial data interoperability and bring about both syntactic and semantic interoperability (i.e., it assists in transferability (reuse) of knowledge and rules). The ontology is created using Protégé [27]—a free, open source, platform-independent environment, or web editor for creating and editing ontologies and knowledge bases implemented using the OWL 2 Web Ontology Language [28]. Our PeSONT framework is a model of spatial/locational items relating to MNCH services and obtained from unstructured texts such as MNCH clinical notes, using a PeNLP parser algorithm [29]. As presented in Fig. 11.1, the context-based (geographic) locations such as patients' residential locations and locations of health facilities within the study area are outputs from PeSONT required for useful decision making and policy formulation. To accomplish the study design, geolocations of available health facilities for Uyo metropolis were retrieved from the Nigeria Health Facility Registry (HFR) ([30]; https://www.hfr.health.gov.ng) while the unstructured Maternal, Neonatal and Child Health (MNCH) input data to the PeNLP parser were retrieved from patients records at St. Luke's General Hospital, Anua, Uyo, Akwa Ibom State (Approval Ref: SLH/ADM/PC/VOL.1/344/21/005). The Health Facility Data in Akwa Ibom State obtained from National Health Registry Database are used as concepts, subconcepts, and instances in the PeSONT. Test cases in the TDDOnto [31] evaluation clearly show these domain concepts. The extracted features (locational items) are then classified based on the underlying reasoning in the health service ontology (PeSONT), from where the geolocations are filtered and visualized.

11.3.1 Concept extraction

One reason for geospatial data mining is to make the extracted knowledge explicit and easily describable. In most cases, semantic search both within proprietary archives and generally accessible repositories uses keywords to locate and retrieve resource. Hence, an end-user needs to correctly provide the predetermined keywords to retrieve expected resources. However, solutions that consider the meaning of terms behind a user query have been proposed. For solutions in the latter category, ontology remains the underlying technology. Using ontology, geospatial entities from unstructured or semistructured natural language text are connected or linked to their meaning derived from the knowledge base [5]. With such linkage, semantic search can follow a top-down approach where the ontology interprets users queries and enriches them with other meaningful terms using ontology concepts and their relations [32]. Furthermore, because geospatial ontologies should be designed to support integration into broader contexts, Claramunt [33] advocates that the explicit representation of abstracted concepts and their relations from reality, formally and allowing numerical notation using symbolic grammars alone is not enough; hence, ontologies should favor interoperability and knowledge sharing between different applications.

11.3.2 Term formalization

Ontology research has moved beyond the strict formalization of geospatial concepts under specific domains to the shared interpretation of meanings across different contexts and the development of lightweight and microontologies tailored toward specific needs [5]. Needs are domain specific and their adoption in recent studies goes beyond their ability to identify entities from natural language text to producing/explaining relations between two entities. An ontology for traditional Ayurvedic medicine was proposed in Gayathri and Kannan [34]. The ontology sought to provide a formal structuring for heterogeneous data on Ayurvedic medicine available in literature. However, structuring such medical information can produce knowledge vital for diagnosis and intervention in therapeutic conditions. Within unstructured and semistructured natural language texts, the same terms are useful to reference similar but different concepts and several concepts can be used to refer to a particular term by different individuals. Ontologies eliminate such limitation and ensure terms describing the same concept are formally represented for appropriate knowledge extraction with a computer-based solution.

11.3.3 Location visualization

Advances in technology is today phasing out paper maps, the same way images are fast dominating text in information presentation and dissemination. For visualization, the proposed PeSONT uses third party software. A visualization of geotagged coordinates is extensively demonstrated in Usip et al. [29]. In [7], extensive geospatial data ontology (GeoDataOnt) for geospatial data integration and sharing is presented; the authors, however, were not successful in direct application of the extracted knowledge from geotext in decision support. PeSONT on the other hand, can extract and visualize geoconcepts. PeSONT adopted the UTM GEO MAP, a simple GPS module designed for getting coordinates in an offline state without internet or cellular access. This module takes advantage of the built-in GPS of handheld devices, and displays in real-time, the latitude, longitude, Universal Transverse Mercator (UTM), MGRS, and all commonly used coordinate reference systems in the world (using EPSG Codes). In future research, IoT-based ontologies such as SSN and SAREF. [35,15] will be used.

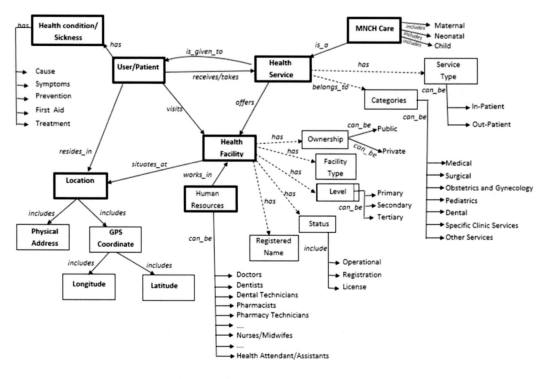

FIGURE 11.2

Health service ontology.

The MNCH care is among the various health services rendered in health facilities and the proposed PeSONT seeks to identify spatial components that are peculiar to this domain. PeSONT is not limited to locations of MNCH patients and facilities only since MNCH services are also available in other health service categories. Hence, Fig. 11.2 presents the health service ontology implementing PeSONT. The developed PeSONT has a total of 531 axioms (logical and declaration), 160 classes being concepts, 155 subclasses, 13 object properties, 1 data property, and 50 individuals. In Fig. 11.2, the major concepts of the ontology include: health_service, patient/user, health_facility, location, human_resource, and health_condition/sickness, while the subconcepts include: Physical_address, GPS_coordinate, Category, Ownership, Registered_name, Level, Status, Facility_Type, Service_Type, etc. The following relations are part of the ontology, takes/receives, works_in, has, offers, belongs_to, visits, includes, resides_in, is_a, situates_at, and can_be. Relations are classified as either object properties or data properties. The individuals in the ontology captured at the ontology development stage use protégé. The Class hierarchy of the developed ontology with the TDDOnto evaluation tab featuring the sample test query is presented in Fig. 11.3. The evaluation tab considers all the axioms of the ontology (e.g., MNCH_Care SubClassOf Health_Service, Maternal SubClassOf MNCH_Care, Neonatal SubClassOf MNCH_Care, Child SubClassOf MNCH_Care, Service_Type SubClassOf Health_Service, Location

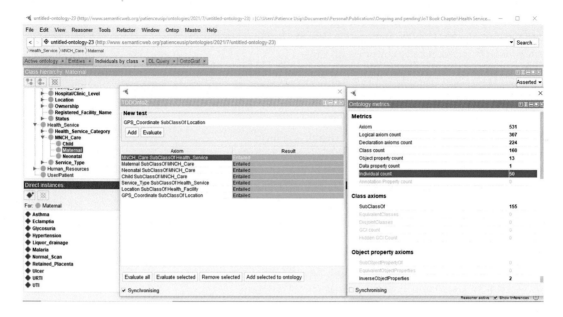

FIGURE 11.3

PeSONT showing class hierarchy and TDDOnto evaluation results.

SubClassOf Health_Facility, GPS_Coordinate SubClassOf Location, etc.) and their results, to validate the health service ontology.

11.4 **PeSONT documentation**

A documentation of the ontology is formally described in this section where the object properties are clearly linked to their domain and range. The version of PeSONT documented in this chapter is available on bioportal and can be accessed with the link: https://bioportal.bioontology.org/ontologies/ PESONT.

Summary: PeSONT is a model that describes health services with related concepts, subconcepts, attributes, and relations among concepts. Inferences of new knowledge are possible using queries.

Introduction: PeSONT classifies linguistic place terms using [16] and uses prepositions of English language for location identification. Its implementation is to further the application of the PeNLP parser on human related problems by classifying place-like terms in natural language and for extracting geolocations.

PeSONT Overview: The proposed ontology has the following concepts/ subconcepts in Table 11.1, and properties in Table 11.2, and named individuals are discussed in Table 11.3.

Cross Reference Description among concepts and properties:
User/Patient: A user/patient visits the Health_Facility in search of Health Services such as MNCH care, which includes maternal, neonatal, and child healthcare.

hasSuperClass: Health_Service
hasSubClass: Health_Condition, Location, Health_Facility, Health_Service
isinDomainOf: visits, receives, resides_in, has
isinRangeOf: is_given_to
Health_Service: This has to do with the services offered in Health_facility for patients with their various Service_Types and categories
hasSuperClass: User/Patient, MNCH_Care
hasSubClass: Service_Type, Categories, Health_Facility
isinDomainOf: hasbelongs_to, Offers, is_given_to
isinRangeOf: is_a, Receives
Health_Facility: The Health_Facility situates_at Location and offers Health_Service
hasSuperClass: User/Patient, Health_Service, Human_Resource
hasSubClass:Status Level, Location, Registered_Name, Facility_Type, Ownership
isinDomainOf: has situates_at
isinRangeOf: works_in, Offers, visits
Location: Location includes Physical_Address and GPS_Coordinates
hasSuperClass: User/Patient, Health_Facility
hasSubClass: Physical_Address, GPS_Coordinates
isinDomainOf: includes
isinRangeOf : resides_in, situates_at
MNCH_Care: MNCH-Care is an example of Health_Service
hasSuperClass: Health_Service
isinDomainOf: is_a
Health_Condition/Sickness: A Patient has Health_Conditions with symptoms, cause. Note that, Health_Service is meant to treat Health_Conditions but they do not have direct links except through the patient
hasSuperClass: User/Patient
hasSubClass: Cause, Treatment, Symptom, Prevention, First_Aid
isinDomainOf: has
isinRangeOf : has
Human_Resource: Human Resource works_in Health_Facility. They have locations but this is not relevant to Health_Service since they are identified under the Health_Facility they serve under. Also, they render Health_Service to patients but they do not have direct link but to patients that visit the Health_Facility. Human resources can_be Doctor, Nurse, Pharmacist, etc.
hasSuperClass: Health_Facility
hasSubClass: Doctor, Dentist, Pharmacist, Pharmacy_Technician, Dental_Technician, Nurse/Midwive, Health Attendant/Assistant
isinDomainOf: works_in, can_be

The description of cross references among the subconcepts will be available as readme.txt file with the published ontology. Table 11.4 provides a comparison of this study with similar state-of-the art tools. We compare the data sets, objectives, NLP methods, evaluation, and geotools adopted and main findings.

Table 11.1 Concepts/subconcepts.

User	Patient	Health_Service
Health_Facility	Location	Human_Resource
MNCH_Care	Service_Category	Service_Type
Medical	Surgical	Dental
Pediatrics	Specific_Clinic_Service	Obstetrics_and_Gynecology
Other_Service	Maternal	Neonatal
Child	Ownership	Facility_Type
Facility_Level	Facility_Status	Registered_Name
Physical_Address	GPS_Coordinate	Longitude
Latitude	Health_Condition	Cause
Prevention	First_Aid	Treatment
Symptoms	Operational	Registration
Licensed	Doctor	Nurse
Midwife	Dentist	Pharmacist
Dental_Technician	Pharmacy_Technician	Health_Attendant

Table 11.2 Object properties.

is_a (data property)	has	belongs_to
can_be	offers	Visits
receives	is_given_to	resides_in
includes	works_in	situates_at
is_offer_by		

Table 11.3 Named individuals (generally stated individuals).

In_Patient	Out_Patient	Primary
Secondary	Tertiary	Etc.
Sample individuals are:		
PM1	St. Luke's Hospital	Anua, Uyo

11.5 Discussion

While other methods exploit unintelligent methods such as direct encoding and geotext/geo images, this study (PeSONT) integrates intelligent reasoning for mining unstructured data. This study however has similarity with state-of-the-art as it exploits the prevailing evaluation tool, the TDDOnto. Currently, we adopt a simple geolocation capturing tool (UTM Geo Map) but will explore IoT-based tools in the future. To the best of our knowledge, none of the state-of-the-art has deployed practical data sets and maintained confidentiality in knowledge extraction with a prototype of the decision support framework. These are achieved in the present study, which offers a cost-effective MNCH decision-support system that will enhance healthcare policy decisions and support disease surveillance measures. The present study builds on a preposition enabled natural language parser, PeNLP [38,29], which integrates intelligent reasoning using a web-based application that mines spatial data from unstructured text. Spatial

Table 11.4 Comparison of PeSONT with similar state of the art tools.

Ref.	Data set	Objective	NLP methods	Evaluation tool	Geo Tool	Findings
[24]	Health Data	SHCO	Direct Encoding	TDDOnto, OOPs	IoT Devices	Smart healthcare service
[7]	Geospatial data	GeoDataOnt	Geotext/ GeoImage	–	–	Data Integration and sharing
[15]	Generic	SSN and SAREF	–	W3C Standardized	Sensors	Ontological requirements and semantic translations
[35]	Generic	SSN and SAREF	–	W3C Standardized	Sensors	Ontological requirements and semantic translations
Proposed Work	Unstructured MNCH Data HFR Database	PeSONT	PeNLP	TDDOnto	UTM Geo Map	Confidentiality in knowledge extraction, data availability and cost-effective decision support system

concepts were identified and extracted from unstructured text using prepositions beyond the IN-ON-AT [36,37] preposition pyramid. Furthermore, Usip et al. [39] further adapted the parser to extract temporal data from natural language text to extract spatiotemporal knowledge from rape news articles for decision support.

Elementary morphologic characteristics of geospatial data [7] such as spatial accuracy and coordinate reference are thoroughly handled in PeSONT via on-site coordinate acquisition for onward geotagging to ensure locational accuracy when its descriptive spatial term is extracted from natural language text and juxtaposed on visualization media. Latitudes and longitudes of MNCH facilities retrieved from the Nigerian Health Facility Registry enable MNCH facility tracking and navigation. Beyond geospatial tagging and mapping for coordinates guided location of geographic objects, Miao et al. [9] proposed a method for geospatial resource discovery. Granted that some geospatial resources are held in organized databases, identification and sharing of resource(s) is tedious and time consuming, as existing search methods use keywords for database query. Different from PeSONT where the spatial resource is extracted from plain text in a natural language without a particular structure, Miao et al. [9] extracts the geospatial resource from metadata, which already contain some predefined format according to ISO standards. Furthermore, while PeSONT uses prepositions as the underlying technique for geospatial term identification, the OGDSSM model uses semantic similarity. Although the ontology by Gayathri and Kannan [34] elaborately represents knowledge in the concerned domain, the implementation of the ontology to the semantic web is not discussed. Their work did not also consider or identify any underlying techniques for "hit" words identification and extraction from the unstructured or semistructured natural language text. These limitations are catered for in the proposed PeSONT framework.

In maintaining proper and elaborate facility description, accurate coordinate determination is crucial for correct facility geocoding—as one of the biggest challenges encountered during on-site geocoding of rural healthcare facilities is accurate coordinate determination. However, the categorization of geospatial health research using a geographic positioning system (GPS), [40], can resolve this challenge. Yasobant et al. [41] agreed that proper spatial representation of clinical data enhances geovisualization for improved healthcare. Geovisualization involves the use of visual aspects of spatial information for building knowledge of the environment using a holistic approach [42]. Given the co-ordinates of a MNCH facility, many persons may be unable to arrive at the location. However, with a map, same persons might get better insights on appropriate routes to take. More so, given a digital map with navigational guidance, the individual will hardly miss the route. Visual directions therefore are immensely important in locating the physical position of objects and facilities if patients or visitors must be interactively guided to given health facilities. For a more accurate on-site geocoding to enable geospatial inclusion and GPS tracking, an ontological representation of clinical data and geolocations is more robust and reliable.

An interesting part of Lee et al. [26] is its utilization of ontology to describe, annotate, and build structured and formalized geocoded information, which enables a 3D visualization and selection of shortest path between two points. The contribution of the authors showed that main and alternative routes to designated MNCH centers can be ontologically represented and geocoded into an online database supporting structured natural language data representation for easy semantic query by patients. Using ontology, geospatial entities from unstructured or semistructured natural language text are connected or linked to their meaning derived from a knowledge base [5]. With this linking, semantic searches can follow a top-down approach where the ontology interprets users queries and enriches them with other meaningful terms using ontology concepts and their relations [32]. Even though geospatial ontologies should support integration into broader contexts, Claramunt [33] advocates that explicitly representing concepts and relations abstracted from reality—formally and allowing numerical notation using symbolic grammars alone is not enough, but ontologies should also favor interoperability and knowledge sharing between different applications. This research therefore ensures that beyond the target domain of MNCH the proposed ontology can be extended to formally represent knowledge in other medical specializations and beyond.

Focusing on MNCH facilities location would specifically help reduce the search time by users to target health facilities during emergency medical situations. Furthermore, an effective spatial healthcare solution requires a correct enumeration of the healthcare facility and a traceable location description. It has been observed that the cornerstone of a viable health system entails the definition of the location of health services in relation to the communities to be served. Whereas this assertion would be valid for new health services being developed for a community, the cornerstone for an already existing community's health service would be proper and elaborate path description with priority and alternative routes based on unique condition.

One reason for geospatial data mining is to make the extracted knowledge explicit and easily describable. In most cases, semantic searches both within proprietary archives and generally accessible repositories use keywords to locate and retrieve resource(s). Hence, a search requires predetermined keywords to retrieve expected resources. However, solutions that consider the meaning of terms behind a user query have been proposed.

While information from national health resource platforms (cf. [17]) could be accepted as authoritative, data held on these platforms are sparse with most of these representing facilities located in cities and suburban areas. The implication is that there is no proximity to healthcare facility for rural dwellers; hence, emergency response healthcare will remain a mirage for rural dwellers. Additionally, the generated database excluded privately owned healthcare facilities on the grounds that the facilities were difficult to audit and that their enumeration and regulation were challenging. The generated database from [17] is however generic and without filters for specialized health facilities, signifying longer search time for health facilities within the database, verbose search space, and difficulty in providing accurate spatial description that is vital for easy navigation to desired health facility.

11.6 Conclusions

Ontologies can contribute immensely to the development of healthcare through the structured knowledge base they generate. They are usually applied to describe the concepts of medical terminologies and the relation between them; hence, enabling shareability of medical knowledge. PeSONT, a general schema for reasoning about the health services offered to MNCH care patients with various health conditions was developed. The locations of both the patient and the health facility were carefully extracted from MNCH clinical notes (unstructured data) using the PeNLP parser and stored in a corpus to aid the retrieval of relevant information for informed decisions by stakeholders (policy makers, healthcare providers, government, etc.). Further works include MNCH data mining using the proposed context-aware ontology for health information discovery and efficient healthcare delivery. The ongoing project is focused on the extension of PeSONT and its application on food recommendations by addressing "eating habits" of patients as a preventive measure for most critical MNCH conditions or sicknesses.

Acknowledgments

This research is funded by the Tertiary Education Trustfund (TETFund) Institution-based Research (IBR) Project grant. The clinical staff at St. Luke's General Hospital, Anua, Uyo, and the postgraduate students who assisted during the data collection phase of this research are also appreciated.

References

[1] Sania Nishtar, Ties Boerma, Sohail Amjad, Ali Yawar Alam, Faraz Khalid, Ihsan ul Haq, Yasir A. Mirza, Pakistan's health system: performance and prospects after the 18th Constitutional Amendment, The Lancet 381 (9884) (2013) 2193–2206.

[2] Saadia Ismail, Majed Alshmari, Khalid Latif, Hafiz Farooq Ahmad, A granular ontology model for maternal and child health information system, Journal of Healthcare Engineering 9519321 (2017) 1–9, https://doi.org/10.1155/2017/9519321.

[3] Mansi Radke, Prarthana Das, Kristin Stock, Christopher B. Jones, Detecting the geospatialness of prepositions from natural language text, in: S. Timpf, C. Schlieder, M. Kattenbeck, B. Ludwig, K. Stewart (Eds.), 14th International Conference on Spatial Information Theory (COSIT 2019), 2019, pp. 11:1–11:8, https://doi.org/10.4230/LIPIcs.COSIT.2019.

[4] Guy Lansley, Michael de Smith, Michael Goodchild, Paul Longley, Big data and geospatial analysis, in: M.J. de Smith, M.F. Goodchild, P.A. Longley (Eds.), Geospatial Analysis: A Comprehensive Guide to Principles, Techniques and Software Tools, 6th edition, The Winchelsea Press, Edinburgh, 2018, pp. 547–570.

[5] Margarita Kokla, Eric Guilbert, A review of geospatial semantic information modeling and elicitation approaches, ISPRS International Journal of Geo-Information 146 (9(3)) (2020) 1–31, https://doi.org/10.3390/ijgi9030146.

[6] Parisa Kordjamshidi, Martijn Van Otterlo, Marie-Francine Moens, Spatial role labeling: towards extraction of spatial relations from natural language, ACM Transactions on Speech and Language Processing 8 (3) (2011) 1–36, https://doi.org/10.1145/2050104.2050105.

[7] Kai Sun, Yunqiang Zhu, Peng Pan, Zhiwei Hou, Dongxu Wang, Weirong Li, Jia Song, Geospatial data ontology: the semantic foundation of geospatial data integration and sharing, Big Earth Data 3 (3) (2019) 269–296, https://doi.org/10.1080/20964471.2019.1661662.

[8] Ifiok J. Udo, Moses E. Ekpenyong, Improving emergency healthcare response using real-time collaborative technology, in: Proceedings of the 4th International Conference on Medical and Health Informatics, 2020, pp. 165–173.

[9] Lizhi Miao, Chengliang Liu, Li Fan, Mei-Po Kwan, An OGC web service geospatial data semantic similarity model for improving geospatial service discovery, Open Geosciences 13 (1) (2021) 245–261.

[10] Arbaz Khan, Maria Vasardani, Stephan Winter, Extracting spatial information from place descriptions, in: ACM SIGSPATIAL COMP'13, November 5, Orlando, FL, USA, 2013, https://doi.org/10.1145/2534848.2534857.

[11] Kiran Adnan, Rehan Akbar, An analytical study of information extraction from unstructured and multidimensional big data, Journal of Big Data 6 (91) (2019) 1–38, https://doi.org/10.1186/s40537-019-0254-8.

[12] Xin Hou, Soh-Kim Ong, Andrew Yeh-Ching Nee, X.T. Zhang, W.J. Liu, GRAONTO: A graph-based approach for automatic construction of domain ontology, Expert Systems with Applications 38 (2011) 11958–11975, https://doi.org/10.1016/j.eswa.2011.03.090.

[13] Clarissa Castellã Xavier, Vera Lúcia Strube De Lima, A semi-automatic method for domain ontology extraction from Portuguese language Wikipedia's categories, in: A.C. da Rocha Costa, R.M. Vicari, F. Tonidandel (Eds.), Advances in Artificial Intelligence – SBIA 2010, SBIA 2010, in: Lecture Notes in Computer Science, Springer, Berlin, 2010, https://doi.org/10.1007/978-3-642-16138-4_2, arXiv:6404.2010.

[14] Zacharias Dermatis, Dimitrios Tsoromokos, Filipos Gozadinos, Athina Lazakidou, The utilization of geographic information system in healthcare, International Journal of Health Research and Innovation 4 (1) (2016) 39–57.

[15] Laura Daniele, Frank den Hartog, Jasper Roes, Created in close interaction with the industry: the smart appliances reference (SAREF) ontology, in: International Workshop Formal Ontologies Meet Industries, Springer, Cham, 2015, pp. 100–112.

[16] Brandon Bennett, Pragya Agarwal, Semantic categories underlying the meaning of 'place', in: International Conference on Spatial Information Theory, Springer, Berlin, Heidelberg, 2007, pp. 78–95.

[17] Joseph Maina, Paul O. Ouma, Peter M. Macharia, Victor A. Alegana, Benard Mitto, Ibrahima Socé Fall, Abdisalan M. Noor, Robert W. Snow, Emelda A. Okiro, A spatial database of health facilities managed by the public health sector in sub-Saharan, Africa Scientific Data 6 (134) (2019) 1–8, https://doi.org/10.1038/s41597-019-0142-2.

[18] Lantao Hu, Qiuli Tong, Zhao Du, Yongqi Liu, Yeming Tang, Location-based service using ontology and collaborative recommendation, in: International Conference on Information Science, Electronics and Electrical Engineering, 2014, pp. 653–656, https://doi.org/10.1109/InfoSEEE.2014.6948195.

[19] Shah Jahan Miah, Huy Quan Vu, John G. Gammack, A location analytics method for the utilisation of geotagged photos in travel marketing decision-making, Journal of Information & Knowledge Management (2018) 1–38, https://doi.org/10.1142/S0219649219500047.

[20] Ioanna Roussaki, Maria Strimpakou, Nikos Kalatzis, Miltos Anagnostou, Carsten Pils, Hybrid context modeling: a location-based scheme using ontologies, in: 4th Annual Conference on Pervasive Computing and Communications, PerCom Workshops, 2006, https://doi.org/10.1109/PERCOMW.2006.65.

[21] Sanju Mishra, Sarika Jain, Ontologies as semantic model in IoT, International Journal of Computers and Applications 42 (3) (2020) 233–243.

[22] Sanju Mishra, Sarika Jain, Towards a semantic knowledge treasure for military intelligence, in: Emerging Technologies in Data Mining and Information Security, Springer, Singapore, 2019, pp. 835–845.

[23] Sanju Mishra Tiwari, Sarika Jain, Ajith Abraham, Smita Shandilya, Secure semantic smart HealthCare (S3HC), Journal of Web Engineering 17 (8) (2018) 617–646.

[24] Sanju Tiwari, Ajith Abraham, Semantic assessment of smart healthcare ontology, International Journal of Web Information Systems 16 (4) (2020) 475–491, https://doi.org/10.1108/IJWIS-05-2020-0027.

[25] Ronak Panchal, Priya Swaminarayan, Sanju Tiwari, Fernando Ortiz-Rodriguez, AISHE-onto: a semantic model for public higher education universities, in: DG. O2021: The 22nd Annual International Conference on Digital Government Research, 2021, pp. 545–547.

[26] Kangjae Lee, Jiyeong Lee, Mei-Po Kwan, Location-based service using ontology-based semantic queries: a study with a focus on indoor activities in a university context, Computers, Environments and Urban Systems 62 (2017) 41–52, https://doi.org/10.1016/j.compenvurbsys.2016.10.009.

[27] Mark A. Musen, The protégé project: a look back and a look forward, AI Matters 1 (4) (2015) 4–12, https://doi.org/10.1145/2757001.2757003.

[28] Pascal Hitzler, Markus Krötzsch, Bijan Parsia, Peter F. Patel-Schneider, Sebastian Rudolph, OWL 2 web ontology language primer, W3C recommendation, 27(1), 123.2009.

[29] Patience Usoro Usip, Moses Effiong Ekpenyong, Funebi Francis Ijebu, Kommomo Jacob Usang, Ifiok James Udo, PeNLP parser, in: Deep Learning in Biomedical and Health Informatics, 2021, https://doi.org/10.1201/9781003161233-8.

[30] Olusesan Ayodeji Makinde, Aderemi Azeez, Samson Bamidele, Akin Oyemakinde, Kolawole Azeez Oyediran, Wura Adebayo, Bolaji Fapohunda, Abimbola Abioye, Stephanie Mullen, Development of a master health facility list in Nigeria, Online Journal of Public Health Informatics 6 (2) (2014) e184, https://doi.org/10.5210/ojphi.v6i2.5287, 1–14.

[31] C. Maria Keet, Agnieszka Ławrynowicz, Test-Driven Development of Ontologies, Extended Semantic Web Conference, LNCS, Springer, 2016.

[32] K. Munir, M.S. Anjum, The use of ontologies for effective knowledge modelling and information retrieval, Applied Computing and Informatics 14 (2017) 116–126, https://doi.org/10.1016/j.aci.2017.07.003.

[33] Christophe Claramunt, Test-driven development of ontologies, ontologies for geospatial information: progress and challenges ahead, Journal of Spatial Information Science 20 (2020) 35–41, https://doi.org/10.5311/JOSIS.2020.20.666.

[34] M. Gayathri, R. Jagadeesh Kannan, Construction of domain ontology for traditional Ayurvedic medicine, in: V.K. Solanki, et al. (Eds.), Intelligent Computing in Engineering, in: Advances in Intelligent Systems and Computing, vol. 1125, Springer Nature Singapore Pte Ltd, 2020, pp. 417–426, https://doi.org/10.1007/978-981-15-2780-7_46.

[35] Michael Compton, Payam Barnaghi, Luis Bermudez, Raul Garcia-Castro, Oscar Corcho, Simon Cox, John Graybeal, M. Hauswirth, C. Henson, A. Herzog, V. Huang, The SSN ontology of the W3C semantic sensor network incubator group, Journal of Web Semantics 17 (2012) 25–32.

[36] Sura Muttlak Nasser, A cognitive-semantic analysis of preposition on: an experimental study at university of Baghdad, Arab World English Journal (AWEJ) 11 (2020).

[37] Asmeza Arjan, Noor Hayati Abdullah, Norwati Roslim, A corpus-based study on English prepositions of place, "In" and "On", English Language Teaching 6 (12) (2013) 167–174.

[38] Patience U. Usip, Moses E. Ekpenyong, James Nwachukwu, A secured preposition-enabled natural language parser for extracting spatial context from unstructured data, in: International Conference on e-Infrastructure and e-Services for Developing Countries, Springer, Cham, 2017, pp. 163–168.

[39] Patience U. Usip, F.F. Ijebu, E.A. Dan, A spatiotemporal knowledge bank from rape news articles for decision support, in: Iberoamerican Knowledge Graphs and Semantic Web Conference, Springer, Cham, 2020, pp. 147–157.

[40] Xun Shi, Mei-Po Kwan, Introduction: geospatial health research and GIS, Annals of GIS 21 (2) (2015) 93–95, https://doi.org/10.1080/19475683.2015.1031204.

[41] Sandul Yasobant, Kranti Suresh Vora, Carl Hughes, Ashish Upadhyay, Dileep V. Mavalankar, Geovisualization: a newer GIS technology for implementing research in health, Journal of Geographic Information System 20 (2015) 20–28, https://doi.org/10.4236/jgis.2015.71002.

[42] Piotr A. Werner, Review of implementation of augmented reality into the georeferenced analogue and digital maps and images, Information 10 (12) (2019) 1–14, https://doi.org/10.3390/info10010012.

IntelliOntoRec: a knowledge infused semiautomatic approach for ontology formulation in healthcare and medical science

12

Gerard Deepak[a,b] **and Deepak Surya S.**[a,b]

[a]*Manipal Institute of Technology Bengaluru, Manipal Academy of Higher Education, Manipal, India*
[b]*National Institute of Technology, Tiruchirappalli, Tiruchirappalli, India*

Contents

12.1 Introduction

The World Wide Web has experienced a significant transformation and continues to do so. Web 1.0, also known as the Web of Documents period, was the first generation of the internet. It had connections between documents. The second-generation Web 2.0, often known as the Social Web, included a connection between user profiles. Web 2.0 is the phrase that refers to a collection of web pages, websites, web services, social networks that are formed by an extensive network of computers. It allows anyone on the interconnected network, i.e., the internet, to access, share, create, download resources and data from the web. Web 2.0 facilitates users to access information via hyperlinks. A hyperlink, also known as a web link, is any object or text that directs toward another resource or document. The internet is made up of hyperlinks that connect billions of sites and data. With the advent of Web 3.0, which emphasizes the internet of objects rather than the internet of computers, there is a call of way to organize and manipulate vast amounts of data so that the computer system could extract the knowledge efficiently and effectively. Ontology is a way of representing and describing domain information as classes and relationships. It encapsulates the domain's different concepts as well as the relationships between them. Ontology provides a framework for connecting the various forms of data on the internet, allowing for information interoperability, interdatabase search, streamlined knowledge generation, and sharing from widely dispersed data sources. Web 3.0, which is still growing, allows the network of all web objects, i.e., the web of data. Knowledge of real-world environments is crucial for intelligence, and it is required for AI entities to demonstrate intelligent behavior. Only if a machine has expertise comprehending knowledge or competence with the data can it respond correctly. Ontology, for instance, could describe ideas, entity relationships, and object categories. These integrated semantics provide significant benefits such as data reasoning and working with diverse data sources, while these standards encourage uniform data formats and web interchange methods, particularly RDF. Web 3.0 refers to the connections between data and entities. It also introduces Semantic Web, making digital material machine-readable, allowing for a more intelligent and linked web. RDF and OWL are the tools that will propel us toward this Web 3.0 goal. To standardize metadata representation, several approaches are used. Ontology is an essential component of the Semantic Web and Web 3.0, as it is also used to generate a knowledge graph. A knowledge graph is a data model in which nodes represent data and connections represent relationships between them. By describing the structure of the information available inside the topic, the ontology sets the basis for acquiring the essential information. As there is a substantial increase in big data now, companies are exploring new approaches to leverage this data to their benefit with machine learning. ML techniques are frequently employed to uncover knowledge to get insight into the problem, ultimately leading toward more effective decision making. The models have constraints associated with them ranging from processing efficiency and lack of transparency and usability since specific systems only operate in particular situations. For example, if an AI system is used to shortlist candidates for a job, neither could it establish a solid reason for its decisions, nor could the algorithm's creator.

Explainable AI is emerging swiftly as we pace toward the Web 3.0. It is considered the third wave of artificial intelligence that enables us to arrive at the solution for a machine learning or deep learning problem that humans can understand and explain. The current AI algorithms are considered a black box as even the algorithm's designer could not reason out why the algorithm has arrived at the particular solution. eXAI tends to enhance the user experience, and people's trust in AI systems as it could provide the explanation model and an explanation interface as its output. XAI's objective is to describe what has already been done, what is being done now, what would be done next, and reveal the evidence on which the actions are based. This is especially significant in fields such as medical, military, finance, and law, where understanding judgments and building faith in algorithms are critical. Explainable AI also enables us to affirm and dispute existing knowledge, produce new assumptions, and unveil biases in the data. This paper presents a novel method to formulate ontologies for various domains belonging to healthcare and medical sectors.

The remainder of the chapter is organized as follows. Section 12.1 includes the introduction. Section 12.2 encompasses the related work. The proposed modeled is presented in the third section of the chapter and the implementation of the proposed approach is presented in the fourth section. The fifth section contains the results. Finally, Section 12.6 concludes the chapter.

12.2 **Related works**

Aguilar et al. [1] propose an ontology, CAMeOnto, designed and developed based on the 5Ws concepts: where, why, what, who, when. The proposed ontology is an introspective middleware for context-aware applications. It has been implemented in various case studies to demonstrate how CAMeOnto operates perfectly and, therefore, can rationale to deduce insights into the context. Fahad et al. [2] explicate on the linking of idea class formulations and offers a method enabling the unification in an autonomous ontological fusion procedure that does not need human involvement. Priya et al. [3] present a novel technique termed HSSM (hybrid semantic similarity measure) based ontology merging based on semantic similarity measures and formal concept analysis. In the realm of ontology engineering, there are several general merging requirements (GMR). Babalou et al. [4] compiled a list of the most frequent GMRs and categorized them into 3 components. Because not all GMRs can be fulfilled concurrently, they offer a mechanism, which enables users to identify the most relevant GMRs for their given job and thereafter finds the greatest acceptable subgroup of GMRs. This result obtained is utilized to either pick a suitable merging technique or parameterize a universal merger mechanism. Chatterjee et al. [5] designed a architecture to enable easy merging of ontologies. Agriculture is characterized as major domain, with numerous subcategories like crops, fertilizer, etc. as add-ons. They also illustrate how scheme software is superior to the well-known Protégé. Nakhla et al. [6] introduce a novel technique to automate database enrichment built on an ontology that models topics through collections of ideas and the semantic links between them. Schoormann et al. [7] highlight how computational linguistics approaches may be used to create domain-specific modeling tools that can deal with these difficulties automatically. They also offer a procedural framework that was created and tested in four sessions, propose resources, techniques, and materials, and remark on common difficulties. Pal et al. [8] provide an all-inclusive method for extracting learning metadata from a educational video. A semiautomatic approach that combines mechanical and computational techniques to retrieve the metadata and evaluate its contents is proposed. Along with establishing a set of particular metadata elements from IEEE LOM, few more extra characteristics are recommended to analyze the appropriateness of a video-based educational object in terms of a learner's customized preferences and compatibility. Xue et al. [9] highlight the ways in which environmental data is collected with the advancements of sensor technology and the emergence of a plethora of sensor ontologies. Their work also presents how the ability to perform semantic interoperation among the sensor ontologies is hindered because of their heterogeneity and how it affects the applications of the sensor ontologies. Rani et al. [10] developed two novel topic modeling approaches, LSI and SVD and Mr.LDA, to facilitate easy learning of topic models. Their primary goal is to identify the statistical link between the document and words to create a subject ontology and ontological network with as little human interaction as possible. The efficacy of the suggested technique is demonstrated by an experimental investigation of constructing a topic ontology and recover the most related topic ontology to the user's query word. Xingjun et al. [11] developed a model using the grey wolf optimization algorithm to provide a unique fuzzy-based solution for data congestion reduction in cloud-based IoT technologies. Franz et al. [12] explore the complexities and limitations in constructing a biological TAXONOMY incorporated into an ontological reasoning framework. Semantic latent analysis was incorporated by Pushpa et al. [13] to measure the ontological relevance. Content-based filtering was used to improve the quality of the proposed framework. These tools have been used to increase the overall efficiency of semantic web content recovery. Nasution et al. [14] have presented an already existing idea of ontology, as well as a mathematically based technique to producing a subject model that supports existence. Tikhamarine et al. [15] conducted a study where the primary goal is to offer an effective hybrid system that combines the optimization algorithm of grey wolves with other AI models. Zhang et al. [16] proposed a minimally supervised system for text categorization with metadata that uses a generative process to explain the relationships between words, records, marks, and metadata. Liu et al. [17] has reviewed the application and development of topic models for bioinformatics, and the article describes the steps involved in creating the application. He stated that while there are no topic models suited for specific biological data, studies on topic modeling in biological data have a long and difficult road ahead, but that topic models are a promising technique for many applications in bioinformatics research. Steyvers et al. [18] has referred to the latent semantic analysis and the three assertions of the LSA method that semantic information can derive from a cooccurrence matrix of the word document, that the reduction in dimensionality is an essential component of such a derivation, and that words and documents are represented in Euclidean space as points. Wallach et al. [19] show that frequently used approaches are unlikely to properly estimate the probability of held-out documents, and present two alternative methods that are both accurate and efficient. In [20–26] several Ontology and Knowledge Centric Models in support of the literature of the proposed framework are depicted.

12.3 Proposed model

12.3.1 Phase 1

Fig. 12.1 depicts the first phase of the architecture for the generation of domain ontologies for specialized medical or specialized healthcare domains. Since the metadata is vast, massive volumes of data are created as obtained straight from the internet, and for each word in the RDF, there is a tremendous demand for metadata classification. Domain indicators are collected from the contents or indexes of a medical textbook to categorize information. Several textbooks are utilized, spanning from radiology to biostatistics and radiology of physics to community medicine, gastro-ontology, dermatology, oncology, and other medical backgrounds. Keywords directly from the textbooks are extracted and utilized as domain indicators. Domain indicators and a preprocessed set of words are fed into the transformers, which classify the metadata under each class reflecting the features and provide the top 50% classified occurrences. A crawling index from a medical textbook is utilized to generate the terminology from which the keywords are extracted using TF-IDF. Since the domain is a high risk like healthcare, a lot of care and attention is required, along with the verification required for ontology modeling. Ontology verification and ontology generation of the healthcare domain are vital because health is one of the scientific domains where high risk is represented. Since it is a vulnerable domain, any case of a miss form or miss formulated ontologies leads to series of consequences. As a result, a semiautomatic approach that incorporates humans in the middle approach for semiautomatic ontology modeling for technique.

Since the domain is the healthcare domain, several domains such as medical science and pharmaceutical are considered core domains for healthcare. So in this approach crawled medical journal, i.e., journal representing medical sciences, the recent 10 years have been crawled directly via the medical journal itself or via google scholar. So the data contained in the medical journal is subjected to preprocessing tokenization, lemmatization, stop word removal, word sense disambiguation, and named entity recognition is performed. Textbooks, e-books of several specializations like radiology, community medicine, gastro ontology, dermatology, oncology, etc., belonging to core medical sciences subjects are incorporated. Their e-books are also used as sources of modeling ontology. So, from the medical

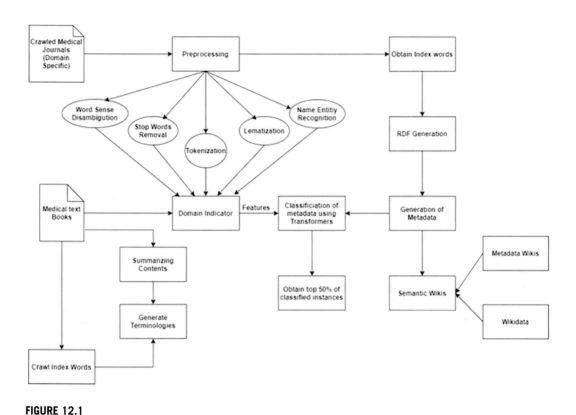

FIGURE 12.1

Phase 1 of the proposed IntelliOntoRec framework.

journal, once preprocessed, the index words are obtained. RDF is generated for all the index words, or the marker keywords used for indexing in the preface of keywords in the medical journal and textbook. The diagram illustrates the first step of the architecture for developing domain ontologies for specialized medical or specialized healthcare domains. Since healthcare is one of the scientific fields where high risk is represented, ontology verification and ontology development in the healthcare sector are critical. It necessitates a great deal of care and attention, in addition to the verification necessary for ontology modeling. Since it is a fragile domain, any misformed, or incorrectly constructed ontologies have a cascade of repercussions. Consequently, a semiautomatic strategy for the ontology modeling technique involves humans in the intermediate approach. Since the topic is healthcare, numerous domains, including medical science and pharmaceutical, are considered essential domains for healthcare. So, in this approach, the last 10 years of crawling medical journals, i.e., journal representing medical sciences, have been crawled directly via the medical journal itself or via Google Scholar. As a result, the data in the medical journal is preprocessed using tokenization, lemmatization, stop word removal, word sensed disambiguation, and named entity recognition. Textbooks and e-books from a variety of specialties, including radiology, community medicine, gastroenterology, dermatology, oncology, and surgery, as well as fundamental medical sciences courses, are included. The e-books are often utilized as ontology modeling resources. The index words are acquired from the medical journal after it has been preprocessed. All index terms or marker keywords used for indexing in the medical journal and textbook introduction are converted to RDF. As a result of preprocessing the crawled medical journal repository, RDF is created in the index words. The OntoCollab is then used to produce RDF. So, XML structure is completed, followed by XML to RDF conversion. The CntoCollab generates RDF from global data as well as user and community-contributed cloud sources. So the XML extracts preprocesses constructs from OntoCollab. Structured XML is created, which is then transformed to RDF. GREDL and RDF distiller are used to generate metadata from RDF. The purpose of creating metadata is to include every possible aspect of online knowledge into the framework. Furthermore, because metadata covers every possible RDF possibility, a large amount of knowledge will be generated, which can be converted into knowledge and further incorporated into the framework to reduce the cognitive gap of the background knowledge between the contents on the World Wide Web and contents incorporated into the framework and cognitive gap is reduced. Although metadata is generated for all RDF, it is a massive quantity of data that requires classification and further preprocessing, refining, and classification before it can be utilized for any further recommendation. In addition, predefined relationships between entities are already accessible in semantic cookies. Thus, entity population and connection, as well as establishing relationships among entities, are computed using semantic wikis. Wikidata and MediaWiki are the two semantic wikis that are employed. The inclusion of semantic wikis guarantees that knowledge density is significantly higher in the framework and that the cognitive gap between global knowledge and knowledge included into the system is much less. As a result of the RDF production of metadata and the inclusion of semantic wikis into the framework, the additive knowledge into the framework is considerably greater, and nearly every kind of entity relationship and their precise placement into the framework are validated and utilized in the proposal.

12.3.2 Transformer architecture

Transformer is a type of neural network architecture that was introduced to overcome several shortcomings of the recurrent neural networks. RNNs are slow to train because of its intense truncated backpropagation through time. They are also not very effective dealing with long sequence inputs due to the problems such as the vanishing gradient and exploding gradients. Vanishing gradient occurs during back propagation when multiplication of small derivative values yields in a very small value that cannot be stored in the memory. Although LSTM neural networks are later introduced to effectively deal with longer sentences by incorporating a memory unit, it is slower than the RNN due to its high complexity. In either of the networks discussed above, the input data must be processed sequentially or serially one after the other. Also, then any hidden state requires some input from its previous state to make operations on its current state. Such sequential or serial processing of inputs does not efficiently exploit the power of modern computer GPUs, which are designed to facilitate parallel processing. One core objective of the transformer architecture is to enable parallel computation for processing sequential data. Transformer adopts a similar type of network architecture as that of LSTMs and RNNs but the units in the input sequence can be passed parallelly. Transformer architecture has two main parts: encoder and decoder. All of the units of the input sequence are passed simultaneously to the transformer and the word embeddings are generated for them consequently. Embedding space is a space that contains the word vectors in such a way that similar words are grouped together spatially. It is used to map a word to a vector. Positional encoders are used in transformers as the same word in different sentences with different meanings and have the same word embeddings. Positional encoders are used to generate a context vector based on the position of the word in the sequence. The input word is preprocessed and converted to its corresponding word embedding. Information from the positional encoders is further added to the embeddings. It is then passed to the encoder block where it goes through a multiheaded attention layer and a feed forward layer. Attention techniques help the model to emphasis more focus on the important parts of the sequence. Attention vector of the ith input unit captures the relevance of the ith unit among other units of the input sequence. Attention with respect to itself is called self-attention and such attention vectors can be generated for each input unit in the sequence. These attention vectors are found in the attention block of the encoder, which is used to capture the contextual relevance among the units in the input the sequence. These attention vectors are then passed through a feed forward neural network. The neural network converts the attention vectors into a form that can be comprehended by the encoder block or the decoder block. Similarly, in the decoder end, the embedding for each word is generated using the input embeddings and positional vectors are added to capture the notion of context of the unit in the sequence. Upon adding the positional information, these vectors are then passed to the decoder block where attention vectors are generated based on self-attention mechanisms. Finally, the attention vectors from the encoder and decoder blocks are passed to the encoder-decoder attention layer to generate an attention vector for each unit in input sequences for the encoder and the decoder blocks. The resulting attention vector captures the relationship of each unit in the input sequence of encoder to each unit in the input sequence of the decoder unit and vice versa. Each attention vector is then passed to a feed forward neural network to make the output layer more comprehensible for the further layers. The next layer is a linear feed forward layer, which passes the data to a SoftMax layer to transform it to a probability distribution. This is how a sequence-to-sequence translation is performed using transformers. In this research study, we implemented transformers using Pytorch to perform classification task.

12.3.3 Phase 2

Fig. 12.2 illustrates the architecture for phase 2 of integrating healthcare domain knowledge with auxiliary associated ontologies. There are several ontologies in various domains such as healthcare, pharmaceutical, general medicine, dermatology, community medicine, nuclear medicine, radiodiagnosis, pediatrics, obstetrics, gynecology, biostatistics, gastroenterology, neurology, nephrology, rheumatology, cardiology, pulmonology, etc. These ontologies are generated automatically using OntoCollab and manually using web Protege from medical text-books. Existing ontologies are included and used for further integration. These eBooks are summarized, and TF-IDF is applied to the contents of the summarized eBooks. Ontologies are also incorporated using ready-made ontologies. Apart from this, 50% of the categorized metadata from step 1 is also incorporated. Semantic wikis, RDF, and metadata are used, and a large knowledge graph is created through reasoning with Shannon's entropy and SemantoSim. Since an enormous number of ontologies are included, a stringent threshold value for SemantoSim and Shannon's entropy is required to only let the most relevant items into the ontology to formulate the knowledge graph. Data integration of ontologies is accomplished using agents, which are modeled using AgentSpeak. Jade is used to calculate the agent's state to compute Shannon's entropy and the SemantoSim measure. Shannon's entropy is used with a minimum step deviation of 0.25 and a SemantoSim threshold value of 0.75. Also, Shannon's entropy and SemantoSim are used as objective functions for the grey wolf search metaheuristic algorithm.

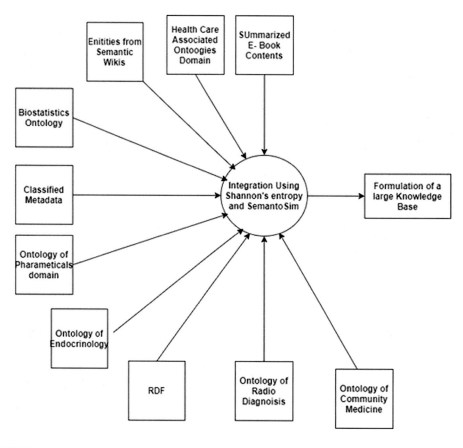

FIGURE 12.2

Phase 2 of the proposed IntelliOntoRec framework.

The grey wolf optimization method is inspired by the grey wolves' foraging behavior and leadership structure. The organizational structure of grey wolves is divided into four divisions: P, Q, R, and S. The grey wolf's hunting process mainly consists of three phases in which optimization is achieved. Seeking for prey, surrounding prey, and striking prey are the three steps. P is assigned to the first optimum solution. The second optimal solution is denoted as Q, while the third solution is denoted as R. These three wolves are followed by the S wolves. The current iteration is specified by t. The equations included in the algorithm are depicted from Eq. (12.1) to Eq. (12.13). The vector of the prey's position is represented by G_q, $G.r1$ and $r2$ in Eqs. (12.3) and (12.4) are vectors with random values between [0, 1]. Using Eqs. (12.8), (12.9), and (12.10), the search agent updates the location of P, Q, and R. Only P, Q, and R are allowed to adjust the position of the prey. Other wolves change their location in the vicinity of the prey at random. The top three best search agent's solutions have been kept, as well as the other search agents' locations will be updated in relation to the best search agent's position. SemantoSim and

Shannon's entropy are calculated as depicted in Eq. (12.12) and Eq. (12.13), respectively.

$$\vec{Y} = |\vec{A}.\vec{G_q} - \vec{G}(t)| \tag{12.1}$$

$$\vec{F}(t+1) = \vec{G_q} - \vec{C}.\vec{Y} \tag{12.2}$$

$$\vec{C} = 2\vec{c}.\vec{r1} - \vec{c} \tag{12.3}$$

$$\vec{A} = 2.\vec{r2} \tag{12.4}$$

$$\vec{Y_\alpha} = |\vec{A_1}.\vec{G_\alpha} - \vec{G}| \tag{12.5}$$

$$\vec{Y_\beta} = |\vec{A_2}.\vec{G_\beta} - \vec{G}| \tag{12.6}$$

$$\vec{Y_\delta} = |\vec{A_3}.\vec{G_\delta} - \vec{G}| \tag{12.7}$$

$$\vec{G_1} = \vec{G_\alpha} - \vec{C_1}.(\vec{Y_\alpha}) \tag{12.8}$$

$$\vec{G_2} = \vec{G_\beta} - \vec{C_2}.(\vec{Y_\beta}) \tag{12.9}$$

$$\vec{G_3} = \vec{G_\delta} - \vec{C_3}.(\vec{Y_\delta}) \tag{12.10}$$

$$\vec{G}(t+1) = \frac{\vec{G_1} + \vec{G_2} + \vec{G_3}}{3} \tag{12.11}$$

$$SemantoSim(a,b) = \frac{pmi(a,b) + p(a,b)\log([p(a,b)])}{[p(a).p(b)] + \log([p(a,b)])} \tag{12.12}$$

$$H(X|Y) = -\sum_{i,j}^{n} P(x_i, y_j)\log\frac{p(x_i, y_j)}{p(y_j)} \tag{12.13}$$

12.3.4 Phase 3

Fig. 12.3 depicts the Phase 3 of the proposed IntelliOntoRec framework, i.e., Ontology Finalization and Review. It is straightforward to incorporate and execute due to its uncomplicated basic storage and computational needs. Also, the continual decrease of parameter space and reduced design alternatives, only two control parameters, along with its ability to avoid local minima, leads to faster convergence. These properties ensure the grey wolf metaheuristic algorithm is very robust and stable. Moreover, since the SemantoSim semantic similarity measure and the Simpson's diversity index are used as objective functions over the grey wolf optimization algorithm, the two parameters of the grey wolf optimizer are linked to SemantoSim and the Simpson's diversity index the control for optimization in choosing the most relevant entity among an environment of highly relevant entities from the knowledge graphs and the domain or clue information conceived by a set of domain experts or similar users. The convergence to optimality in selecting the most appropriate solution from a set of feasible solutions is quite distinguishable and reliable when grey wolf metaheuristics is applied. The reason why other optimization metaheuristics are not used owes to the fact that some of them either have multiple parameters or multiple stage objective functions, or multiple crossover functions. Also, other metaheuristic algorithms with two parameters can be feasibly applied, but we chose grey wolf as its intermediate steps are pretty reliable with low computer requirements.

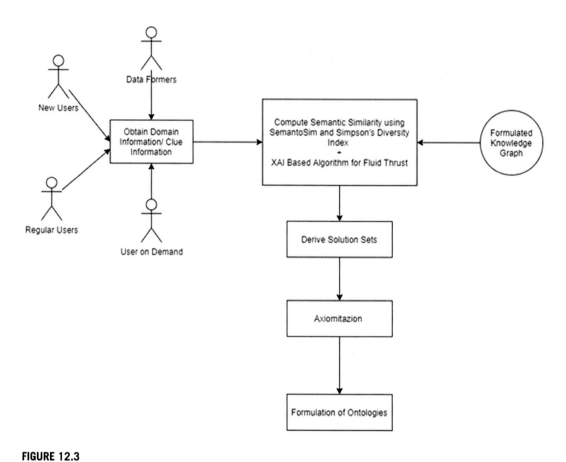

FIGURE 12.3

Phase 3 of the proposed IntelliOntoRec framework i.e., Ontology Finalization and Review.

An empirical analysis based on the structure of the grey wolf optimization algorithm, its computational requirements, the feasibility of rendering solution sets from initial and intermediate solutions, the feasibility of parameter tuning, the feasibility of parameter linking, a parametric coalition of the objective function with parametric tunability, linking of external objective functions into the native objective functions of the grey wolf metaheuristics is analyzed based on its implications and applicability across various problem domains. The derivation of solution sets and their intermediate steps were also analyzed empirically. Moreover, quantitative evaluation of the efficiency, scalability, and convergence has not been conducted. The problem focuses on integrating knowledge from heterogenous sources across a specialized domain and achieving a semiautomatic model with a human-in-the-middle approach formalized. Due to this, the focus of the quantitative evaluation is not on grey wolf optimization but on the entire framework that has been proposed.

The necessity for ontology finalization and evaluation is mostly owing to healthcare being a high-risk domain requiring medical ontologies. Any departure from the realities of the ontology might have disastrous consequences. This is a human-in-the-middle method in which human intervention is necessary for ontology verification. Manual verification of ontologies may be time consuming, but it is required. At any moment, any number of domain experts may be engaged and communicate with experts from other specialized fields. In addition to normal users, data farmers, knowledge engineers, and domain specialists might work together to collect domain information or true terms from users. The semantic similarity is computed based on SemantoSim and Simpson's diversity Index under explainable AI algorithm based on fluid thrust.

12.4 **Implementation**

The implementation of the proposed approach has been done in computer with i7 processor and 16GB RAM. A modularity approach is followed in which most of the phases is implemented using latest version of Python. The phase for Ontology aggregation and knowledge-base generation was implemented using JAVA as it required the use of agents and also the RDF XML Ontology format class. A module view controller pattern is incorporated, and the centralized knowledge bases are used. A customized Java crawler is used for crawling the medical journals by using the results of Google scholar and other specialized medical journals. Journal articles from 2008 to 2020 were all crawled under each domain and are used for experimentation. The preprocessing was carried out using the Python's natural language toolkit. Tokenization is performed by using customized Landspace character tokenizer and lemmatization is done using the WordNet lemmatizer. Regular expression removal is used to remove the stop words and Stanford Names Entity Recognizer is used for Name Entity recognition. Word Sense Disambiguation is carried out based on Corpus Lesk algorithm using the Semantic Concordance. The RDF is generated using the OntoCollab and where the WWW open community forums, community contributed ontologies, Linked Open Data Cloud, etc. are accessed. GREDL and RDF distiller are used to generate metadata from RDF. Wikidata API has been used to get the wikidata and SPARQL endpoints are written to gain access to the media wikis. Journals and articles from 1985 to 2020 have been archived under several broad domain areas via Google Scholar using a customized web crawler. AgentSpeak has been used for agent modeling and Pytorch is used for transformer implementation.

In this study, ontology belonging to the domain radiodiagnosis, pharmaceuticals, general medicine, dermatology, community medicine, nuclear medicine, pediatrics, obstetrics and gynecology, biostatistics, gastroenterology, neurology, nephrology, cardiology, rheumatology, pulmonology, otorhinolaryngology, ophthalmology, geriatric medicine, orthopedics, and pathology are considered. Table 12.1 shows the number of classes and individuals found in ontologies of the domains considered in the study. Algorithm 1 illustrates the algorithm for the proposed framework, IntelliOntoRec.

The parameter settings are utterly implicit for the grey wolf optimization as the concept of agent-driven computation based on goal setting, goal achievement through the computability of the intermediate states have been put forth. The agent is written using AgentSpeak and JADE with a state of computing the semantic relatedness without compromising on the diversity, and the behavior of the agent is to filter out the most appropriate entities among a set of relevant entities. To achieve this, thresholding is the only parametric setting that applies to both the extrinsic objective functions: Simpson's Diversity Index and SemantoSim measure. A threshold of 0.75 has been set for the SemantoSim measure, a stringent threshold setting that allows only the most relevant entities into the framework to conceive knowledge. The very stringent value for threshold is considered because the domain is susceptible, critical, specialized, and scientific. Similarly, a set deviation of 0.25 is chosen for Simpson's

Table 12.1 Details of concepts and individuals in the initial seed domain ontologies used.

Domain ontologies	No. of concepts	No. of individuals
Pharmaceuticals	785	3123
General Medicine	2437	6684
Dermatology	3676	7812
Community Medicine	2481	4933
Nuclear Medicine	2431	9812
Pediatrics	1412	6771
Obstetrics and Gynecology	997	7611
Biostatistics	689	3872
Gastroenterology	1714	5102
Neurology	2481	4861
Nephrology	1419	3308
Cardiology	996	4112
Rheumatology	1687	5009
Pulmonology	1412	6134
Otorhinolaryngology	3372	7108
Ophthalmology	3919	8117
Geriatric Medicine	1716	2106
Orthopedics	1422	3040
Pathology	1799	3812

diversity index for the same reasons. The i7 processor with 16 GB RAM is highlighted because it is the minimum requirement for the integrating agents and grey wolf optimization algorithm in an environment of such computability requirements. However, it is also an indication that high-powered GPUs or other fancy collaborative computing environments powered by large-scale GPUs are not required.

12.5 Results

The proposed IntelliOntoRec's performance has been assessed employing precision, recall, accuracy, F-measure, FDR, and n-DCG as evaluation metrics. Precision, recall, accuracy, and F-measure essentially belong to the class of retrieval and recommendation system. As a result, these measures are best related when retrieval or recommendation is performed. Since ontologies are suggested in this case, this is one of the best performance evaluation metrics that may be integrated if the ground truth of the ontologies is maintained or validated. As a result, precision, recall, accuracy, and the f measure have been used to calculate the relevance of the results. The FDR, which stands for False Discovery Rate, is related with the error rate and quantifies the number of false positives recommended by the algorithm. As a result, the FDR False Discovery Rate is employed. In this case, n-DCG computes the diversity of the suggested results; the diversity in the recommended ontologies is utilized as a performance metric. In addition to these performance measures, because ontologies are suggested, the reuse and reference rate for each ontology is utilized as a prospective metric.

Algorithm 1 The proposed algorithm of the IntelliOntoRec framework.

Input: Crawled Medical Journals

Output: Ontologies belonging to various medical domains.

Start

Step 1: Crawled medical Ontologies are preprocessed. Preprocessing includes Tokenization, Lemmatization, Named Entity Recognition, Word Sense Disambiguation and Stop Words Removal.

Step 2: Obtain index words

Step 3: Generate RDF

Step 4: Generate Metadata

Step 5: Crawl index words to generate terminologies and obtain summarized content from medical textbooks to incorporate to the Domain Ontologies.

Step 6: Incorporate wiki data and media wiki into semantic wikis

Step 7: Classification using Transformer on the features extracted from the domain Ontologies upon preprocessing and enriched metadata.

Step 8: Integration of domain knowledge in health care with associated ontologies

Step 9: From medical textbooks, ontologies are created automatically with OntoCollab and manually with Web Protege.

Step 10: Half of the classified metadata from step 1 is also included.

Step 11: Using Semantic Wikis, RDF, and metadata, a huge knowledge network is generated by reasoning using Shannon's entropy with a minimal step deviation of 0.25 and SemantoSim with a threshold value of 0.75.

Step 12: The grey wolf search metaheuristic algorithm is employed with Shannon's entropy and SemantoSim as objective functions.

    Randomly launch the population of the Grey Wolves

    Set c, C, and A

    Estimate the fitness for each agent X α X_{α}

    Y_{α} = Entropy, T1X β X_{β}

    Y_{β} = Morisita's overlap index, T2

    X δ X_{δ} Y_{δ} = Normalized Point Wise Mutual Information, T3

      While (t t x < Maximum number of iterations)

        For each search agent

          Modify the location for T

        end for;

        Modify ca a c, A A C, and AC C

        Evaluate the fitness of all T

        Modify X α X_{α} Y_{α}, X β X_{β} Y_{β}, and X δ X_{δ} Y_{δ}

        x = x + 1

      end while;

    return X α X_{α};

Step 13: In addition to regular users, data collectors, knowledge engineers, and domain professionals also collaborate to gather domain information or actual words from users.

Step 14: Under an explainable AI system based on fluid thrust, semantic similarity is determined using SemantoSim and Simpson's diversity Index.

Step 15: New Ontologies belonging to various domains in medical and healthcare sector are formulated.

End

HFOM [13] does not perform exceptionally well in terms of precision, recall, accuracy, FDR but performs extensively well in n-DCG values because it is hybridized in nature. It includes latent semantic analysis and content-based filtering as it is hybridized. However, HFOM's content-based filtering is less effective than the suggested technique. The synonym extraction agent, latent semantic analysis for ontology topic modeling, ensures that the n-DCG is as high as possible, but the content-based filtering is not as effective as it could be. However, Dynamic Logic Formulation, Dynamic Logic induced axiomatization, and domain knowledge acquisition via interdomain thesaurus ensure that n-DCG is quite strong, whereas validating concepts based on content-based filtering are rather poor. If hybridization is taken into account to enhance content-based filtering, the HFOM would have performed far better than the other systems. Onto Yield is an ontology recommendation system that uses the SemantoSim measure to match ontologies. SemantoSim is a superior metric than others as it alone is not insufficient to incorporate relevance to results. There is also no lateral knowledge. This approach's knowledge density is extremely low in comparison to other techniques. Thus, there is room for development through increasing the incorporation of auxiliary information into the framework. As a result, OntoYield [22] does not perform as expected. Still, incorporating OWL to XML, derivation of RDF and making this knowledge RDF centric, and induction of description logics to the initially matching concepts and individuals ensures that this yields one of the high-quality ontologies. Still, precision, recall, accuracy, and F-Measure has always a scope for improvement. It has a very low n-DCG of 0.91.

Improvement of n-DCG can be made with the improvement of other measures that can be done by incorporating auxiliary knowledge from several heterogeneous sources into the framework. SAMO is a semiautomatic approach of merging and producing ontologies using web Protégé. It includes only the class reliance properties in Protégé by using PROMPT, where classes and individuals are merged only based on the class type, individual type, and the relationship and the data property associated with them. Quantifier restrictions and similarity alone are used, and finally, consistency checking has been done using the merger available in Protégé. So, the merger, although it works well it yields very low precision, accuracy and recall, and high FDR mainly for the reason that traditional and naïve similarity algorithms and also there is no concrete reasoning mechanisms in terms of term similarity or concept similarity, which should eternally be associated with the terms which incorporate ontologies into the approach. As a result, SAMO has a large scope for improvement. SAMO has a very high n-DCG value mainly because it focuses on topic modeling schemes.

STOL [20] incorporates LSI (Latent Semantic Indexing), Singular Vector Decomposition, Latent Allocation for topic modeling to facilitate ontology learning. So, text to onto nil and OIE problems are resolved. In this approach, the main core strategies employed are fuzzy coclustering and fuzzy scale type 1 and 2 to facilitate semantic compliant retrieval. Topic and Word deflections are computed using an ontology similarity and ranking approach. It is a full semantic approach with a very high n-DCG value as it incorporates three distinct topic modeling schemes. The model records low recall and accuracy value as it uses a fuzzy based coclustering algorithm. However, the results could be enhanced by incorporating powerful semantic reasoning schemes to compute the ontological similarity.

SAMO [21] is a semiautomatic method for merging and generating ontologies using web Protégé. It includes only the class reliance properties in PROMPT Protégé, where classes and people are merged according to the class type, person type, relationship, and data property linked with them. Quantifier constraints and similarity alone are utilized, and lastly, consistency testing is performed using the Protégé merger. So, while the merger works well, it yields very low precision, accuracy, and recall, as well as a high FDR, owing to traditional and naive similarity algorithms, as well as the lack of concrete reasoning mechanisms in terms of term similarity or concept similarity, which should always be associated with the terms that incorporate ontologies into the approach. As a result, SAMO has a lot of room for development. SAMO has a very high n-DCG value, owing to its emphasis on topic modeling methods. To assist ontology learning, STOL integrates LSI (Latent Semantic Indexing), Singular Vector Decomposition, and Latent Allocation for topic modeling. This approach uses fuzzy coclustering and fuzzy scales of types 1 and 2 for semantic compliance retrieval. Topic and word deflection are computed using ontologies, and the ontology similarity ranking approach has been used. So, it is an entire semantic approach with a very high n-DCG value because three different topic modeling schemes are incorporated. While the recall and precision is low because a fuzzy-based coclustering approach is used wherein this can be powerfully incorporated using semantic reasoning schemes with other semantic similarity schemes that are much powerful in ontological similarity. This ensures the precision, accuracy, recall, and F-Measure could be high with a low FDR.

The proposed approach is a highly specialized approach for high-risk domains like medical science and healthcare. Healthcare is a part of pharmaceutical and medical science ontology. A full cover approach has been proposed by incorporating full content circles from medical journals, medical eBooks, medical research papers, and medical domain experts. The proposed approach comprises an intuitive approach and a human in the middle approach. For the automatic approach, Transformer, a deep learning algorithm, is used to facilitate learning. Metadata is generated along with media wiki data, and wiki data is incorporated to include auxiliary knowledge. The initial knowledge graph is derived from the various domains such as healthcare, pharmaceutical, general medicine, dermatology, community medicine, nuclear medicine, radiodiagnosis, pediatrics, obstetrics, gynecology, biostatistics, gastroenterology, neurology, nephrology, rheumatology, cardiology, pulmonology, etc. These domains get the content of the eBooks that are summarized and the preface or the index of the eBooks and also from several research articles that are crawled from Google Scholar comprised of the same keywords in the eBooks. As a result, the ontology is initially modeled, and in turn, a knowledge graph has been modeled by combining the terms of vocabularies from all the sources. A generation of metadata ensures that the hidden contents relevant to the domain from the world wide web are also uncovered, and a large amount of metadata from the web structured linked open data of the semantic web is incorporated in the form of indexes. So, as a result, the metadata generated ensures that every possible content under the full cover search is infused into the system. Incorporation of Shannon's entropy and SemantoSim with 0.75 thresholds and Shannon's entropy with a step deviation of 0.25 is considered. However, the integration is done via an agent, which is modeled using the AGENTSPEAK and JADE. It is evident that the agents employed along with the SemantoSim and Shannon's entropy to filter the contents that are deviating from the step size of Shannon's entropy and the threshold of SemantoSim and finalizing the content in the form of Hash map separating them based on the key-value pairs by keeping the semantic similarity alone as the value even for the instances that are incorporated based on Shannon's entropy and SemantoSim. The axiomatization is done using an axiomatization agent where several relationships based on "is a part of," "as a part of" are taken into consideration, and axioms are infused into the terms to formulate the ontology. The proposed approach facilitates learning using Transformer, which uncovers the hidden instances, concepts, and individuals. Wikidata is also incorporated to enhance the density of the entities along with the media wiki and incorporation of Shannon's entropy, and SemantoSim makes the proposed approach IntelliOntoRec highly powerful. The proposed model has heterogeneous sources of information like eBooks, journals, and seed ontologies representing every domain. The approach has several levels of users as it is a human in the middle approach, including data farmers, regular users, domain experts, new users, naïve users, and knowledge engineers. These users, especially the domain experts, are mainly used to interpret and review the modeled ontology. The approach does not include just one domain expert but rather multiple domain experts. Based on the instant reviews made by the top n domain experts, the solution is considered, and ontologies are formulated annually. Upon manual formulation, Simpson's diversity index, SemantoSim, is further used through the XAI-based fluid

Table 12.2 Performance comparison of the proposed approach IntelliOntoRec model.

Search technique	Average precision	Average recall	Average accuracy	Average F-measure	FDR	nDCG
HFOM	87.46	90.38	88.92	88.89	0.12	0.94
OntoYield	92.43	94.69	93.56	93.54	0.07	0.93
STOL	91.17	93.63	92.40	92.38	0.08	0.94
SAMO	81.12	83.71	82.41	82.39	0.18	0.87
Proposed IntelliOntoRec	97.71	99.42	98.56	98.55	0.02	0.98

thrust algorithm to derive the solution sets, which is further passed for axiomatization and formulation of ontologies. The reason for using an XAI-based fluid thrust algorithm with SemantoSim measure is to incorporate only the highly relevant Ontologies among a wide variety of ontologies finalized by the domain experts.

Thus, we could conclude that the proposed IntelliOntoRec is the best-in-class approach for formulating ontologies.

$$\text{Precision}\% = \frac{\text{No. of Relevant and Accepted Ontologies into the system}}{\text{Total No of Ontologies Accepted by the system}} \quad (12.14)$$

$$\text{Recall}\% = \frac{\text{No. of Relevant and Accepted Ontologies into the system}}{\text{Total No of Relevant Ontologies Authored}} \quad (12.15)$$

$$\text{F} - \text{Measure}\% = \frac{2 * \text{Precision}\% * \text{Recall}\%}{\text{Precision}\% + \text{Recall}\%} \quad (12.16)$$

$$\text{FDR} = 1 - \text{Precision} \quad (12.17)$$

The Precision, Recall, F-Measure, and FDR can be calculated as illustrated in Eq. (12.14), Eq. (12.15), Eq. (12.16), and Eq. (12.17), respectively. The average precision value of the proposed method, IntelliOntoRec is greater than HFOM by 10.25, OntoYield by 5.28%, STOL by 6.54%, and SAMO by 16.59%. The average recall percentage of IntelliOntoRec is more remarkable than HFOM by 9.04%, OntoYield by 4.73%, STOL by 5.79%, and SAMO by 15.71%. The accuracy percentage of the proposed approach is higher than HFOM by 9.65%, OntoYield by 5.02%, STOL by 6.17%, and SAMO by 16.18%. The average F-Measure of IntelliOntoRec is higher than HFOM by 9.66%, OntoYield by 5.06%, STOL by 6.18%, and SAMO by 16.17%. The FDR of the proposed approach is lesser than HFOM by 0.10, OntoYield by 0.05, STOL by 0.07, and SAMO by 0.17. The average nDCG score of the proposed IntelliOntoRec is higher than HFOM by 0.04, OntoYield by 0.05, STOL by 0.04, and SAMO by 0.11. Table 12.2 compares the performance of the proposed approach, IntelliOntoRec with other baseline approaches considered in the study. Table 12.3 shows the performance metrics for each domain.

The system conceives ontologies belonging to various domains. The domains furnished by the proposed model include radiodiagnosis, pharmaceuticals, general medicine, dermatology, community medicine, nuclear medicine, pediatrics, obstetrics and gynecology, biostatistics, gastroenterology, neurology, nephrology, cardiology, rheumatology, pulmonology, otorhinolaryngology, ophthalmology, geriatric medicine, orthopedics, pathology. The average precision percentage obtained for the domains mentioned in the previous line are 96.32, 95.81, 97.32, 94.89, 98.39, 94.48, 99.81, 96.87, 95.99, 97.63,

Table 12.3 Performance metrics for the individual ontologies furnished by the system.

Domain ontologies	Average precision %	Average recall %	Average accuracy %	Average F-measure %	FDR	nDCG
Radiodiagnosis	96.32	98.32	97.32	97.30	0.03	0.97
Pharmaceuticals	95.81	97.32	96.56	96.55	0.04	0.98
General Medicine	97.32	99.33	98.32	98.31	0.02	0.96
Dermatology	94.89	96.21	95.55	95.54	0.05	0.97
Community Medicine	98.39	98.27	98.33	98.32	0.01	0.99
Nuclear Medicine	94.48	96.36	95.42	95.41	0.05	0.96
Pediatrics	99.81	99.93	99.87	99.86	0.001	0.95
Obstetrics and Gynecology	96.87	98.31	97.59	97.58	0.03	0.97
Biostatistics	95.99	97.37	96.68	96.67	0.04	0.98
Gastroenterology	97.63	99.36	98.49	98.48	0.02	0.96
Neurology	98.36	99.07	98.71	98.71	0.01	0.97
Nephrology	97.23	99.03	98.13	98.12	0.02	0.96
Cardiology	96.21	97.83	97.02	97.01	0.03	0.95
Rheumatology	94.12	96.37	95.24	95.23	0.05	0.95
Pulmonology	94.96	96.39	95.67	95.66	0.05	0.99
Otorhinolaryngology	96.81	99.07	97.94	97.92	0.03	0.94
Ophthalmology	97.33	99.87	98.60	98.58	0.02	0.97
Geriatric Medicine	95.39	97.34	96.36	96.35	0.04	0.97
Orthopedics	96.32	98.71	97.51	97.50	0.03	0.96
Pathology	95.71	97.87	96.79	96.77	0.04	0.95

98.36, 97.23, 96.21, 94.12, 94.96, 96.81, 97.33, 95.39, 96.32, 95.71, respectively. The average accuracy percentage for the above mentioned domains are 98.32, 97.32, 99.33, 96.21, 98.27, 96.36, 99.93, 98.31, 97.37, 99.36, 99.07, 99.03, 97.83, 96.37, 96.39, 99.07, 99.87, 97.34, 98.71, 97.87, respectively.

The accuracy in percentage, F-Measure, FDR, nDCG for radiodiagnosis is 97.32, 97.30972462, 0.0368, 0.97; pharmaceuticals is 96.565, 96.55909698, 0.0419, 0.98; general medicine is 98.325, 98.31472769, 0.0268, 0.96; dermatology is 95.55, 95.54544113, 0.0511, 0.97; community medicine is 98.33, 98.32996339, 0.0161, 0.99; nuclear medicine is 95.42, 95.41073989, 0.0552, 0.96; pediatrics is 99.87, 99.86996395, 0.0019, 0.95; obstetrics and gynecology is 97.59, 97.58468798, 0.0313, 0.97; biostatistics is 96.68, 96.67507551, 0.0401, 0.98; gastroenterology is 98.495, 98.48740342, 0.0237, 0.96; neurology is 98.715, 98.71372334, 0.0164, 0.97; nephrology is 98.13, 98.12174564, 0.0277, 0.96; cardiology is 97.02, 97.01323748, 0.0379, 0.95; rheumatology is 95.245, 95.2317119, 0.0588, 0.95; pulmonology is 95.675, 95.66965665, 0.0504, 0.99; otorhinolaryngology is 97.94, 97.92696243, 0.0319, 0.94; ophthalmology is 98.6, 98.58364199, 0.0267, 0.97; geriatric medicine is 96.365, 96.35513516, 0.0461, 0.97; orthopedics is 97.515, 97.50035584, 0.0368, 0.96; pathology is 96.79, 96.77794917, 0.0429, 0.95, respectively.

Fig. 12.4 depicts the variation of precision in percentage versus the number of recommendations generated by the system. It is evident from the graph that the suggested approach, IntelliOntoRec, has better average precision than any other baseline model. The average precision decreases gradually

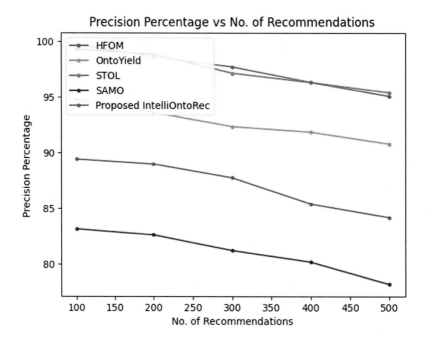

FIGURE 12.4

Precision distribution graph.

with the number of recommendations for all models considered in this study. It is also evident that the proposed approach performance is better than any other baseline approaches considered in the study as it has the highest average precision value than HFOM, OntoYield, STOL, and SAMO.

Table 12.4 presents the information about the numerous ontologies on various domains used in this study. Ontology belonging to the domain radiodiagnosis, pharmaceuticals, general medicine, dermatology, community medicine, nuclear medicine, pediatrics, obstetrics and gynecology, biostatistics, gastroenterology, neurology, nephrology, cardiology, rheumatology, pulmonology, otorhinolaryngology, ophthalmology, geriatric medicine, orthopedics, pathology has 22805, 28916, 43817, 44001, 58914, 41113, 68181, 38106, 27804, 51106, 40316, 28110, 36174, 38761, 44328, 50201, 44176, 33044, 51121, 27476 classes, respectively. The ontologies of the domains mentioned in the previous sentence have 76714, 87681, 120271, 138101, 111786, 101171, 102021, 76381, 86378, 96874, 78122, 81175, 104023, 82674, 71144, 108746, 102121, 85714, 96832, 81674 individuals, respectively.

12.6 Conclusion

In this chapter, an efficient method, the knowledge infused semiautomatic approach, is put forward for ontology formulation in healthcare and medical science domains. The outcomes obtained prove the adequacy of the proposed approach, InelliOntoRec for intended purposes. It achieves a precision

Table 12.4 Details of concepts and individuals in the final ontologies generated.

Domain ontologies	No. of concepts	No. of individuals
Radiodiagnosis	22805	76714
Pharmaceuticals	28916	87681
General Medicine	43817	120271
Dermatology	44001	138101
Community Medicine	58914	111786
Nuclear Medicine	41113	101171
Pediatrics	68181	102021
Obstetrics and Gynecology	38106	76381
Biostatistics	27804	86378
Gastroenterology	51106	96874
Neurology	40316	78122
Nephrology	36174	81175
Cardiology	38761	104023
Rheumatology	44328	82674
Pulmonology	50201	71144
Otorhinolaryngology	44176	108746
Ophthalmology	33044	102121
Geriatric Medicine	44176	85714
Orthopedics	33044	96832
Pathology	51121	81674

of 97.71%, average recall of 99.42%, average accuracy of 98.57%, F-Measure of 98.58%, FDR of 0.023, and nDCG of 0.98 implying that it is by far the most effective solution. The proposed model outperforms the baseline approaches considered in this study significantly to formulate ontologies for its intended applications in the medical and healthcare related domains. The prospective future includes developing more efficient and accurate hybrid statistical and heuristic models incorporating diversity indexes.

References

[1] J. Aguilar, M. Jerez, T. Rodríguez, CAMeOnto: context awareness meta ontology modeling, Applied Computing and Informatics 14 (2) (2018) 202–213.

[2] M. Fahad, Merging of axiomatic definitions of concepts in the complex OWL ontologies, Artificial Intelligence Review 47 (2) (2017) 181–215.

[3] M. Priya, A.K. Ch, A novel method for merging academic social network ontologies using formal concept analysis and hybrid semantic similarity measure, Library Hi Tech. (2019).

[4] S. Babalou, B. König-Ries, GMRs: reconciliation of generic merge requirements in ontology integration, in: SEMANTICS Posters&Demos, 2019.

[5] N. Chatterjee, N. Kaushik, D. Gupta, R. Bhatia, Ontology merging: a practical perspective, in: International Conference on Information and Communication Technology for Intelligent Systems, Springer, Cham, 2017,

March, pp. 136–145.

[6] Z. Nakhla, K. Nouira, Automatic approach to enrich databases using ontology: application in medical domain, Procedia Computer Science 112 (2017) 387–396.

[7] T. Schoormann, D. Behrens, U. Heid, R. Knackstedt, Semi-automatic development of modelling techniques with computational linguistics methods – a procedure model and its application, in: International Conference on Business Information Systems, Springer, Cham, 2017, June, pp. 194–206.

[8] S. Pal, P.K.D. Pramanik, T. Majumdar, P. Choudhury, A semi-automatic metadata extraction model and method for video-based e-learning contents, Education and Information Technologies 24 (6) (2019) 3243–3268.

[9] X. Xue, J. Chen, A preference-based multi-objective evolutionary algorithm for semiautomatic sensor ontology matching, International Journal of Swarm Intelligence Research (IJSIR) 9 (2) (2018) 1–14.

[10] M. Rani, A.K. Dhar, O.P. Vyas, Semi-automatic terminology ontology learning based on topic modeling, Engineering Applications of Artificial Intelligence 63 (2017) 108–125.

[11] L. Xingjun, S. Zhiwei, C. Hongping, B.O. Mohammed, A new fuzzy-based method for load balancing in the cloud-based Internet of things using a grey wolf optimization algorithm, International Journal of Communication Systems 33 (8) (2020 May 25) e4370.

[12] N.M. Franz, Biological taxonomy and ontology development: scope and limitations, Biodiversity informatics 7 (1) (2010).

[13] C.N. Pushpa, G. Deepak, J. Thriveni, K.R. Venugopal, A hybridized framework for ontology modeling incorporating latent semantic analysis and content-based filtering, International Journal of Computer Applications 150 (11) (2016).

[14] M.K.M. Nasution, Ontology, Journal of Physics Conference Series 1116 (2) (2018 December) 022030, IOP Publishing.

[15] Y. Tikhamarine, D. Souag-Gamane, A.N. Ahmed, O. Kisi, A. El-Shafie, Improving artificial intelligence models accuracy for monthly streamflow forecasting using grey Wolf optimization (GWO) algorithm, Journal of Hydrology 582 (2020 Mar 1) 124435.

[16] Y. Zhang, Y. Meng, J. Huang, F.F. Xu, X. Wang, J. Han, Minimally supervised categorization of text with metadata, in: Proceedings of the 43rd International ACM SIGIR Conference on Research and Development in Information Retrieval, 2020 Jul 25, pp. 1231–1240.

[17] L. Liu, L. Tang, W. Dong, et al., An overview of topic modeling and its current applications in bioinformatics, SpringerPlus 5 (2016) 1608.

[18] M. Steyvers, T. Griffiths, Probabilistic topic models, in: Handbook of Latent Semantic Analysis, Psychology Press, 2007, pp. 439–460.

[19] H.M. Wallach, I. Murray, R. Salakhutdinov, D. Mimno, Evaluation methods for topic models, in: Proceedings of the 26th Annual International Conference on Machine Learning, 2009, June, pp. 1105–1112.

[20] M. Rani, A.K. Dhar, O.P. Vyas, Semi-automatic terminology ontology learning based on topic modeling, Engineering Applications of Artificial Intelligence 63 (2017) 108–125.

[21] A. Ameen, K.U.R. Khan, B.P. Rani, Semi-automatic merging of ontologies using protégé, International Journal of Computer Applications 85 (12) (2014).

[22] G.L. Giri, G. Deepak, S.H. Manjula, K.R. Venugopal, OntoYield: a semantic approach for context-based ontology recommendation based on structure preservation, in: Proceedings of International Conference on Computational Intelligence and Data Engineering, Springer, Singapore, 2018, pp. 265–275.

[23] D. Surya, G. Deepak, A. Santhanavijayan, KSTAR: a knowledge based approach for socially relevant term aggregation for web page recommendation, in: International Conference on Digital Technologies and Applications, Springer, Cham, 2021, January, pp. 555–564.

[24] D. Surya, G. Deepak, A. Santhanavijayan, QFRDBF: query facet recommendation using knowledge centric DBSCAN and firefly optimization, in: International Conference on Digital Technologies and Applications, Springer, Cham, 2021, January, pp. 801–811.

[25] D. Surya, G. Deepak, A. Santhanavijayan, Ontology-based knowledge description model for climate change, in: International Conference on Intelligent Systems Design and Applications, Springer, Cham, 2020, December, pp. 1124–1133.

[26] M. Arulmozhivarman, G. Deepak, OWLW: ontology focused user centric architecture for web service recommendation based on LSTM and whale optimization, in: European, Asian, Middle Eastern, North African Conference on Management & Information Systems, Springer, Cham, 2021, March, pp. 334–344.

Index

Printed in the United States
by Baker & Taylor Publisher Services